Supply Chain				People			Technology		Competitive Benchmarking			
Supplier Responsibilities	Supplier Comm.	Supplier Management Mechanisms	Supplier Roles	Job Designs	Skills	Motivation Systems	Internal	Suppliers	Coseat Now	Coseat Future	Competitor A	Competitor B
-3	-3	-9	-9		3	-3	3	-3	16 months	8 months	14 months	10 months
-3	-9	-3	-3		9	-3	3	-3	230 ppm	23 ppm	300 ppm	200 ppm
-3	-3	-3	-3		3	-3	3	-3	250	190	200	150
	-3				3				$75	$53	$100	$100
		-3	-3						8%	6%	5-10%	5-10%
-9	-9	-3	-3					-3	10%	7%	8%	7%
-66	-90	-75	-75	NA	54	-30	30	-42				
			Mostly contractual, a few parental									
Poor relations with tooling suppliers, most suppliers involved late	Low frequency one way with parts & some component makers before production	Qualification is only mechanisms used consistently, early involvement with 2 suppliers	Even highly involved supliers are at best parental	NA in assessment	Highly skilled workforce; engineers up to date on tech. issues; willing, energetic staff; training not systematic	Rewards tend to focus on functional org. at expense of product goals; some goal conflicts; appraisal inconsisent; some rewards viewed as unfair	Slow CAD system, IGES for translation, no links from CAD to CAM	Most have primitive technology				
Strong supplier involvement early		Supplier with early involvement usually gets the contract, price agreements			Heavy commitment to training and skill development	Strong PDA and QSR processes, measure everything	Wide array of tools					
Strong supplier involvement throughout design, more design responsibility at supplier	Generally high levels of communication much face to face at Chrysler Tech Center	Early price commitments based on target pricing	Many at mature, moving toward partner	Traditional roles maintained			Central CAD database using single CAD system (CATIA)	All req. to use CATIA for CAD - some do but many use translation internally or thru service				
Full service supplier-develop complete valve system; now work w/their suppliers to validate process		Early commitment from customer, but no contract until late-"strategic alliance" not "partnership"			40 hours training per year for all employees							
							VIVID sys. [illegible]	Exchange [illegible]				
Strong emphasis on early involvement in design, suppliers responsible for product test		Strategic suppliers member of team early, pre-source to them	Many in mature roles				[illegible]	[illegible]				
Battery supplier involved early in new technology development			Battery supplier in mature or partner role				[illegible] effective	[illegible] effectively used to link with other systems				
			Developing full-service suppliers such as Eaton-much like mature or partner roles				Ford uses a single CAD system within major functions	Attempts to require suppliers to use same CAD system, but has not been totally successful - some IGES used				

Concurrent Engineering Effectiveness: Integrating Product Development Across Organizations

by

Mitchell Fleischer
Center for Electronic Commerce
Industrial Technology Institute

and

Jeffrey K. Liker
Department of Industrial and Operations Engineering
University of Michigan

Hanser Gardner Publications
Cincinnati

Fleischer, Mitchell,
Concurrent engineering effectiveness : integrating product development across organizations / by Mitchell Fleischer and Jeffrey K. Liker.
p. 518 15.25 × 22.85 cm.
Includes bibliographical references and index.
ISBN 1-56990-231-3
1. New products. 2. Concurrent engineering. 3. Production engineering. 4. Production managment. I. Liker, Jeffrey K. II. Title.
TS170.F54 1997
658.5'75--dc21 97-18614
CIP

Hanser Gardner Publications
6915 Valley Avenue
Cincinnati, OH 45244-3029

2 3 4 5 6 01 00

Dedication

To Carol for her love and support over the years.
Mitchell Fleischer

To Deb, Jesse, and Emma, who bring joy and meaning to my life.
Jeffrey Liker

TABLE OF CONTENTS

PREFACE

Concurrent Engineering Effectiveness: Integrating Product Development Across Organizations provides you with an understanding of the non-technical issues involved in establishing a company-wide context for concurrent engineering (CE). This includes the work process, organization, people, and supply chain issues involved with CE. But understanding is not all we hope to provide. The book also includes a structured methodology for using that understanding to plan and implement changes that will ensure your company is able to do CE consistently over the long run, rather than just as a one-shot experiment.

This book is the result of more than 12 years of research and consulting experience working with companies trying to improve their product development processes. We started down this path in 1984 with an interest in how to help manufacturing companies make better use of computer-aided design (CAD) technologies. The more we studied this, the more we realized we needed to understand the context in which that technology was being used; so we began to refocus our efforts on product development and concurrent engineering (and, as it later came to be called, Integrated Product/Process Development or IPPD).

We started out doing surveys and interviews about CAD in six large manufacturing companies in the automotive, aerospace, and heavy appliance industries. Since then we have worked extensively with a wide range of firms in the automotive industry, as well as a number of firms in a scattering of other industries, including defense contractors. The first author (Fleischer) has focused much of his effort on working with smaller firms and suppliers to the large automotive manufacturers as part of his work with the Center for Electronic Commerce at the Industrial Technology Institute. The second author (Liker) has focused much of his effort on studying Japanese product development and manufacturing practices, especially within the context of the Japan Technology Management Program at the University of Michigan.

We quickly learned that product development was one of the keys to manufacturing improvement. The original design of the product has huge downstream effects on manufacturing, sales, service, and disposal. We learned that products and processes not only could be designed concurrently, they *needed* to be. We also learned that while suppliers always had an important role to play in the product development process, in CE that role was magnified enormously. We came to see CE as an enormous problem of integration — across functional boundaries within an organization and across company boundaries in a supply chain. The problems of how to create links across these different kinds of boundaries were not very different, although the solutions might be very different.

A second thread running through this work is our experience with Socio-Technical Systems (STS) design. In STS design, the technical system is designed concurrently with the social system. Since we had both been trained in STS before we even knew about concurrent engineering, much of CE came to us almost intuitively. We have applied the principles of STS to our work in CE to great benefit.

A final thread is the importance of building tools to empower manufacturing companies to change themselves. Our experience as both researchers and consultants has suggested that too many companies place too much reliance on consultants to help them analyze and solve their problems. We have long hoped to be able to make many of the concepts and tools used by experienced consultants available directly for staff in a manufacturing firm. Part I of this book is intended to make the *concepts* of work process, organization, and people system design in CE accessible to a wider audience. Part II is intended to make the consulting *tools* for work process, organization, and people system design available for use by a wider audience.

The Audience for this Book

Concurrent Engineering Effectiveness is for you if you work in a manufacturing firm and have some responsibility for planning how your firm will do CE. This book is also for you if you currently consult (or plan to consult) with firms that want to learn how to do CE, or how to do CE better. The book does not cover the technical aspects of CE. Instead it focuses on the work process, organization, people systems, and supply chain issues that are necessary to do CE, and on how those systems interface with the technology used in CE.

Concurrent Engineering Effectiveness is also suitable for the growing number of graduate-level courses on product development management and concurrent engineering. Whether your course is in mechanical engineering, industrial engineering, operations management, or cross-listed across departments, this book can provide a framework for students to think about how to design a product development organization to concurrently design products and processes. It will also be useful for small groups of students assigned a project to work through how to manage themselves as a team.

Acknowledgments

This book is based on an earlier project we worked on together called the Cross-Organizational STEP Adoption Tool (COSAT). COSAT is a hypermedia computer software program that includes much of the material in this book, but is accessible only on a personal computer. One of the problems with COSAT was that it was difficult for people to grasp what it was about — they couldn't easily browse the material to find out if they wanted to get involved with what obviously could be a dense and difficult subject. Several people who used COSAT suggested that the whole thing would be easier to use, at least initially, if the material were in book form. You see the result. In the process of writing the book, however, we discovered that there were a lot of things missing from COSAT — as software, there was a level of detail we simply couldn't cover. Hence most of the written material in COSAT was completely redone for this book. Many people contributed to COSAT, and many of their ideas are contained in this book, even if many of their words were changed. Those people are: Thomas Phelps, Stanley Przybylinski, Michael Wood, Lida Orta-Anes, Lorraine Hendrickson, Mark Brown, Tihamer Toth-Fejel, Judith Robinson, Steve Clark, and William Neal. We thank them for their work on COSAT and their ideas that carried on to this work. COSAT was funded by the U.S. Navy Manufacturing Technologies Program, with award #70NANB2H1720 through the National Institute of Standards and Technology (NIST).

After COSAT was complete, Leo Plonsky and Steve Linder of the Navy

Mantech program had the wisdom to understand that a book like this would make disseminating ideas about the work process, organization, and people systems in CE much easier. So they provided the additional funding to get it written with award #70NANB5H0025, also through NIST. We thank them and hope that their efforts to encourage CE and the use of STEP are aided by our work. Mary Mitchell at NIST was the project officer. We thank her for her encouragement and many useful suggestions.

Many of the concepts brought to this book by the second author (Liker) were based directly on his research funded under the Japan Technology Management Program by the Air Force Office of Scientific Research. The chapters on supplier management, the references throughout the book to Japanese approaches, and the case study on Toyota (by Durward Sobek) all gre directly out of that work.

Ann Majchrzak of the University of Southern California provided much of the inspiration for the methodology and tools in Part II. Her work with the first author on High Integration of Technology Organization and People (HITOP) provided critical insights about how to develop tools for analysis and change that could be used by people who were not professionals in the field (Majchrzak, Fleischer, and Roitman, 1991). This would be a very different book without her influence.

Several people at the Industrial Technology Institute also helped get this book written. Jack White helped enormously by providing the time and encouragement to get it done, not to mention quite a few good ideas. Tom Phelps helped to write Chapter 7. Bill Hetzner provided a large number of very insightful comments and suggestions on the manuscript. Lindsay McPherson provided copyediting, and Caron Wiesner provided administrative support. Jim Shearer did wonderful work on the figures and the book cover. We thank them all.

Lou Tornatzky (now at the Southern Technology Council) got us started on our original work on CAD. We were inspired by his dedication to the development of practical tools and methods that would make a difference. In many ways he has been a mentor to both of us.

Finally, we want to thank the numerous people from the many manufacturing companies who provided us with the background that fills the case studies and other examples in this book. Many of them gave freely of their time to participate in research projects in hopes of advancing the state of knowledge about product development. Others were working with us in hopes of advancing their companies and careers. Still others were participants in the Concurrent Engineering User Group, run by Professor Frank Hull of Fordham University. John McKenzie from that group served as a reader and provided many helpful suggestions. We especially thank Professor Hull for providing the first author with the opportunity to participate as a full-fledged member of the group.

PART I

•

Organization and Process in Concurrent Engineering

CHAPTER 1

•

Product Development Matters

Thirty-odd years ago I began to counsel that you should build organized abandonment into your system. It follows the old line that it makes more sense for you to make obsolete your own products than to wait for the competitor to do it. But this is very hard for organizations to do. The internal resistance is great. They have to be forced. Remember the Edsel? After eighteen months the Ford Motor Company announced that it was abandoning the Edsel. I think we all roared with laughter. We had already abandoned the Edsel. The Ford Motor Company just took a hell of a long time to accept it. • Peter Drucker

THE DEATH OF THE CASH COW

The list of major manufacturing firms which have undergone "near-death" experiences over the past 20 years is remarkable: General Motors, Ford, Chrysler, IBM, Digital Equipment, Xerox, Polaroid, and many other equally well-known names. And, of course, there are many more formerly well-known companies that are simply history, such as Cray Research, Kaypro Computers, and American Motors. These companies have something in common: they failed to bring new products to market quickly enough, and they paid a huge price for it.

- IBM developed the architecture for the System/360-370 mainframe in 1964, and continued to focus on selling it well into the 1980s. While this enabled IBM to dominate the computer business during this period, they were unable to understand how to control the personal computer market.
- Digital Equipment made its fortune in mini-computers based on its proprietary VAX architecture. It was unable to successfully

develop new products beyond the VAX and has been in steady decline for the past decade.

The lesson? It's not enough to simply develop an outstanding, market-leading product. Competition in most markets is now so intense that no company can afford to rest on its laurels. *Rapid, cost-effective product development is essential for any company that wants to be in business ten years from now*.

Not all new products are revolutionary. It makes perfect sense to build a range of new products based on a single platform or architecture. But those products need to stay fresh and keep up with constantly changing customer needs. When the platform no longer serves those needs it must change. Consider the following examples.

- Sony very effectively exploited the technology of the Walkman through a combination of fundamental and incremental innovation which has kept them as a market leader in this technology since 1979 (Sanderson, 1992). "Tiger teams," which are separate from day-to-day design and manufacture, conduct aggressive technological development. The basic platforms then provide a mechanical and electronic core for many variants designed to meet particular market niches. Marketing/industrial design teams in Japan and major regional markets develop new product concepts attuned to different market segments (e.g., the "My First Sony" line of children's products). Teams of design and manufacturing engineers develop new models off of the five core designs that make up the bulk of Sony's product introductions. Sony has used a similar strategy very effectively in developing other product lines such as portable CD players and the 8-mm camcorder.
- Intel has exploited its "xx86" microprocessor architecture through a steady progression of increasingly powerful chips. Intel introduced its 8086 processor for personal computers in 1978. About every three to four years it has introduced a new version of the chip which is not only substantially faster, but also cheaper on a cost/performance basis. So far, it has managed to keep one step ahead of its direct competition (makers of xx86 clones) as well as competing architectures, such as the PowerPC. The speed of introduction actually increased over this time period — between the 8086 and the 80286, 4 years passed, while between the Pentium and the Pentium Pro, only 2 years went by.
- In the 1970s and 1980s, GM, Ford, and Chrysler worked on a schedule of bringing out totally new models about every 10-12 years, while their Japanese competitors were bringing out new models

much more often. Between 1970 and 1995, the Japanese share of the U.S. automobile market went from 3.6 percent to 22.8 percent.[1] Even today, Ford went 10 years between total remakes of the Taurus as compared to four years per revision for the Honda Accord (Figure 1.1). It also costs Ford about $300 more to build a vehicle than Honda (Harbour and Associates, 1996).

The lesson? New products generate more money than old ones. If you can introduce new products more quickly than your competitors, you'll make more money than they do. But there are some exceptions to this rule. You have to introduce new products which *meet customer needs*, and you must do it cost-effectively.

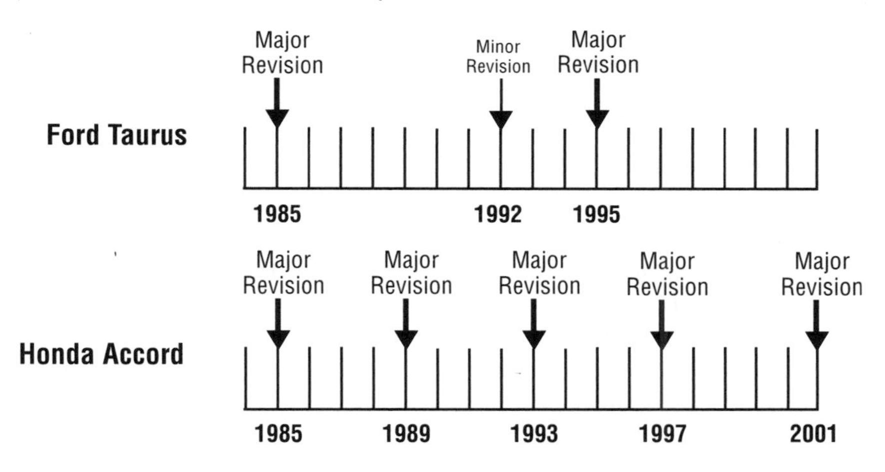

Figure 1.1 • Honda vs. Ford Product Development

- In the first half of the 1990s, Chrysler virtually recreated their entire vehicle line, introducing the LH, the Neon, the Jeep Grand Cherokee, and the fully redesigned minivan. They did it quickly (about as fast as the best of the Japanese), designed in the features consumers wanted at good value, and did it at relatively low cost. The only problem was that quality suffered — to the extent that sales of even these great new products were hurt.
- Ford introduced a new "world car" (called the Mondeo in Europe, and Ford Contour/Mercury Mystique in the U.S.) in 1995 that received excellent reviews and won several awards. However, it cost $6 billion to develop and introduce (in contrast to $900 million Chrysler used to develop its comparable Cirrus vehicle). Some experts wonder if Ford will ever be able to make money on the Mondeo.

[1] Calculated based on data from "American Auto Centennial 1896-1996," *Automotive News*, April 24, 1996.

- Apple made a breakthrough in the market when it came out with the Apple Macintosh® computer. Apple got about a 10 year "ease of use" headstart. Yet because Apple failed to develop the right products customers wanted at the right cost quickly enough, Microsoft had time to build market share and then quickly dominate with Windows 95®. Apple also spent too much money on development. In 1995 Apple spent 6 percent of revenue on R&D compared to 2 percent by rival Compaq (*USA Today*, 1996); and Compaq seemed to do a better job of giving their customers what they wanted at the right time.

The lesson? You have to do everything right when you introduce new products or you stand a significant chance of failure. How can you ensure you'll do everything right?

THE PRODUCT DEVELOPMENT PROCESS: A KEY TO SUCCESS

Most manufacturers recognize the contribution their manufacturing process makes to the quality and cost of their products and their ability to deliver on time to customers. The same is true of their Product Development (PD) process — it too has a major impact on cost, quality, and timing. The only difference is that the PD process has a much *greater* impact on those success factors than does the manufacturing process. The reason for this is simple: the design (which is the result of the PD process) is the ultimate influence on how the product is made, its cost, and how easy it is to make. The well-known graph in Figure 1.2 shows the effect the PD process has on engineering changes, and hence the cost

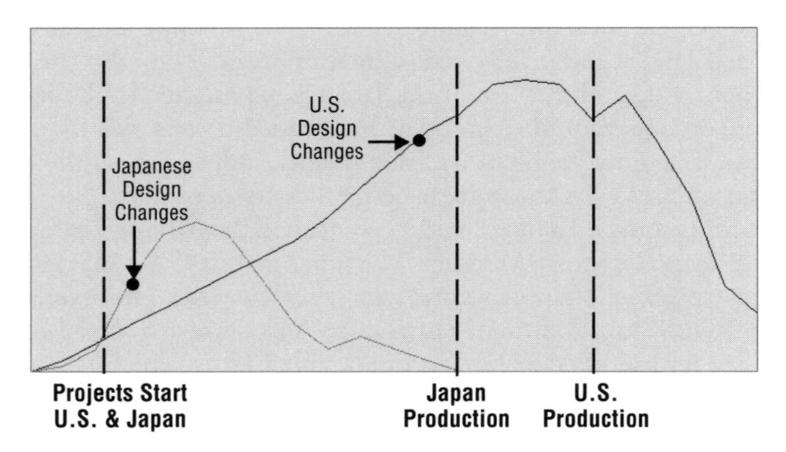

Figure 1.2 • US vs. Japanese Patterns of Design Changes

and timing of the product (Barkan, 1992). Japanese automakers focused more time on making design revisions early on in the prototype stage so the bugs were worked out by the start of production. The U.S., by contrast, invested less time in exploring alternatives early in the cycle and thus made major changes just before production, continued making costly engineering changes after start of production, and the overall cycle was slower and more costly.

The design has other effects. For example, tolerances are established as part of the design. Unnecessarily high tolerances can make the product more costly and time consuming to produce than necessary. On the other hand, tolerances that are too loose may mean quality will suffer. Designers who do not consider the interrelationships of tolerances of mating parts, so-called stacking tolerances, may find individual parts that are within tolerance but do not fit together effectively. Tolerances *sound* like a manufacturing issue, but they start life as a *design* issue.

Design also affects manufacturability. Designers are imaginative people — they can design anything.[2] Making that thing is another story. As resourceful as many manufacturing engineers may be, there are many things which are too costly to make as designed or which may be impossible to make at any cost. Every time a design is produced which is difficult or impossible to make, the result is a costly engineering change and a delay in getting to market.

The design of an automotive body is one of the longest-lead time items and the second most costly part of a vehicle design program. While it would seem that most of the time and cost would be involved with styling, in fact most of it is downstream when the large dies used to stamp body panels are designed and manufactured. In the 1980s, U.S. automakers, trying to ensure high quality, went to great lengths to be sure that the individual parts stamped from these dies accurately matched the specifications developed by design engineers. Dies that missed these specifications were modified at great time and expense. Yet even when individual parts met the specification they often did not fit together right. Metal deforms in unpredictable ways, so it is difficult to predict the precise shape of a part until you try stamping parts and assembling them. Despite all their efforts to be more and more precise, the U.S. automakers were unable to match the Japanese in the quality of their body panels.

An investigation of the Japanese approach (which not only produced better quality but also took less time) found they used a method called "functional build." Die makers do their best to meet the specifications and

[2] Some M.C. Escher prints come to mind here!

try out the dies making actual parts. They then try to fit these parts together and modify the dies (those easiest to rework) to make the parts fit together better and meet the design intent, even if that means deviating from the initial specifications on individual parts. They work together with design engineers to agree on an acceptable stamped body and then design engineers may change the specifications to fit the stamped body that fits together well.[3]

If design is so important, how can we make sure it's done right? The mechanism used inside a company to control design is the Product Development Process. A PD Process is a structured series of decisions supported by information. The culmination of these decisions is a completed product design. There are two types of decisions made during the PD Process. The first addresses the management of the design process itself: how the design problem will be decomposed into smaller, more tractable parts, how the work will be divided up among the design team, when phases are complete, etc. The second type of decision addresses the design specifications and the product itself, its appearance, dimension, materials, etc. In effect, the PD Process ensures that good decisions of both kinds are made in a timely and coordinated manner.

How can the PD Process help you to introduce new products more quickly and at lower cost? The key is increased coordination and integration across functions in your company. By "functions" we mean the many different specialty groups that all companies have, such as manufacturing engineering, production, design engineering, purchasing, etc. The biggest time and money wasters in any process are usually in the links between parts (or functions) of the process (Womack and Jones, 1996).

One of the classic places to look for waste in any manufacturing system is in work in process (WIP). WIP of course only occurs between process steps. WIP is not only expensive inventory sitting idle, it means that a process may be running out of control long before it is discovered, producing waste instead of good parts. The same principle is true for a product development process. Many product development efforts involve substantial amounts of waiting time — e.g., waiting for an engineering analysis to be done, waiting for prototypes to be built, or waiting to get comments back from another function.

A related type of waste in manufacturing is the waste of overbuilding—that is, building material ahead of when it is needed and stockpiling it. An example of this in product development is when an engineering analyst performs many different analyses in anticipation of design

[3] For further discussion of the functional build approach, see P. Hammett, W. Hancock, and J. Baron (1995).

engineers needing the analysis. In a study of design organizations we found it was common for engineering analysis groups to be separated from design engineers. They often would work on an analysis project for a development program in isolation from the design engineers, and thus were trying to "push" the analysis results onto the design engineers rather than the analysis being pulled by immediate questions and needs of the design engineer. Thus, much of the analysis was wasted and never used to make design decisions—it was the wrong analysis at the wrong time (Liker, Fleischer and Arnsdorf, 1992).

In defense shipbuilding, it is common for design contracts to be given separate from manufacturing contracts. Different design houses can even bid on different aspects of the design, e.g., electrical, plumbing, HVAC (heating, ventilation, air conditioning). In one shipbuilding example, the development of a new mine sweeper, three different design houses got contracts for electrical, plumbing, and HVAC and handed off their designs to the shipbuilding firm that was building the first ship. The result was utter chaos in the building stage. For example, a plumber would attempt to install pipes only to find there was ductwork where the pipes were to go. The shipbuilding firm had 100 design engineers whose full-time job was making thousands of engineering changes to resolve these conflicts—essentially redesigning the ship as it was being built. They instituted a procedure where a design engineering group was stationed out in the yard and was "red lining" blue prints (marking them up with red pens) right in the yard, as the ship was being built, to bypass the usual, lengthy bureaucracy for making engineering changes (Liker, 1988).

By far the most well-known approach to improving integration between functions in the PD Process is known as Concurrent Engineering (CE) or Integrated Product/Process Development (IPPD).

CONCURRENT ENGINEERING

Concurrent Engineering means that there is a tight link between all participants in the product development process, such that they can perform much of their work at about the same time. For many people, CE or IPPD implies only the link between design engineering and manufacturing. Indeed, these are the two areas to which CE is most commonly applied. However, the same logic which says design and manufacturing should work concurrently, also says that styling or industrial design should do so as well, as should development of sales strategies, mechanisms for delivering service after the sale, and even methods for disposal of the product when it is all used up. In other words, *CE covers the entire life cycle of the product.*

Taken to its logical conclusion, CE would mean everything would be happening all at once. Of course this is never the case. A more accurate description of the reality of CE is that activities *overlap*. For example, work on developing downstream systems, such as manufacturing, begins before work on developing the product itself is complete (Figure 1.3). This is in direct contrast to more traditional approaches to product development in which the process proceeds sequentially, i.e., each phase, with some minor exceptions, waits for its predecessor to finish before starting (Figure 1.4).

Task Name	Jan	Feb	Mar	April	May	June	July	Aug	Sept	Oct	Nov	Dec
Identify Customer Needs												
Develop Concept												
Get Market Reaction												
Design Prototype												
Build/Test Prototype												
Final Design												
Design Tooling												
Make Tooling												
Install Tooling/Machinery												
Purchase Parts/Materials												
Production Launch												

Figure 1.3 • Overlapping or Concurrent Product Development Process

Task Name	Jan	Feb	Mar	April	May	June	July	Aug	Sept	Oct	Nov	Dec
Identify Customer Needs												
Develop Concept												
Get Market Reaction												
Design Prototype												
Build/Test Prototype												
Final Design												
Design Tooling												
Make Tooling												
Install Tooling/Machinery												
Purchase Parts/Materials												
Production Launch												

Figure 1.4 • Sequential or Waterfall Product Development Process

Sequential design (often called the *waterfall* method) has long been the traditional process taught to American design students (Shigley and Mischke, 1989). And indeed, the logic of a sequential design process at first seems impeccable. After all, the design of the product will take many wrong turns, and involve many changes, how can you possibly design the manufacturing process until you know what the product will *really* look like? There are three arguments against sequential design.

1. ***You can't finalize the design of the product until you know how it will be made.*** As we showed earlier in this chapter, manufacturability issues are as important to the design of the product as knowing performance specifications. While design for manufacturability analysis and participation by manufacturing personnel on a design team can help a great deal, many problems will not be discovered until some work has begun on the manufacturing system. This is particularly true if your design is pushing the limits of producibility in some way (e.g., closer tolerances, higher production rates). Thus, some product design problems will not be discovered until work is significantly along on developing the manufacturing system or process plan. In other words, the sequential method is inherently flawed — design is always an iterative process that covers the entire life cycle, not just the design of the product.
2. ***Sequential design takes too long.*** It seems obvious that doing tasks sequentially will take longer than doing them in parallel. A key benefit of critical path analysis for project planning is that it forces you to consciously decide what needs to happen in sequence and what can happen in parallel. Doing more things at the same time would seem to be naturally faster than waiting to complete each one before starting the next. A less obvious benefit of working in parallel is the shorter feedback loops that can help you adjust each task based on feedback from the other. If we wait for the product to be "completed" before starting to design the manufacturing system, and then find a manufacturability error, we have to go back to the beginning and wait for the product to be redesigned (an "engineering change") before we can start all over again (or at least resume work) on the manufacturing system. This involves large amounts of waiting time and large (and hence expensive) engineering changes.
3. ***Ultimately, a sequential process results in a poor design.*** Quite simply, management and the market won't wait for a sequential process to be completed. Some deadline will be reached (either a formal deadline or some executive's frustration level) and the product will be released, even if all its problems haven't been solved. The customer will then find the flaws which were missed, provide feedback to the designers, and more (and very expensive) changes will be made. The costs of changes at this point in the process are incredibly expensive both in cash terms and in terms of customer satisfaction and later sales.

This is of course no mere theoretical argument. As James Womack and his colleagues showed in their book *The Machine That Changed the World* (Womack, Jones, and Roos, 1990), this is one of the primary reasons for the problems faced by the U.S. automotive industry in the 1980's. The reader with an interest in a more technical discussion, which served as the basis for some of Womack's work, should see *Product Development Performance* by Kim Clark and Takahiro Fujimoto (1991). Clark and Fujimoto studied 29 "clean sheet" development projects reaching the market between 1983 and 1987, and found that a totally new Japanese car required 1.2 million engineering hours and took 43 months from the initial concept to building the first car. By contrast, the U.S. and European projects both took over 60 months and over 3 million hours. Thus, the U.S. and European projects took about a third more time and almost 3 times the engineering hours.

Clark and Fujimoto attributed much of this gap to the use of concurrent engineering by Japanese automakers—design and manufacturing worked together with overlapping tasks. Since the original study, U.S. automakers have all worked hard to adopt concurrent engineering methods, and as a result have dramatically reduced lead time and engineering hours to the point where the gap in lead time has narrowed a great deal.

Chrysler, the most advanced of the Big 3 in concurrent engineering, took only 42 months to develop the LH line (see case history in the Appendix) which included a brand new powertrain, and 29 months to develop the Neon (from clay model freeze to start of production). In the same time Toyota has gotten down to 27 months for a major redesign of a vehicle (e.g., the Camry). In a follow-up study Clark and his associates (1995) found the gap in engineering lead time had narrowed a great deal, but there was still a large gap in engineering hours. The Japanese actually got slightly less efficient, moving up to 1.3 million engineering hours, the U.S. went down to 2.3 million hours, and the Europeans got worse — moving up to 3.4 million hours.

All stakeholders have a fair chance of influencing the design to meet their needs in concurrent engineering. That includes such groups as customers (usually through sales and marketing), manufacturing, service, and shipping. In complex products such as vehicles, major appliances, or aircraft, this involvement would include the supply chain, as well as groups within the manufacturer itself.

Design is inherently a process of making decisions while balancing the many requirements that affect the design. Tradeoffs must always be made. The point of CE is that those tradeoffs are addressed consciously, during the design process, before a lot of money and time are spent putting the manufacturing capability in place. The closer the product is to actual production, the more expensive a change will be. CE provides an important means to catch problems early in the process.

If CE saves time and money, and improves product quality to the customer, it sounds like a pretty good thing. The problem is, as always, *how do you do it?* Understanding "how to do" CE will require that we discuss three topics.

1. ***The Work of CE.*** The activities which take place as CE is performed, and how these activities are controlled. On a more general level, this is the work and business process of product development. A detailed discussion of this can be found in Chapter 3.
2. ***The Context for CE.*** The internal organization, the skills and attitudes of the people doing CE, their technology and tools, and the ways in which they link to customers and suppliers (Figure 1.5). A detailed discussion of this can be found in Chapters 4 through 7.

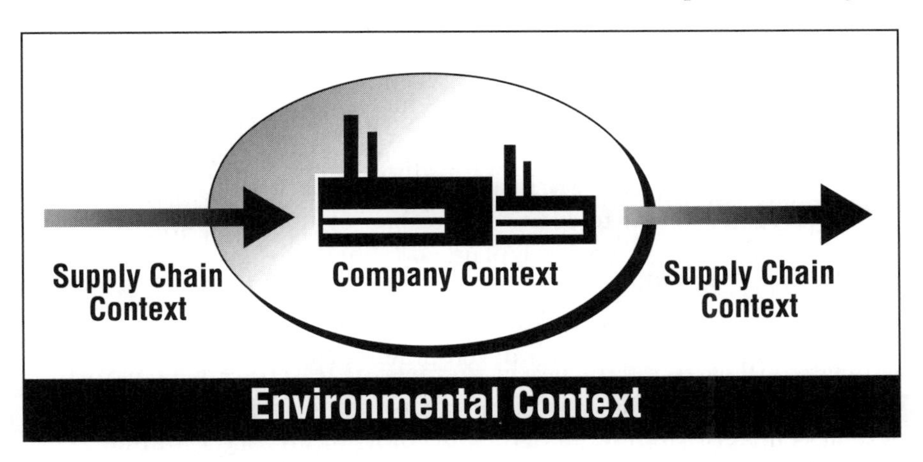

Figure 1.5 • The Context for CE

3. ***The Process for Getting to CE.*** CE is a significant change not only in design process, but in the way designers and others think about their work. Neither a small amount of training nor a simple mandate from the chief executive is going to result in effective, long-term change. A comprehensive process is needed to change the way people do design and to make that change stick. A brief discussion of this is in Chapter 9, and a more detailed discussion is in Chapters 10 through 13.

When we bring up these three topics, we are sometimes asked by technical managers: "Why should I worry about 'context' and 'change process?' If I get the 'Work of CE' in place then we're doing CE. And if we're doing CE, why mess around with all this other stuff?" The answer is that by simply putting the elements of CE in place without considering the context or change process, two things are likely to happen.

- ***You will not get the full benefits of CE.*** For example, if you use CAD/CAM software, but fail to improve communication between design and manufacturing, you will get some benefits, but you will miss a lot more. A study of 53 small tool and dies firms that tried to use their customers' CAD data for CAM found the data they received needed to be extensively reworked before it could be used for CAM (Fleischer et al., 1991). In almost half the jobs, lines did not meet at corners, or lines crossed at corners, and the tooling supplier had to fix the data. In 25 percent of the cases, the suppliers *completely recreated the data set.* Further analysis found this was not a technical problem, but a problem of communication between the design engineers of the customer and the tooling engineers in the tooling companies—the design engineers did not understand what level of data accuracy was required by the tooling engineers.
- ***You will fail to sustain the improvements you've made.*** Even if you make some gains, you will be unable to keep them up. Consider the Ford Taurus, a world famous example of CE completed in 1985. Ford failed to learn the lessons of the Taurus, and subsequent Ford products were plagued by high costs and late product launches. By failing to consider the contextual issues (such as changing the organization and people systems), Ford failed to sustain its learning about CE in later development projects. The case of Texas Instruments Defense Systems and Electronics Group, discussed later in this chapter, is another good example. A small concurrent engineering team did an exceptional job of developing a new missile decoy in record time, under budget, at low cost. But this group worked largely in isolation from the dominant functional organization. Members of the concurrent engineering team were viewed as renegades and many felt that they were being punished by their functional managers for participating in the project.

The point is, by paying attention to the context and the process of change you ensure the *long-term viability* of CE in your company. If your focus is only on doing CE for a single project, then long-term viability won't matter much to you.

Designing Across Organizations

One of the central elements of CE is the Supply Chain Context (Figure 1.5). A trend in many industries is the increased outsourcing of parts and components, and the involvement of those suppliers in the development of new products. Traditionally, suppliers of parts and components were treated much like the manufacturing function was treated — engineering gave them a design and said "Go make it." In CE, manufacturing is involved in the design of the product. By implication, this means suppliers also need to be involved in the design of the product.

An extension of this trend is that suppliers are increasingly being asked to do their own design of the parts and components they make. Product development is moving from being a highly centralized activity to being a distributed one. Design is taking place across organizations, across company boundaries, across functional boundaries, and across status boundaries. We think this trend is so important that we've made it a central theme of this book.

The Benefits of Concurrent Engineering

Concurrent engineering provides many benefits to a manufacturing company. Conceptually, CE should reduce time to market, improve quality, and reduce costs. However, quantifying the benefits of CE has proven to be a difficult task. There are several reasons for this.

- ***Confusing metrics***. Different companies use different metrics for measuring performance, and even a measure that appears the same across companies, such as lead time, may actually be measured differently. For example, some companies start the clock when the concept has been approved by top management, while others start the clock at the start of the conceptual development stage, and when conceptual development really begins is often a mystery.
- ***Project variety***. Different projects vary widely in lead time, manufacturing costs, scope, how good or bad they were before CE was used, etc. So a 20 percent improvement in some metric like productivity may mean very different things for a complete car as compared to a bearing. Equally, it makes no sense to compare product development time for a totally new product with one that is a modification from a previous generation.
- ***Implementation varies***. Not all companies are equally effective in implementing CE. A simple average of benefits achieved combines those who did a bad job with those who did a good job, thus reducing estimates of the effects of CE.

- ***What's in a name?*** Different companies define concurrent engineering differently; so when reports are given of the benefits of CE, we do not know what they have actually adopted under the name of "concurrent engineering."

Having said this, can we say anything at all about the benefits of CE? We believe we can, with caution, draw some conclusions. We can say that there have been many documented case studies that suggest enormous benefits of CE. For example, the well-known Institute for Defense Analysis study on CE (Wenner et al., 1988) reported benefits on a variety of measures, including product development cycle time (40-60 percent reduction), manufacturing costs (30-40 percent reduction), engineering changes (more than 50 percent reduction), and scrap and rework (up to 75 percent reduction). Many case studies have appeared in the popular business press. For example, *IEEE Spectrum* ran a Special Report on CE in the July 1991 issue. They reported the following.

- John Deere & Co. used CE to cut 30 percent off the cost of developing new construction equipment and 60 percent off development time.
- Hewlett-Packard decided to improve the quality of its products by 1000 percent in five years—and succeeded. For example, on one product, the development team for a 54600 oscilloscope developed the product from idea to finished product in one-third the time it took for similar projects in the past and were able to bring the product to market at a significantly reduced cost, competitive with inexpensive Asian-manufactured units.
- AT&T adopted CE and halved the time needed to develop a 5ESS electronic switching system.

These and examples like them suggest that CE is worth taking seriously. Generally the projects involve cross-functional teams who undertake a heroic task, signing on to goals never before achieved by their company, and then they succeed. Sometimes formal CE tools are used, but often the group process seems to be the key success factor. You have probably seen many such examples through internal company presentations or at public seminars. One cannot help but be impressed.

There is a danger, however, of overinterpreting these case studies and assuming that we know what it is about CE that caused the improvements. What were the specific factors that led HP to be successful, for example? Was it teamwork? Was it the outside pressure of formidable competition from Asia? Was it the setting of specific strategic objectives?

One attempt to formally test the benefits of CE was a survey of manufacturing engineers in 74 Fortune 100 firms making a wide range of products (Liker, Collins, and Hull, 1995). Manufacturing engineers were asked to think of a specific development project and answer questions about the process of development and the outcomes for that project compared to a similar previous project. The focus was on the manufacturing issues of tooling lead time and production ramp-up time, as well as the impact of the changes in design approach on manufacturing costs.

This study found that most of the CE practices examined were beneficial, reducing the time it took to tool the factory and ramp-up to full production, as well as reducing manufacturing costs. But the results were rather complex. Some CE practices were beneficial across the board, but most were mainly beneficial for *new versions* of a product, as opposed to minor tweaking of an existing product. For example, CE education and training was beneficial for the full range, from minor modifications of existing designs to fundamentally new product designs. CAD/CAM also led to reduced lead time and manufacturing costs for both major and minor design changes. Practices that were particularly beneficial for new product designs, and less useful for modified designs, included the use of organizational methods (such as cross-functional teams) to integrate design and manufacturing, and establishing strong communication links with suppliers and customers. What was particularly *harmful* to manufacturability outcomes was excessive bureaucracy that provided specialized job descriptions and standard operating procedures in great detail, and created a tall organizational hierarchy, and this was particularly harmful for relatively new designs. The use of design reviews and design standards was equally beneficial for modified and new designs.

As part of our background research in preparing for this book, we conducted a set of case studies (in the Appendix). We particularly sought out companies that were reputed to be at the leading edge of concurrent engineering, and asked them to give us a specific example of a very successful design project. As others before us, we discovered some outstanding examples. All these examples revolved around cross-functional teams that got together and magically seemed to accomplish what others in their company before them seemed unable to accomplish. A summary of the accomplishments of several of these projects follows.

1. In 1992 the Defense Systems and Electronics Group of Texas Instruments won the Malcolm Baldridge Award, in part because of its strengths in using concurrent engineering teams that included both customers and suppliers. The project that most successfully

used concurrent engineering teams involved the development of the GEN-X decoy. They were able to design this new product in 1/5 the time it previously took for a similar device, cost was reduced by a factor of 4 compared to a previous version of the product, and part count was reduced by half. This was all achieved even though 90 percent of the design was changed from the previous version and no off-the-shelf parts were used.

2. Whirlpool, anticipating increased competition, and needing to get break-even points down and remain profitable in the face of increased government regulation of appliances, began to aggressively pursue concurrent engineering in the 1990s. This included the use of design for manufacturability methods, cross-functional teams, early supplier involvement, and a structured development process. A number of projects were successfully performed with this new structure. For example, on the design of a plastic tub for a "next generation washer," the team was able to generate good parts at the first prototype stage, something not done before, and saved a substantial amount of development time as well.
3. Chrysler completely reorganized its engineering organization in 1988 from a functional organization to "platform teams." The platform teams include all the engineering functions necessary to design a complete vehicle, along with a large number of manufacturing representatives. The first vehicles developed using this new team approach were code named the LH line (e.g., Dodge Intrepid and Chrysler Concorde). They were developed in about 3 1/2 years, as compared to about 5 years for previous vehicles, and they met or exceeded all cost and performance targets.
4. Eaton Engine Components supplies most of Ford's engine valves in the U.S. In 1986 they were asked to take significantly more responsibility, this time for the entire valve train, and to design as well as build it. By working hand-in-hand with Ford on this development process, Eaton was able to achieve a cpk[4] of 2 when it came to shipping parts (which they never even thought was possible), reduce the cost of the valve train to Ford through design innovations, and do all this in less time than previous development projects. Moreover, some of the new capabilities and technologies developed led to Eaton getting additional business from other customers.

No one that we talked to on any of these projects has any doubts about the power of concurrent engineering. They know what it did for

[4] cpk is a measure of manufacturing process capability — lower is better.

them on these projects and what it can do. But even in these successful cases there were clear, unexploited opportunities to do more, and some concerns about whether the approaches used would be sustained in the future.

The Problem of Sustaining CE

The companies we visited all had gone far in implementing CE and had success stories to prove it. But consider the following.

1. The accolades the Defense Systems and Electronics Group of Texas Instruments received for its concurrent engineering program were largely based on a single project and a structured methodology for the design process called Integrated Product Development (IPD). Yet we learned that the GEN-X decoy was the first case in which this division had successfully used concurrent engineering, and several years later was still the only case. Other projects were in process, and were beginning to use IPD, but it was not clear if any were as dramatic examples of CE success as the GEN-X. In the meantime, the GEN-X decoy was soon to go into production and the team that developed it was being dissolved and reintegrated into the functional organization. There was fear that once they got back to their "home organization" they would go back to business as usual, and not get to function as a team again in the way they had for GEN-X.
2. Whirlpool took nearly a decade of struggling to improve its design process before moving to cross-functional teams in a serious way. At a recent visit the teams still seemed to be a fragile temporary structure that dissolved once a project ended. It was not clear whether this structure could be sustained in each of the product groups, particularly if a new manager were chosen who was not completely supportive of the approach.
3. Chrysler has had great success in selling the LH line which was a significant part of the revival of the company, which was on the brink of collapse. But this has been despite the manufactured quality of the cars rather than because of it. The line of vehicles has been plagued with "things gone wrong" in the J.D. Power surveys, and were bad enough that *Consumer Reports* dropped these cars from its recommended list.
4. Eaton Engine Components had a very successful experience in the case of the V8 engine developed for the Ford Romeo engine plant. Eaton was a beneficiary of Ford's new approach to managing suppliers on that project. Yet, several years into production of

the Romeo engine there were signs that Ford's engine division was slipping back into old supplier management practices. For example, Eaton had been given responsibility for developing and supplying the entire valve train subsystem and invested much money in developing this *as a system*. After several years of production, Ford began to explore alternative, cheaper manufacturing sources for *individual items* in the subsystem. This meant Eaton would lose some of its ability to make money on the original investment in development. Ford's actions risk moving the cooperative, win-win relationship back to a more traditional relationship in which Eaton will be less willing to make heroic efforts for Ford.

At the heart of the problems in implementing CE in the companies we visited is a basic but powerful lesson: *full and lasting benefits of CE depend on changing the system*, not just an isolated piece of the system. This does not necessarily mean that every company must make a complete, wholesale change of every aspect of its culture, structure, and standard operating procedures to get benefits from CE. It does suggest that simply getting a group together from different departments and assigning them responsibility for a development program will not lead to full and lasting benefits, unless it happens that the organizational context is already well suited to concurrent engineering. Let us consider that for each of our four examples.

1. ***Texas Instruments***. The Defense Systems and Electronics Group of TI made their change in the GEN-X development process in large part out of desperation. In 1987 DSEG had developed an earlier version of GEN-X under a design contract which met the functional requirements, but was too heavy and expensive for practical use. They knew they would lose the contract unless they developed a new version in much less time and at a markedly lower cost to manufacture. This was a highly visible project inside of TI and, thus, the project manager was given a good deal of latitude to run the project using "best practice" methods. The project had in fact run for nine months using a traditional top-down structure and there had been relatively little progress made. Fortunately the project leader decided a new, participative team approach was needed. Even so, the team struggled for some time before arriving at the structure that ultimately made the project a success. Meanwhile, DSEG's strong functional organization continued to operate, and team members, who were assigned only part-time to the GEN-X team, were pulled between devoting themselves to GEN-X and getting trapped in the bureaucracy of their

functional group. GEN-X was like an island in which different rules operated, surrounded by the hostile environment of DSEG's functional organization.

2. ***Whirlpool.*** Whirlpool, like DSEG, never really made any major organization structure change. Fortunately, Whirlpool has a relatively small design organization. The St. Joseph organization is only responsible for the design of laundry products and dishwashers. Only about 200 engineers work there. The functional organization was never very strong, and the head of each product group (e.g., Director of Dryer Engineering) has a direct reporting link to engineers assigned to their group. Thus, once management got serious about cross-functional teams, the organization structure barriers were not as severe as they were for DSEG. On the other hand, it took many years of small steps to reach the point of serious use of cross-functional teams. And even after several years the team structure was not formalized but seemed to be a hit and miss process that depended heavily on the personalities of the managers.
3. ***Chrysler***. Chrysler was the most ambitious of our cases in the degree to which it changed its entire system. Right now it has gone farther than any other car company in adopting the platform team concept as the primary organizational structure for vehicle development. Many manufacturing representatives are assigned full time to the team. So why did the quality problems in the LH and later vehicles surface? We cannot be certain, but we believe that the primary problem was that the manufacturing organization had not gone through the same degree of change as the design organization. The manufacturing organization did not have the quality systems in place to support the kind of quality output we see at companies like Toyota (e.g., true just-in-time, standardized work, systems designed to automatically stop when a quality problem occurs). It is possible that without these systems manufacturing is less able to make the best suggestions on how to improve the design. In short, only the design part of the system had changed a great deal, and what may have been a good design was handed off to an outdated manufacturing organization. When this became clear to Chrysler they began to aggressively work on various aspects of lean manufacturing in the mid-1990's under the name of the Chrysler Operating System. Hopefully this will bring their manufacturing operation up to the same level of excellence as their design organization.

4. ***Eaton.*** Eaton Engine Components had a successful experience with the Ford V8 engine developed for the Romeo engine plant, both in design and manufacture. This was a unique and highly visible program for Ford, a make or break program for the engine division. They needed to meet aggressive targets or future engine business would be contracted out to Japanese competitors. There was an effort to do everything best-in-class, from concurrent engineering to the use of self-directed teams and world class quality methods on the shopfloor. The slippage backward at Ford suggested this change was fragile and isolated in the specialty organization put together to develop and make the Romeo engine. As in DSEG, the part of engine division devoted to this one program had advanced much farther than the rest of the organization.

These cases suggest two things. First, one should be suspicious of taking "success stories" at face value. Each of these cases has been well publicized as successes, for good reason, but when one actually talks to the participants and digs deeper it is clear that each of these companies had only implemented CE to a limited degree. Second, we believe that implementation can be smoother, faster, and have greater staying power if the change is carefully planned, considering the work process and broader organizational context right from the start. Perhaps these early pioneers of CE needed to go through their struggles to achieve what they did. And indeed, all companies will go through their own struggles. But we believe companies can learn from the successes, and failures, of their own early efforts and those of other companies if a careful and *intentional* planning process is used.

CE and Other Manufacturing Initiatives

Industry is inundated these days with new initiatives clamoring for attention. Just consider the following list:

- Concurrent Engineering
- Just-In-Time (JIT)
- Total Quality Management (TQM)
- Lean Manufacturing
- Agile Manufacturing
- Business Process Reengineering
- Continuous Improvement
- Kaizen
- Downsizing

There are several ways to view these kinds of initiatives.

- *Program of the month*. An executive learns about a new initiative and gets very excited about it. People are trained and start to try to implement it, but before they get a chance to gain the benefits, the executive is already off on the next initiative and the cycle starts all over. The main product of this kind of situation is high income for consultants combined with low morale and high cynicism for people in the company.
- *Fundamentalist religion*. The executive treats the initiative like a new religion, and any book written by the initiative's spokesperson like a set of fundamental truths which must be adhered to without question. The main reason for taking this approach seems to be a desire to achieve "*the answer*" without a lot of thought or work. The result is an inflexible application of the initiative without consideration for the unique conditions at the company — usually leading to no gain, since the initiative is applied with no real understanding of its intent, limitations, or necessary modification.
- *Use what works for you and move on*. The executive takes the time to really understand the principles behind the initiative and then applies those principles to his/her operation. If something doesn't work after a reasonable trial, that part gets dropped. If something does work, then it becomes routine practice. When the company has gotten as much benefit as it can from one initiative, it stops treating it as a separate initiative and moves on to the next set of exciting new ideas. The result of this is that the company continues to learn and grow.

Obviously, we think it better to take a very flexible approach to any of these major initiatives. For one thing, there is tremendous overlap among them. For example, Lean Manufacturing probably implies the use of CE, TQM, continuous improvement, and JIT. This suggests a strong need to integrate initiatives such as this when more than one is being attempted. We will briefly explore the connections between CE and some of the initiatives listed above.

CE and Business Process Reengineering. Business Process Reengineering (BPR) is "the fundamental rethinking and radical redesign of business process to achieve dramatic improvements in critical, contemporary measures of performance, such as cost, quality, service, and speed" (Hammer and Champy, 1993). The emphasis in BPR is on the *business process*, as opposed to functions. A business process is a thread running through your business which provides value to your customer. Product development is usually considered to be a business process; order fulfillment is another example. If you reengineer your product development business process, the end that you would have in mind would typically be concurrent engineering. Thus, BPR can be the "means" to the "end" of CE. Earlier, when we said you needed *a process for getting to CE*, that process could be BPR. And, indeed, the whole second half of this book will describe a process for doing something much like that. It's important to make the distinction between the typical practice of BPR and our approach. Most BPR efforts fall into two categories. The first is a primarily technical approach which changes the business process and supporting information technology, but which leaves the rest of the context alone. This often has some benefits, but may fail to meet its objectives because the rest of the infrastructure has been left behind. The second category is "downsizing" in the guise of BPR. Substantial numbers of people are cut from the organization and the business process is left to itself to reorganize. This approach has devastating effects on morale (and in many cases on performance) and is a primary reason for BPR getting a "bad name" over the past few years.

CE and Lean Manufacturing. Lean Manufacturing is an approach to manufacturing in which you attempt to minimize everything, from work in process, to part count, to labor hours. "Less of everything" not only inherently reduces cost, it increases pressure to get things done more quickly as well as correctly (no slack means you can't waste anything). Invented by Toyota and popularized in the U.S. by *The Machine that Changed the World*, Lean Manufacturing is currently the dominant paradigm in manufacturing — guiding many of the improvements now taking place around the world. Unfortunately, Lean Manufacturing has tended to focus on the shop floor, as opposed to product development. It's just as critical to make the product development process lean as it is to make the manufacturing system lean (Womack and Jones, 1996). Even Toyota, the oft-cited paragon of a manufacturing company, has been much more effective at systematizing their manufacturing systems (witness the famed Toyota Production System) than their product development processes. We believe CE embodies Lean Manufacturing by its emphasis on reducing time and waste in the entire system of product development. In this book, we propose ways to systematize product development more completely so that it can become a more complete part of a Lean Manufacturing operation.

CE and Agile Manufacturing. Agile Manufacturing is an approach to manufacturing in which groups of companies (enterprises) flexibly join together as needed to bring out new products on demand in any lot size (Iacocca Institute, 1991). It's the result of flexible technology which allows for economic production in lot sizes of one, innovative management structures which permit instant reconfiguration of the enterprise, and a skilled base of knowledgeable workers. Today, Agile Manufacturing is more of a vision than a reality; tomorrow, we can expect enterprises to become more agile. However, whether they ever fulfill the complete vision remains a question. Regardless, CE is a critical element of agility. This is partly because adoption of Lean Manufacturing is almost certainly a prerequisite for agility, and CE is clearly a part of Lean. Moreover, CE between customers and suppliers, i.e., including the entire supply chain, is central to the process of flexibly developing new products across company lines. As practiced, CE is primarily a within-company exercise. However, as agility becomes increasingly important, we believe it will expand to include large parts of the supply chain in a full "lean enterprise" (Womack and Jones, 1996).

Overview of the Book

In our reading we have found that there are many books and lots of training available which describe the work and tools of CE. We have referred to some of these materials earlier in this chapter. In any case, as social scientists, we have little to teach people who develop products for a living about the mechanics of doing their jobs. What we do know a great deal about is the *context* in which CE resides, and the *process* for making CE happen. The rest of this book will be devoted to explaining those two topics.

In Part I, we work through in detail a model for the context in which CE takes place. Chapter 2 provides an overview of the whole model. Chapter 3 focuses on the Work Processes of CE; Chapter 4 describes the internal organizational context for CE; Chapter 5 discusses people issues; Chapter 6 describes the role of the supply chain along with mechanisms for linking with these external organizations; and Chapter 7 discusses the technological context for CE. For each of those topics we will discuss the *choices* which need to be made in order to provide a suitable context for CE, as well as best practices and lessons learned. The final chapter in Part I provides an opportunity for you to check yourself against the model we have presented here using a tool we call the *CE Profile*. This detailed discussion provides conceptual frameworks and models that are the foundation for the analysis and design methodology we present in Part II.

In Part II, we provide directions for working through a process of changing from your current product development process to concurrent engineering. Chapter 9 provides an overview of the structured process and describes the benefits of using it. Chapters 10 through 13 describe the four stages in the change process: Scope, Assessment, Design, and Implementation. In each chapter we provide instructions and tools for working through the process. These tools use the categories and measures discussed in Part I. The Appendix includes a set of descriptive case studies which demonstrate the best practices in CE.

CHAPTER 2

•

A Context Model for Concurrent Engineering

"Everyone in engineering design, purchase of materials, testing materials, and testing performance of a product has a customer, namely the manager (e.g., a plant manager) that must make, with the material purchased, the thing that was designed. Why not get acquainted with the customer? Why not spend time in the factory, see the problems, and hear about them?" • Dr. W. Edwards Deming

This book provides a practical approach to help companies develop an *appropriate* concurrent engineering process and supporting organization. The key word here is "appropriate." In Chapter 1 we stated our belief that there is no single best way to do CE. Each company (or division of a company) has a unique set of circumstances and resources which keep a single prescription from working the same way for everyone. In contrast, we believe that only through a rigorous process of analysis and design can each company develop its own best path to CE, or, indeed, its path to anything else worth doing well. Our purpose in this book is to provide the background and the tools to make that rigorous process much easier and faster.

That process of analysis and design begins with understanding, and the easiest way to help someone understand something is to give them a model. A model is a simplification of reality that makes that reality easier to understand. It does this by decomposing some large, complex "thing" into smaller, more easily understood categories. This aids understanding by simplification, since the model focuses only on the most important categories and leaves out many others which might confuse the main issues. The model can then help guide you as you work to assess where you are today and develop a vision of where you want to go. It is important that the model is general enough to give you the flexibility to tailor the design to your unique circumstances, but specific enough that it provides some useful guidance.

Our model of CE is consistent with our design philosophy in that it is based on a contingency view of organizations and organization design (Daft, 1995; Aldrich, 1979; Thompson, 1967). By contingency we simply mean the right organization design depends on your circumstances. The circumstances you must design for include characteristics of the product you are designing, the "design task," as well as characteristics of the broader competitive, business, and cultural environment faced externally by your firm. Therefore, we must provide tools to help characterize design tasks, as well as a model to characterize the external environment and some guidance on how the design task and environment influence the kind of CE process and organization best for you. Some readers may recognize our model as an "open-systems" view of organizations (Checkland, 1981; Senge, 1990). An open-systems view recognizes that any organization is a "system" which has parts that influence each other, and the organization influences and is influenced by the broader environmental context.

We call our model a "contextual model." We begin with the design task and then examine the broader "context" in which CE work takes place. This is in contrast to most models of CE which focus primarily on technical approaches to doing the work of CE, such as tools for Design of Experiments.[1] Instead, we have assumed you can learn how to do that from the many books and courses that are already available on these CE tools. We have focused on the organization and human concerns which provide the context for that work.

As a simplification of reality, our model will focus on only a few elements of the whole picture, those which evidence shows are the most important for making the critical decisions about the context for CE. In this chapter, after presenting our entire model, we will provide a brief discussion of the "contingencies" that should be considered when choosing an appropriate CE process and organization, namely the characteristics of your environment and tasks.

Why Another Model?

Our shelves are filled with books about CE, each of which has its own model. Why create yet another one? In looking at the existing models

[1] See, for example, G. Taguchi (1987). Another example which attempts to codify design as a set of mathematical axioms is N.P. Suh (1990).

of CE, we find that most are not really models, but rather lists of processes or tools or activities which seem to be associated with effective CE. For example, Syan and Menon (1994) list eight "key features that can be identified as essential elements for successful CE implementation:"

1. multidisciplinary teams
2. cross-discipline communication and coordination
3. quality management methods
4. computer simulations of products and processes
5. integration of databases, application tools and user interfaces
6. education programs for employees at all levels
7. attitude from employees of ownership toward processes
8. commitment to continual improvement.

Himmelfarb (1992) lists 15 elements:

1. formal planning
2. front-end project planning
3. multifunctional teams
4. empowering the teams
5. supporting the teams with adequate equipment
6. senior management support
7. freezing features and specifications as early as possible
8. reuse of components and processes
9. eliminate top down decisions
10. continuity of team leadership
11. minimize bureaucracy
12. allocate time for the project
13. assure production piloting
14. collocate team members
15. seed the next projects.

There is nothing wrong with these lists; they are all "true" in some sense. But they are unorganized and include items of varying specificity. For example, "formal planning" is at a much more general level than "reuse of components and processes." Given such a mixed bag of items we could easily brainstorm another 20 or 30 items that would belong equally well on the list. But certainly a long, exhaustive list would not help us design a CE process. In contrast, we want a simple, organized model which describes the context in which CE takes place and which will serve as a guide to action: "What must change if I want my company to do CE effectively?" We also do not want to assume there is a single list of prescriptions that are right for all circumstances. We want to identify the circumstances under which different CE features are more or less important. Our model starts with the work process of CE — the design task. It then builds out from there to the context of that work, first inside the company and then outside.

In later chapters we will discuss in great detail the work process and the other elements of the model which make up the context of CE. We will then provide methods for analyzing each of these in your firm. Our intent in this chapter is to introduce the model and its elements and define key terms.

The Work Process of CE

The "Work Process of CE" is really the process of product development. The term *product development* includes the development of the product and systems which surround it over the entire product life cycle, from initial conception to systems for disposal. The Work Process can be broken down into four elements: activities, the order in which activities take place, the flow of information among those activities, and the ways in which activities are controlled. The reader familiar with process flow modeling methods, such as IDEF0, will immediately recognize these elements. We define them briefly as follows.

- ***Activities*** – the core work that adds value for the customer or which supports that core work.
- ***Flow of information and objects between activities*** – the information and things that move between activities as they are performed.
- ***The order and timing of activities*** – these are important because some activities can occur in parallel while others can only begin when another activity is complete. By modeling and planning the order and timing of activities, we can ensure they take place when desired.
- ***Control mechanisms*** – these are used to ensure that activities remain aligned and that the project remains on target.

The Context for CE

Having defined the work process, we need to consider the broader context that supports (or fails to support) this work. One can think of the layers of an onion where the inside of the onion is the work process and we then work our way out to the outer layers. The context in which the work process of CE takes place can be divided into three layers or "contexts" (see Figure 2.1). We will briefly describe each context and then delve into greater detail about each.

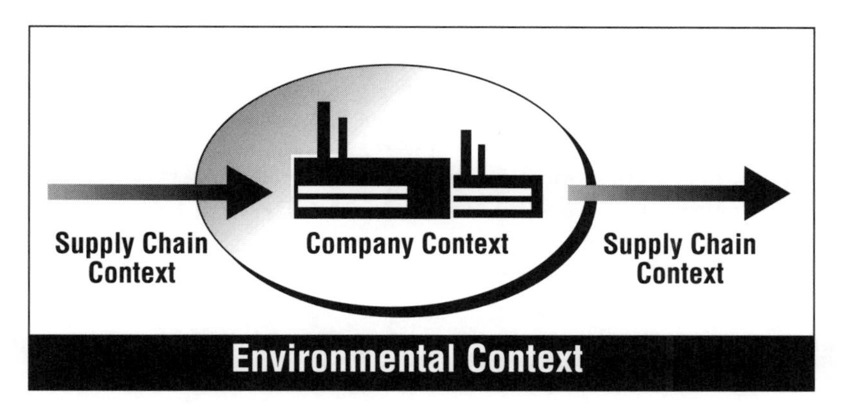

Figure 2.1 • Three Contexts for CE

1. *Internal Company Context.* The most immediate context for CE is the company in which product development is taking place (we'll sometimes refer to this as "your company"). This includes the company's strategy, the structure of the organization, the people in product development and support, and the technology they use. Defining what is inside and what is outside is always somewhat arbitrary. In some cases it would make sense to consider the legal definition of a corporation, while in other cases the corporation may be so large and diverse that it will be more useful to consider a particular division or business unit as the "Company."
2. *Supply Chain Context.* Every manufacturing company has customers and suppliers (and every company is a customer and is a supplier). Customers and suppliers have critical roles to play in the development of new products. Moreover, both suppliers and most customers[2] have their own internal company contexts which must work in concert with yours for an effective concurrent engineering process.
3. *Environmental Context.* The work of both the company which is developing a product and its customers and suppliers takes place in an even larger environment of people and organizations which includes the larger marketplace for the product, developers of new technology for both product and manufacturing process, an economy which may be growing or shrinking, government regulators, and many others.

We will discuss each of these three components in turn. Again, our purpose here is to introduce terms and provide a big picture view of the

[2] The only exception is customers who are individual end users. However, the overwhelming majority of customers for manufacturing companies are in fact other companies.

model. In later chapters we will use this model as a framework to guide the analysis and design of the CE process and organization.

The Internal Company Context

Looking inside the company, there are four major parts or "contexts" which influence, and are influenced by, the CE Work Process (Figure 2.2). The links between CE Work Process and the other elements are shown with double-headed arrows to illustrate these relationships. As a simple example, if the work process requires that you produce 3-D solid models of your parts, then the technology required will include a suitable CAD system. The influence may go the other way; for example, the fact that relatively inexpensive software is now available to produce finite element analyses (FEA) may affect the work process, since it may now be possible to perform more of these analyses, while at one time it may have been too costly to do so. In subsequent chapters we will discuss these connections in greater detail.

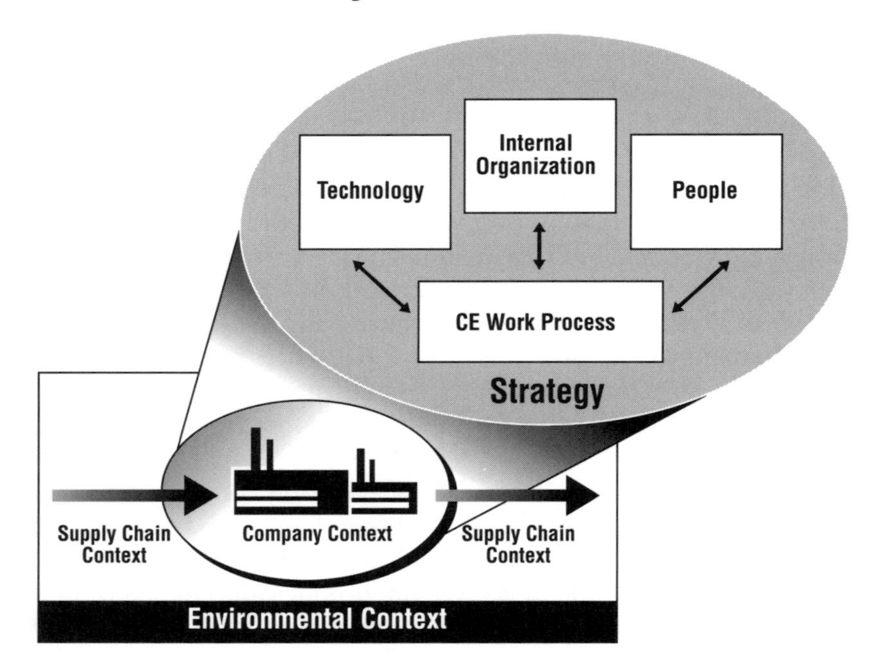

Figure 2.2 • Company Context for CE

The company strategy is shown outside of the work process, technology, internal organization, and people systems as it is a broader part of the context for all of these. Strategy is a set of management decisions and a planning framework that helps to tie together the other aspects of the company with the external environment. We define each of the four parts of the company context as follows:

1. ***People.*** This includes skills, selection, job design, reward systems, motivation. People are central to the context because they do the work, and their support of CE is dependent on how their work is organized, their skills, and their motivation to do it. Concurrent engineering may require new technical skills (e.g., designing using CAD solid modeling systems) and new social skills (e.g., working in cross-functional teams). The skills and motivation of the existing workforce will influence the ability to adopt CE practices and the way CE is defined may influence future hiring.
2. ***Internal Organization.*** This includes reporting relationships, project management structure, job descriptions, communication, coordination mechanisms, and performance measurement systems. The internal organization is important because it is the social infrastructure for getting work done in complex systems. In a traditional organization the focus is on staffing individual functions, and on the parts of the system. In CE the focus is on monitoring, communication, and coordination mechanisms to ensure that the various parts of a design project actually work together.
3. ***Technology.*** This includes communication technology, design technology, and program management technology. Technology is important because it provides the physical infrastructure for getting work done. For example, CE is a much easier process if data can be quickly and easily shared between different functions in the organization and with suppliers.
4. ***Strategy.*** As defined by Michael Porter (1980), a competitive strategy is ". . . a broad formula for how a business is going to compete, what its goals should be, and what policies will be needed to carry out those goals." Figure 2.2 shows strategy outside of, but supporting, work process, technology, organization, and people, since a well-implemented strategy will in some ways drive all of those company elements. For example, a strategy to be a technology leader will require a different approach to CE than a strategy to be a fast follower of technology proven in the marketplace by others. A strategy of building incrementally on a stable product line, such as many mature companies follow (e.g., for home appliances), will require a different CE approach than a strategy of creating new markets with breakthrough technologies (e.g., consider the 3M Company).

THE SUPPLY CHAIN CONTEXT

The Supply Chain Context for CE includes the customers and suppliers outside your company. Customers need to provide input to your product development activities through market research, formal requirements, and feedback about existing products. Suppliers also need to be involved as members of the product development team. If we are really interested in giving manufacturing a voice in product development, and suppliers are manufacturing our parts, then they need to be involved in product development. Many companies are adopting the Japanese practice of turning over responsibility for the design of manufactured components to outside suppliers.

Interestingly, since your suppliers are also manufacturing companies, they too have the same context elements as you do[3] (Figure 2.3). The ability to form links between members of a supply chain is at least partly a function of the "match" between your suppliers' context elements and your own. For example, if you and your supplier have different CAD systems, then you may have difficulty sending CAD files back and forth as a means of communication. As a less obvious example, consider a situation in which you have highly paid design engineers communicating with a supplier's relatively low paid (and low status) manufacturing engineers — there may be assumptions made about the manufacturing engineers' ability to criticize decisions made by the design engineer. There are also many aspects of traditional contracting (e.g., going out after the design is complete for competitive bids based on lowest price) that make it very difficult to work cooperatively with suppliers in the early phases of design. Just as the company context elements must sup-

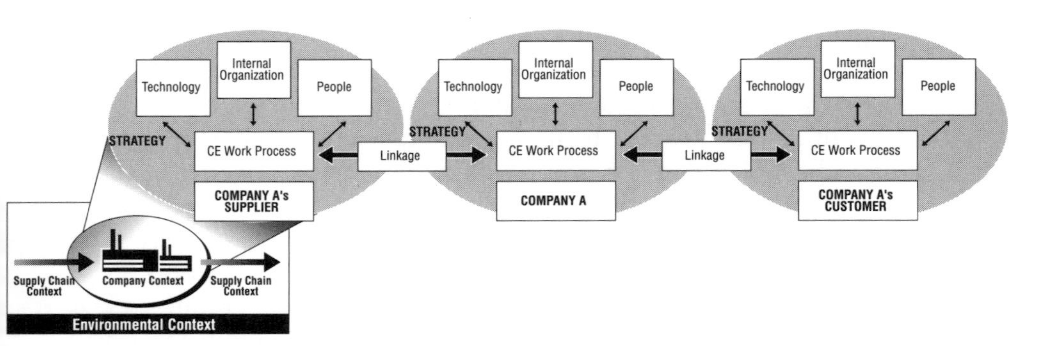

Figure 2.3 • Supply Chain Context for CE[4]

[3] If your customer is also a manufacturing company, that is, if your company is an intermediate supplier, this applies to them as well.

[4] It should go without saying, but we'll say it anyway — most manufacturing companies have multiple suppliers and customers. We show one of each so the picture will be legible.

port doing CE, the supply chain context elements must support it as well.

THE ENVIRONMENTAL CONTEXT

The Environmental Context in which CE takes place includes "everything" outside of the supply chain context which has an impact on that chain (Figure 2.4). Note that we have defined it as the environment of the whole chain. Obviously each company in the chain has an environment which is a subset of the chain's environment. Thus, the specific environment for each member of the supply chain will be somewhat different. Nonetheless, as a system, we need to be concerned with the whole supply chain's environment. We can divide this Environmental Context into eight parts.[5]

1. ***Industry.*** This includes the competitors and the level of competitiveness in the industry. As an example, imagine a supply chain in which Company A makes automotive seats. Company A's customer is an automotive manufacturer, and its supplier provides fabric for the seats. The "industry" involved here is the automotive seat industry, since that is the focus of the supply chain. However, the larger automotive OEM industry and the fabric industry are also relevant, since they have a considerable impact on this supply chain.
2. ***Human Resources.*** This includes the labor market, presence and influence of unions, skills available in the work force, and the availability of education and training in the community. Each company's human resource environment will be somewhat different (especially if they are located in different places). In fact people in different parts of the same company (e.g., different departments, roles, and responsibilities) can experience very different human resource environments.
3. ***Financial Resources.*** This is primarily concerned with the availability of financing, including characteristics of the stock market, banks, and private investors. While these will appear to be relatively constant (at least within a single country or region), the financial resource environment reacts differently to different companies. Thus, one member of the chain may find it easy to obtain financial resources, while another may have great difficulty.
4. ***Technology.*** This includes vendors of production equipment and design tools, and developers of new technology and materials. We distinguish between vendors of production equipment/design tools

[5] This is based on a model proposed by Daft (1995). We have modified it slightly, by pulling suppliers and customers into the Supply Chain Context.

and members of the product supply chain, since vendors of equipment and design tools typically engage in a relatively small number of transactions with their customer, while part and component suppliers act as ongoing members of the extended enterprise for making a product over its life cycle.

5. ***Economic Conditions.*** Here we include the unemployment rate, rate of growth in the economy, inflation rate, and interest rates. This interacts with other elements of the environment, including financial resources and human resources.
6. ***Government.*** This includes laws and regulations, taxes, and political concerns. A major concern for most product development efforts is compliance with various regulations, including health, safety, and energy efficiency laws.
7. ***Socio-cultural.*** This includes the values, beliefs, education, and work ethic in both the community and the work force. This interacts strongly with the human resource component of the environment, as well as with the industry component, since there is considerable commonality within an industry in terms of the socio-cultural elements.
8. ***International.*** This includes participation in international markets, exchange rates, foreign customs, and international competition. This is primarily a concern for product development if products are being developed for export markets. Many companies which never before considered international issues are now doing so as the market for many products becomes more global and less national in base.

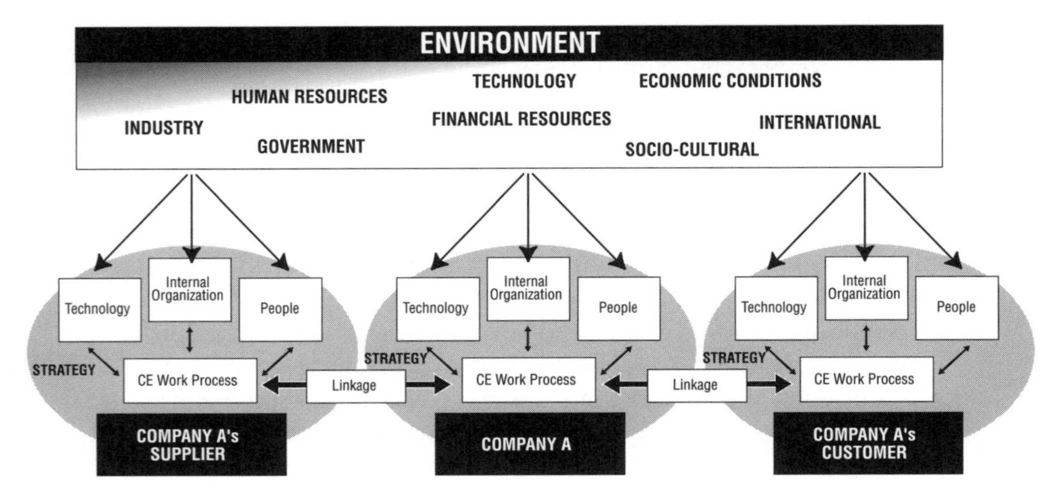

Figure 2.4 • Environmental Context for CE

As Figure 2.4 shows, characteristics of the environment have an impact on choices which must be made in technology, people, internal organization, and strategy. Figure 2.4 shows the arrows from environment going through the strategy "bubble" because the strategy is one way the company communicates its desired response to the environment.

Some of the effects of the environment on the company context are obvious. For example, new production technology developments affect what technologies the company will use to make its products. Equally, characteristics of the available work force (in the environment) affect skills and training needs of the workers in the company. Other impacts are less obvious. For example, unstable economic conditions might lead a company to find ways to be more flexible in its product development capability — able to develop a range of high-end products when economic conditions are prosperous, and low-end products in an economic downturn. The ways in which it becomes more flexible might include something like changes in work processes to shorten lead time by increasing concurrency among functions.

These less obvious effects of the environment are of considerable interest here because many of them have great impact on both the product development process and its company and supply chain contexts. A great deal of research has shown that one of the most critical of these issues is *environmental uncertainty* which cuts across specific changes in environmental sectors.

CHARACTERIZING YOUR ENVIRONMENT

Environmental uncertainty is the extent to which the environment around your company is unpredictable. Environments can vary a great deal in the uncertainty they produce. An obvious example is the way the market for automobiles changed between 1967 and 1995. In 1967 the U.S. auto market contained four domestic manufacturers (one of which, American Motors, was about to become extinct) and a small number of cars coming in from European and Japanese manufacturers such as VW and Toyota. The market operated as a classic oligopoly, with near total control by the largest competitor, GM. The market was relatively predictable, with overall sales following the ebb and tide of the overall economy, and a relatively stable market share held by the top three automakers (GM, Ford, and Chrysler). By 1995 the market had become much more unstable and unpredictable. Overall sales continued to track with the overall economy, but market share was much more volatile, thanks to the introduction of numerous new competitors, mostly from Asia. The number of models had exploded, from 148 car and light truck models on the market in 1967, to 261 in 1996.[6] This is not only due to the increased preference for light trucks; the number of domestic passenger car models doubled from 40 to 80 in that time period. In addition, the speed of development accelerated, from five to seven years, then to about three years or less today.

As another example, consider the difference between the automotive industry and the computer industry today. A typical product life (i.e., the length of time for which a particular model can be sold) for an automobile is 4-6 years. In contrast, the life for a specific personal computer model may be as short as a few months, after which it may be replaced by another model which is both more powerful and less costly! Designers in the computer industry can be much less certain about both customer needs and technological changes than can designers in the automotive industry, hence we would say that their environmental uncertainty is much higher.

So far, we have treated environmental uncertainty as a single concept. But, in fact, it is composed of several elements, all of which may combine to create an overall level of uncertainty. We can divide environmental uncertainty into three components.

1. ***Predictability of Change.*** How easy is it to know what changes are coming up? In some sense, change in the computer industry is relatively predictable, since it is easy to predict that the cost of a given amount of processor power will halve about every 18 months. On the other hand, the implications of that increase in power are much harder to forecast. For example, few people in the early 1970s would have predicted that most office workers would have an enormously powerful computer on their desk and that vast numbers of these workers would be using their computer for the (then) equivalent of typing and making diagrams and drawings. Equally, there are many predictions about the future of the Internet, but the only thing certain about any of them is that most will be wrong. Environments change in your ability to predict them. The environment for automobile sales was highly predictable in the 1950s and 1960s; the 1973 and 1978 oil crises, combined with the growth in Japan's auto industry, changed the equation significantly.
2. ***Rate of Change.*** How fast is change taking place? Faster change, even if you can predict it (as in the computer industry), still makes the overall environment less certain. Imagine a computer company which introduces new models every few months with both higher processing power and more features being offered for the same or lower price. The life of a designer in such a company must be ter-

[6] Based on an analysis of tables from *Automotive News Almanac* in 1968 and 1996.

ribly nerve-wracking — will we succeed at building this new model and will it succeed in the marketplace? How will the market turn this month? What new twist will a competitor introduce that drives the market in a new direction? What new twist can we introduce that will do the same next time around? The pace at which these changes are introduced makes it much more difficult to know what to do and when to do it.

3. ***Environmental Complexity.*** How many different aspects of the environment are changing that will impact product development in some way? A more complex environment (i.e., one that has more parts to be considered) results in more uncertainty about how to approach it. For example, an airplane design team must deal with a very complex environment, including airline customers throughout the world, end-user customers (the flying public) throughout the world, government regulators in over 100 different countries, environmental groups, suppliers for a variety of components (such as engines and wings), the oil industry (which must make and supply fuel), and many others. In contrast, a designer of electric toasters deals with a much less complex environment since the toaster is probably going to be sold in only one country, by retailers who will never open the box except to set up a display, and with relatively little need to worry about government regulations or environmental concerns (at least as compared to an airplane).

Concurrent engineering is one response to high levels of environmental uncertainty. CE reduces the time it takes to develop and produce new products, speeding the firm's response to environmental changes. CE also reduces the cost of new product development and therefore improves the firm's ability to respond to changes in the environment. CE gets the right people together to consider the implications of key environmental influences on product development. Finally, CE helps to ensure that new product quality is as high or higher than previous products, making it more likely that customer requirements will be met immediately, rather than at some later time when those requirements may already have changed. Through its emphasis on more structured processes, CE also helps to make sure that the voice of the customer is heard, even if there is not a specific emphasis on additional market research.

What are the implications of high environmental uncertainty? There are three primary implications. First, the organization may expend much greater energy seeking, sorting, and using information about the environment. If it makes sense that the organization wishes to reduce uncertainty as much as possible, then the primary way to do this is to work harder at getting the means to improve predictability. A common example of this is to do more market research so you have a better idea of what customers want. Second, the organization may seek to improve its control over the environment, through advertising, or lobbying government officials. Finally, the organization may seek to increase its internal flexibility, making it easier to respond when the environment does something unexpected. CE is an approach to increasing flexibility. In Chapter 4 we will discuss what it means to increase internal flexibility — specifically what it means for organization design.

Most importantly from our point of view, the nature of the environment has a profound effect on how your concurrent engineering process should be structured, who should be involved when, how you should design your organization, etc. For example, if the environment is highly uncertain, in order to respond quickly to a diverse set of demands you will probably need a higher level of concurrency and more two-way communication between different parts of your organization.

In later chapters we will provide tools to help you measure environmental uncertainty and methods to respond to different levels of uncertainty.

Characterizing Your Task Requirements

There are two main drivers that will affect the appropriate way to design your internal company context and your linkages with your customers and suppliers. We have already discussed one—the external environment. This comes from the outside as a set of resources, pressures, and demands that will shape the operation of your concurrent engineering process. The second driver comes from your work processes and pushes out to the technology, internal organization, people, and strategy.

There are many ways that your work process will affect the broader context. For example, the flow of tasks will influence who needs to be involved, and when in the process. The level of concurrence will influence the degree of two-way communication and joint problem solving needed. In addition to these specific effects, there are two concepts analogous to environmental uncertainty. These are task uncertainty and task interdependence.

Task Uncertainty

Task uncertainty is the extent to which the way to perform your task is unpredictable. "Task" in this case is defined as the work performed by someone in an individual job. Just as environments can vary in the degree of uncertainty they produce, so do tasks. Compare the task of a product development group in a technology intensive company (e.g., developing a new spacecraft) with that of a maintenance group in a traditional, "low-tech" factory. Clearly the product development group is much less clear about how to perform its task under most circumstances.

As with environmental uncertainty, task uncertainty has multiple components.

1. ***Analyzability.*** This is how easy it is to analyze or program the task. A task that is easily analyzable can be broken down into a series of standard activities or steps. A classic example of analyzable work is the tasks on an assembly line which have been thoroughly analyzed and programmed; some are so totally programmed that they can be performed by a robot. In contrast, most "creative" work is less analyzable. For example, while we can list a series of tasks to be performed in product development, many of them are very difficult to describe in detail, much less actually program them. In most such cases, any attempt to describe the task in detail ultimately ends in a "black box" which cannot be fully explained. The more analyzable a task is, the more certain one can be about how to go about doing it and how long it will take, and hence the lower its uncertainty.
2. ***Variety.*** This is defined as how often new and unexpected events happen in the course of doing the task. An example of an unexpected event might be a problem which has to be resolved. Even if the solution to the problem can be looked up in a manual, if it is not part of the routine of the task, then it is considered unexpected, and contributes to variety.

What are the implications of task uncertainty? There are two. First, the more uncertain the task is, the greater the effort required by the organization to find ways to get people together to solve problems. This suggests a greater reliance on teams and other mechanisms designed to get people from different perspectives to communicate. Second, the more uncertain the task, the more effort required to find and collect information and to make it accessible when needed. In Chapter 4 we will discuss the implications of this on organization design and specifically on the design of cross-functional teams. In Chapter 11 we will provide tools and methods to help you assess task uncertainty in a way that will help you with the team design issue.

Task Interdependence

Task interdependence is the extent to which a performer of a task is dependent on another individual or department for information or material in order to perform his/her task. The most obvious example of interdependence can be seen in the classic assembly line, in which any given work station depends on receiving a work piece from the previous station upstream, as well as on receiving parts and material from stores, and information from manufacturing engineering about what to do with the material received. Another example would be a product designer who depends on having information about manufacturing capability from the factory and information from testing about material life before he/she can determine tolerances.

While the examples in the previous paragraph all demonstrate interdependence, they are not all the same. In fact, they describe different types of interdependence. There are three types of task interdependence.

1. ***Pooled interdependence.*** This means that units don't actually interact in performance of the task, but rather that they are dependent on the same "pooled" sources of information or material, to which they all contribute and from which they all draw. The most basic example of this is when all organizational units contribute to corporate profits — all units benefit and depend on each other to contribute to profits. Another example might be when design groups of three independent product lines depend on a single CAD support group. These design groups may not have to share product technologies, but do have to share the pooled CAD support.
2. ***Sequential interdependence.*** This means one task depends on another to be completed before it can begin. The "downstream" operation requires information or material from the "upstream" operation. Again, the classic assembly line comes to mind as an example. A less obvious example is in the traditional, "over the wall" product development process. Traditionally, engineering is sequentially interdependent with marketing for customer information; manufacturing is sequentially interdependent with engineering for the product design.
3. ***Reciprocal interdependence.*** This means that two (or more) tasks must mutually inform each other. One way to think of this is that the first task requires inputs from the second, and the second also requires inputs from the first. An example might be a concurrent engineering process in which design engineering and manufacturing engineering work together to help each other develop their designs for both the product and the manufacturing process.

The three types of task interdependence can be thought of as being in order of increasing degree of interdependence. As one moves up toward higher levels of interdependence, more information or material must pass between tasks as they take place and, as a consequence, more effort will be required during performance for interdependence to be successful. Sequential interdependence is more intensive than pooled, while reciprocal is more intensive than sequential. While most organizational units clearly have some level of interdependence with others for some or even most of their tasks, for the most part that interdependence is pooled and therefore of relatively little consequence. On the other hand, many tasks are, in principle, reciprocally interdependent, but the organizational units don't act that way. For example, if you have reciprocally interdependent tasks performed by organizational units which are located far apart and have no means to communicate other than mail and telephone, then they will find it much more difficult to succeed at their tasks.

What are the implications of task interdependence? In general, higher levels of task interdependence require greater intensity and quantity of communication between the performers of those tasks. This often results in placing such performers together in a single structure (such as a department or a team), physical collocation, or the creation of intensive communication and coordination mechanisms designed to facilitate sharing of information. In Chapter 4 we will discuss the implications of task interdependence for organization design and in particular on the design of cross-functional teams. In Chapter 6, we will discuss the implications of task interdependence on job design.

CE AND CORPORATE STRATEGY

In many ways, a company's first response to its environment is its competitive strategy. Porter (1980, p. xvi) and others have defined three fundamental types of competitive strategies for businesses.

1. *Low Cost* – ensuring that the company's products will always cost less than those of competitors. This requires using approaches which will increase production efficiency and reduce overhead costs.
2. *Differentiation* – ensuring that the company's products have unique characteristics which differentiate them from those of competitors in ways desirable to the customer. This might require advanced technology, creative design, or an unusual devotion to some aspect of the product, such as delivering certain levels of service that no competitor is providing.

3. *Focus* – addressing a particular market segment based on cost or differentiation.

CE can be used for any of these fundamental types, although it may not be appropriate for some specific cases. For example, CE can be very useful in a low cost strategy since it helps to ensure that the efficiency of production has been considered during design. This makes it much more likely that the most efficient production methods will be used. CE also tends to reduce engineering changes late in the product development process, which can result in a considerable cost savings.

CE can also be useful in a differentiation strategy, if the specific strategy involves rapid development of new products to keep ahead of the competition. This is because CE has such a strong impact on time to develop new products.

On the other hand, a strategy of focus on high levels of personalized customer service would probably not benefit very much from the use of CE, since product development (even the development of new services) is not much of an issue. Equally, a strategy based on focus in some geographic region would not benefit from CE.

Our point here is that strategy matters a lot, but its impact on our model is mostly only in two areas. First, it serves as a "gate." This means that if your strategy requires improvements in time to market, cost or (to a lesser extent) quality, then you should seriously consider using CE. Once you have made that determination, the specific strategy has relatively little direct effect on how you will proceed to plan and implement CE. Second, strategy has an indirect effect on the approach you will take to CE. This is because strategy affects both environmental and task uncertainty. It affects environmental uncertainty because one element of your strategy will be to choose a position in the market, and that affects how you interact with the environment. For example, if your strategy is to be a "technology follower," one who watches the competition carefully for new technical developments, but responds quickly, then your environmental uncertainty will be very high — because it can change quickly and unexpectedly. On the other hand, if your strategy is to always be the first to market with new technologies, you probably spend a lot of time gathering information about new developments from universities and other researchers — hence you will only rarely be surprised by a new technical development in the environment. The point is that we actually choose how the environment affects us — not completely, but our strategy is a way to do this.

SUMMARY

In this chapter we have introduced a model of concurrent engineering which emphasizes the context in which CE takes place. That model shows the basic context, which includes strategy, organization, technology, and people, as being strongly influenced by the work process of CE and by the environment in which CE takes place. We described a key mechanism by which the environment affects the context of CE, namely environmental uncertainty. We also discussed two ways to characterize your work processes: task uncertainty and task interdependence. In the next chapter we will move on to the first step you need to take in order to assess these task characteristics, specifically the development of a model of your work processes. We will follow that with detailed discussions of several of the context elements, starting with the internal organization in Chapter 4, moving on to the supply chain in Chapter 5, people in Chapter 6, and technology in Chapter 7. In Part II, we will provide you with detailed methods and tools for measuring the parts of the model we have described here, and a process for using those methods and tools so you can effectively make the transition to CE.

CHAPTER 3

•

WORK PROCESSES

The clock, not the steam engine, is the key machine of the modern industrial age. • Lewis Mumford

PURPOSE OF THIS CHAPTER

What is sometimes missed in discussions of concurrent engineering is that the work itself must be designed. In this case, that work is the product development process. Design work, like some other forms of knowledge work, is inherently ambiguous. It is not clear at the outset how to define tasks, which order they should be in, or how these tasks should be related to each other. Any engineering organization that designs things has somehow decided what tasks will be done, and then does some things in series and others in parallel, but this does not mean that process is the most efficient or effective. In fact, it is safe to assume there is plenty of room for improvement.

Any good book on business process reengineering (Rummler and Brache, 1995; Andrews and Stalik, 1994) will tell you that after identifying your strategy and customer needs you should focus on the core work processes and then design your organization to support the core work.[1] We tend to agree with this philosophy, but with a caution. To some extent the analysis and design of your core work process can serve as the foundation for the later work you will do on designing the organization and human resource systems.

Often companies skip this step and leap right to a particular organizational solution, such as cross-functional teams. However, this may not

[1] If you have exposure to the sociotechnical systems (STS) school, you will recognize that, in STS terms, you start with an analysis of your technical system, then analyze how the social system supports or fails to support the desired technical system. See, for example, W. Pasmore and J.J. Sherwood (1978).

lead you to an improved work process. The existing work process probably evolved in a way that makes sense for sequential design, but not for a concurrent approach. Thus, the teams will just execute sequential design with lots more communication. This may solve some of your problems, but you are not likely to get the fundamental improvements in quality, lead time, and cost that you are after.

The caution we have is that there is no "perfect" work process any more than there is perfect technology or a perfect organization system. In some sense, all of the CE elements ought to be designed together, concurrently. Our process actually suggests that you view the design of the CE elements as an interative, recursive process — one in which you design each element at a high level and then return to provide feedback from the design of the later levels before you move on to greater detail. That said, you have to start somewhere, and we find that the work process is a better place than most.

To design the core CE work, you need a way to *model* the work process. That is, you need to represent the activities, their order, the flow of information between them, and how they are controlled. This chapter is about how to model your work processes.

There are a number of different potential uses for your work process model:

- ***Assessing your work process.*** The model is an aid to understanding, in the form of an assessment. Thus, you can use a modeling approach to understand the work processes you use today, in order to assess their strengths and weaknesses.
- ***An input to organizational assessment and design.*** You need to understand your current work process to some degree to help guide your organizational assessment. You should begin work on your new work process first, and then start to design your organization to support the work.
- ***Designing a new work process.*** You can design a new work process that will be a significant improvement over current practice.
- ***A project management tool.*** Each project should be managed based on a detailed work process model appropriate for that project.

We would like to be able to present you with a single modeling approach that is best for all purposes. Unfortunately, we do not believe there is one best approach, despite what some reengineering books might argue. There is a wide variety of methods for modeling work processes, some of which will be familiar to many readers. In some cases, you will be familiar with these methods as project management tools, although they also have applicability as more general modeling approaches. However, each method has different strengths and weaknesses in this new application.

This chapter was a difficult one for us to write, in part because there is such a wide variety of modeling methods available, and also because different readers will have different levels of experience with the methods. Should the chapter be a primer on all the various modeling methods? That did not seem right, since there are so many reengineering and project management books available which already do a fine job of this. On the other hand, should we presume all of our readers are already familiar with these methods? That seemed like a bad assumption since our intended audience was not limited to project managers. We concluded that our main purpose should be to discuss the strengths and weaknesses of the most well-known modeling methods *as tools for analysis and design*. Ultimately you will probably use more than a single method. In fact, at some point you may use all of the methods we discuss. As with all the tools in this book, you need to decide what makes sense for your particular situation, and perhaps experiment with several approaches.

We will discuss a set of common modeling approaches that range from very simple to relatively complex. For each method we will briefly review the basics. If you are already familiar with these methods you will probably want to skip over these sections. We then discuss each method's strengths and weaknesses for the analysis and design of a product development process.

What are Work Processes?

The "work processes" of product development are the set of activities and the connections between those activities that are used to develop a product and the processes for making it. They include:

- activities
- flow of information and physical objects between activities
- ordering and timing of activities
- control mechanisms.

Each of these elements needs to be included in whatever model of the work process you construct (Figure 3.1). We will briefly define each of these elements.

Activities

This is the actual work that adds value to the product, e.g., creating specifications, collecting information on customer wants, making a sketch of a mechanical part, creating a 3-D model, building a prototype.

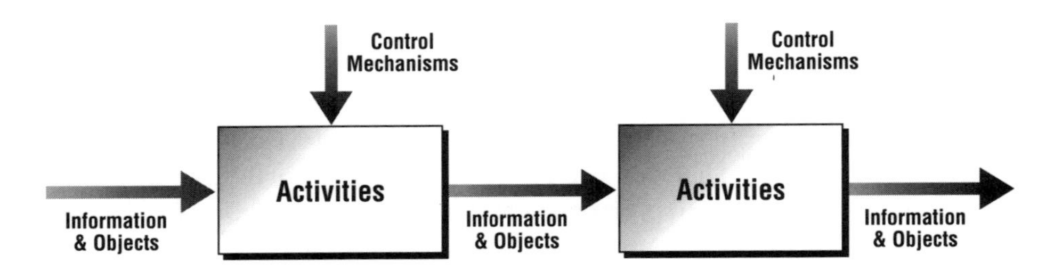

Figure 3.1 • Elements of the Work Process

FLOW OF INFORMATION BETWEEN ACTIVITIES

A few activities can proceed independently, while most require information from other activities. For example, you can't build a mechanical prototype without part dimensions. On the other hand, you can begin to build the prototype even if you only have incomplete information about dimensions. Delays in the flow of information may cause delays in downstream activities. The flow of information can be modeled and planned. Both technologies and social structures can be established to facilitate the planned flow. As an example, consider the flow of objects and information associated with the task of making a prototype part for Elementary Engine Parts, Inc. (Figure 3.2), a fictional producer of small engine parts.

In this example, the prototype is made in another company, Fred's Prototypes, which is specially organized for making parts in small lot sizes. The prototype maker might receive any or all of the following: a CAD model in the form of a computer file (perhaps delivered by floppy disk or over a network), a physical drawing, a physical model (e.g., made out of wood), written instructions describing design intent, verbal instructions received by telephone or in person, a purchase order to perform the work for a certain amount of money, and materials for making the prototype. After completing the work, the prototype maker might send out the physical prototype, written or verbal information about why certain things were done the way they were, and a confirmation to his billing department to invoice the customer. He might also send out questions to the customer about details on the drawing he doesn't understand.

In the above example, there are a great many objects and pieces of information flowing back and forth. At some point in the analysis and design process, you will need to capture this complexity. For example, unless we understand the details of information flowing around, we probably will fail to set up the necessary systems to support all of the differ-

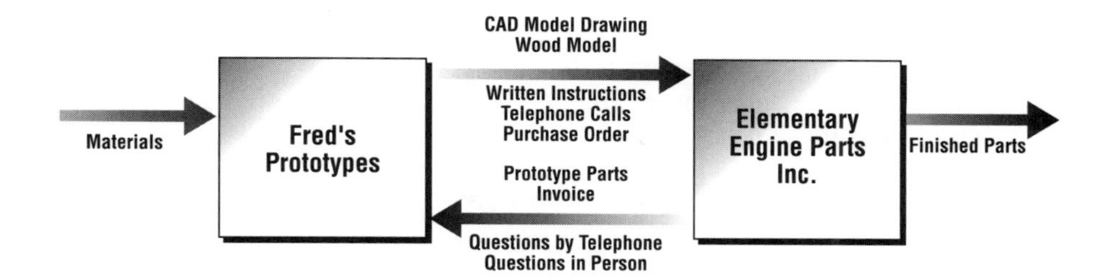

Figure 3.2 • Flow of Parts and Information

ent forms of information flow. Equally, we can observe the flow of objects or information, and may well find flows in which there is little or no value added. We can then work to eliminate or at least reduce these.

The Order and Timing of Activities

Some activities can occur in parallel, while others can only begin when another activity is complete. In the case of design, this is generally because they must wait on information from the prior activity. In some cases, the wait is due to capacity constraints, e.g., we have only a limited design staff and prefer to have designers work on complex parts with long build times before they start on simpler parts. By modeling and planning the order and timing of activities, we are in a better position to monitor and ensure they take place when desired.

Control Mechanisms

These are the mechanisms used to ensure that activities remain aligned and that the project remains on target (e.g., on schedule, within budget). Most companies will have some mechanism in place to ensure the project will meet its goals, but few also work to make sure activities remain aligned. By aligned, we mean how well activities fit together. For example, can we be sure that design specifications determined by product engineers consider the process capabilities of equipment specified by manufacturing engineers? Can we ensure proper coordination between our engineers and outside suppliers? Alignment is critical for CE to be done effectively and is often only considered after the fact.

Modeling activities and the flows between them only tells you what *should* be happening. The control processes are your mechanisms for making sure they *do* happen. Examples of control processes include reporting and review activities. Thus, the prototype maker in the example above might be required to provide weekly reports to his customer about progress. A design project might have reviews at specified points in the process to ensure coordination among tasks. Many formal project management methods have been devised to provide this form of control. There are, of course, many methods other than control *processes* for controlling the flow of objects and information. For example, the organization structure, teaming, communication, and standards are just a few of the possible methods used to control flow. In this chapter we will focus on control processes. The other methods for control will be covered in Chapter 4.

A complete model of work processes needs to address all four of these elements. None of the commonly used modeling methods does a really complete job of providing such a model. We'll now look at some of those methods and discuss their strengths and weaknesses.

Approaches to Modeling Work Processes

In this section we provide an overview and comparison among four major approaches to modeling work processes. We have not attempted to provide detailed instructions on how to use each method, since there is ample information available on this elsewhere. The structure for this discussion will be to briefly describe the basic approach used in each method, discuss the strengths and weaknesses of each, and provide a big picture comparison of them all. Finally, we will move on to a discussion of the criteria for selecting methods, i.e., the circumstances under which you might choose to use one or the other. The four approaches to work flow modeling we will discuss are:

- descriptive
- schedule-focused
- flowchart
- phases and gates.

The Descriptive Approach

In this approach, individual job and task descriptions (either formal or implicit), or standard operating procedures (SOP), are used to describe the product development work process. For example, a design engineer's job description might specify that the individual:

- discuss specifications with the customer
- develop 3-D drawings for a product which meets the customer specifications
- coordinate prototype development and testing
- ensure that product specifications meet all applicable federal and state standards.

While a job description is usually done at a fairly high level of abstraction, individual task descriptions or SOPs may be quite detailed; as, for example, when a testing procedure is described in complete detail.

This approach obviously includes some description of core tasks. However, the description of the flow of physical objects and information between tasks is often incomplete and may well be implicit, rather than formally described. The processes for control may not be included at all, unless some control mechanism happens to be part of an individual task or SOP. In some cases, such as those involving quality or safety procedures, the flow of objects or information may be described in great detail. But, on the whole, the descriptive approach tends to miss most of the flow of both objects and information. And there is generally no timing discussed.

Nearly all companies have job and task descriptions and/or SOPs in some form. However, in many companies, these descriptions are incomplete, obsolete, or simply ignored. Often, they form a type of running joke around the company. For very small companies and those which are just beginning to be involved in product development (either their own or their customers'), the descriptive approach may be the only one used. At such companies, if the descriptions are incomplete or out of date, this may mean there is no model at all of the product development work process.

Strengths and Weaknesses of the Descriptive Approach. In fact, the descriptive approach is not really a model of the work process. Some of the SOPs or job descriptions may include statements that imply a certain work process, but this is apt to be by accident. It is probably because these kinds of formal procedures and rules are so static that they often sit unused. They tell you little about how to plan or execute a project. More often than not they get in the way, placing general constraints on what people can do without regard to particular circumstances. For these reasons we are mainly presenting the descriptive approach as an undesirable baseline against which to illustrate the advantages of other approaches.

Descriptive Approach Example.[2] *A small company we will call Attachments Inc. makes clips for attaching pieces of rubber conveyor belts. They also make machines for attaching the clips to the belts. Attachments, Inc. has been in pretty much the same business for over 100 years. The last major innovation in the business was over 50 years ago, and in the interval the product engineering capability of the company virtually disappeared. Until recently, the only new product ca-*

[2] This is a real example, only slightly modified to disguise the company involved.

pability the company had was to develop minor variations in clips that are slightly easier to insert or provide greater strength or lower weight. Each new variation in clip required some variation in the machines which attach the clips. The company had three sources of these product variations: the sales manager, a product designer, and one of the manufacturing engineers. There was no formal process for product development. Both the product designer and the manufacturing engineer had product development written into either their job description or their personal goals. While the sales manager had no formal sanction for being involved, he was the major source of new product ideas, and he worked quite closely with the others to bring new products to market in line with his promises to the customer. (We will continue this case example later in the chapter.)

Note that for its environment, Attachments, Inc. was quite successful. There was no need for further elaboration of the process. As we will see later in this chapter, as Attachments, Inc.'s environment changed, this was no longer the case.

SCHEDULE-FOCUSED APPROACH

The schedule-focused approach is primarily concerned with the timing (start and end) of various tasks. In this approach, what is nominally a project management tool can act as a model of the product development process. A Gantt chart (Figure 3.3) shows tasks, with an emphasis on showing overlap among tasks. The Gantt chart also shows milestones for the project, and can be adapted to show dependencies and the critical path.

Strengths and Weaknesses of Schedule-Focused Approach. A Gantt chart is often used as a model of the product development process

July: 9 16 23 30 | August: 6 13 20 27 | September: 3 10 17 24 | October: 1 8 15 22 29 | November: 5 12 19 26 | December: 3 10 17 24 31

ID	Task Name
1	Obtain customer orders
2	Create project plan
3	Design Parts
4	Design Finished
5	Purchase raw materials
6	Manufacture parts
7	Assemble components
8	Package and ship finished parts
9	Project Complete Report
10	

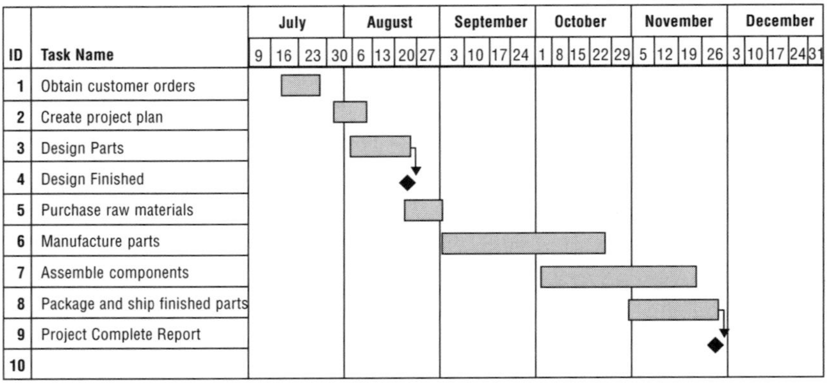

Figure 3.3 • Sample Gantt Chart

when concurrent engineering is being planned since it shows the overlap among tasks so well. While most project managers will use a Gantt chart at some point in their work, if only as a planning tool, it is also useful for modeling a more generic process. Often a fairly detailed Gantt chart can fit on one legal size page. This can be carried around and at a glance you can tell whether you are ahead or behind on individual tasks. However, it is of limited use for analysis and design of a generic work process. Its ability to show the degree of overlap of different tasks is useful. It also can provide a good first cut at assigning generic timing to activities. For all but the most simple processes, where one can mentally picture the precedence relationships among tasks, this should only be used in conjunction with some type of flowchart.

Schedule-Focused Approach Example. *A schedule-focused approach to product development is more common than one might expect, and not just in small companies. For example, Chrysler has no formal in-depth product development process, in contrast with its U.S. brethren, GM and Ford, who have very elaborate, structured processes. If you ask Chrysler engineers to describe their process, they can describe a process, but it is mainly a description of what they, as individuals, happen to do — and each engineer will describe a different process. For the most part, they focus on the schedule, which is very firmly established at the start of a vehicle program. For Chrysler, the schedule is a series of review points at which various forms of prototypes must be ready. Everyone just sort of "knows" what they have to do in order to get ready for a given prototype on time. There is some discomfort within the company about this lack of structure and, as we write this, an attempt is being made to develop a more formal process. Nonetheless, Chrysler has been quite successful to date without a formal process.*

It might be useful to think of the first vehicles designed by Chrysler under the platform team approach as experiments. There was an advantage to managing each one differently to learn what works and what doesn't work. This learning was mainly passed on through word of mouth from project manager to project manager. If Chrysler wants to capture this learning and continuously improve, we believe they may need to more formally document the work process and use this as a standard against which to make new improvements.

The Flowchart Approach

In the flowchart approach, tasks are described with a particular focus on the flows between them. In the simplest form of flowcharting, tasks will simply have inputs and outputs to describe the flows. In more complex forms, additional information is provided about timing or items used in the task, such as tools and other resources. We will discuss three methods in this approach: simple flowcharting, PERT, and IDEF0.

Simple Flowcharts. A simple flowchart only shows the tasks and describes each task's inputs and outputs. Figure 3.4 just shows the simplest form, in which the output of one task is the input for the next. The method can easily be adapted to include additional inputs and outputs for each task. Some additional inputs can come from "outside the system" in the sense that an input might arrive from someplace that isn't a task in the model, e.g., the task producing the prototype might receive materials from an outside supplier which isn't shown in the model. A figure such as this would typically be supplemented by some text description of the details of the task being performed, as well as some description of the inputs and outputs.

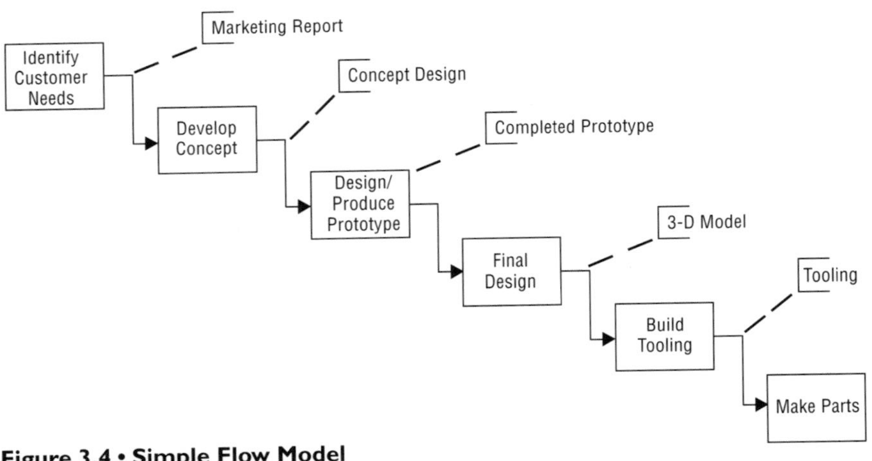

Figure 3.4 • Simple Flow Model

This type of model provides four primary advantages over the job/ task description approach.

1. It formally describes the links between tasks. While this is often provided implicitly in the descriptive approach, it is usually incomplete.
2. It provides information about the order in which tasks are to be performed.
3. It ensures that each task is described. In the descriptive approach, many tasks are left implicit. While this provides plenty of flexibility for job holders, it defies analysis and makes improvement difficult.

4. It provides a graphical view of the process that greatly aids understanding.

Simple Flowchart Example. *The simple, descriptive process described in the above-mentioned Attachments, Inc. example worked reasonably well for decades. In the last few years, however, a new type of vulcanized belt has appeared which threatens to all but eliminate the company's primary product line. The company is working hard to get into new types of business which take advantage of their core competencies in making attachment devices. They hired several new product designers and have created several new lines of business. However, they have been plagued by late deliveries and products for which the market is too small. In response to these problems, they have started to use simple flowchart models (Figure 3.5) to design a formal product development process. One advantage of the process is that it helps to ensure the use of market research information by the new product decision making process. It also ensures that development projects stay on track so they are more likely to be delivered on time. Note that in Figure 3.5, the flowchart elements show both the activity and who is responsible for it. This demonstrates the flexibility of the method.*

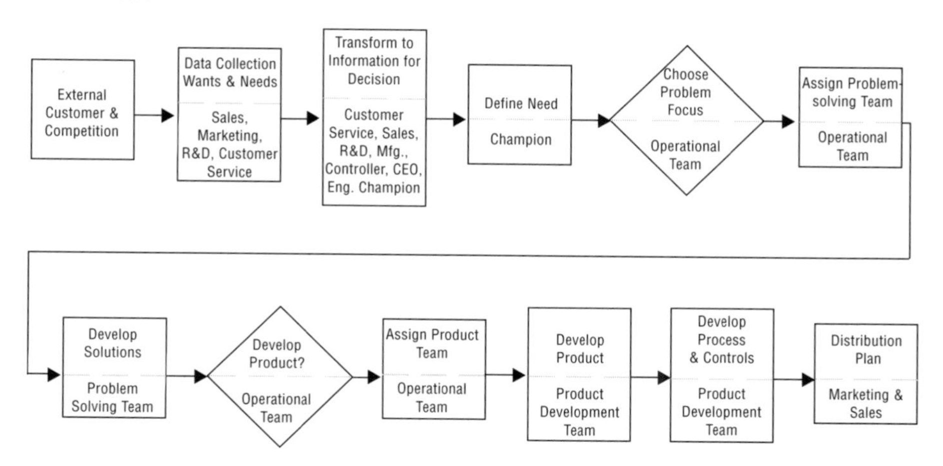

Figure 3.5 • Attachments, Inc. Simple Flowchart

Program Evaluation and Review Technique (PERT). PERT is a method which combines simple flowcharting with timing information. As shown in Figure 3.6, the PERT diagram shows the connections between tasks, the length of time, and the start and stop time for each task. It does not explicitly describe the tasks or their inputs and outputs, but it is easy to create additions to the chart, as might be added to a simple flowchart. It is important to recognize that the arrows in the

PERT chart only show timing dependencies, not information or resource dependencies. Thus, the arrow coming out of the end of the task "Build and Test Prototype" which goes into the start of the task "Create Final Design" in Figure 3.6 only says that creating the final design must logically follow building and testing the prototype. It says nothing about what is coming out of one task and into the other. The timing information shows that work on the final design is expected to start before work on the prototype is complete.

Another advantage to a PERT chart is that it can provide a look at the critical path. The critical path is the path between tasks which, if delayed, would immediately delay the end of the project. Tasks not on the critical path may be very important, but their completion is less time-critical than those on the critical path. Information about the critical path warns management to pay closer attention to tasks on it, since those tasks are much more time-sensitive.

IDEF0 Analysis. Integrated Definition (IDEF) is a structured analysis and design method for graphical and textual descriptions of activities, activity relationships, information, and information relationships used

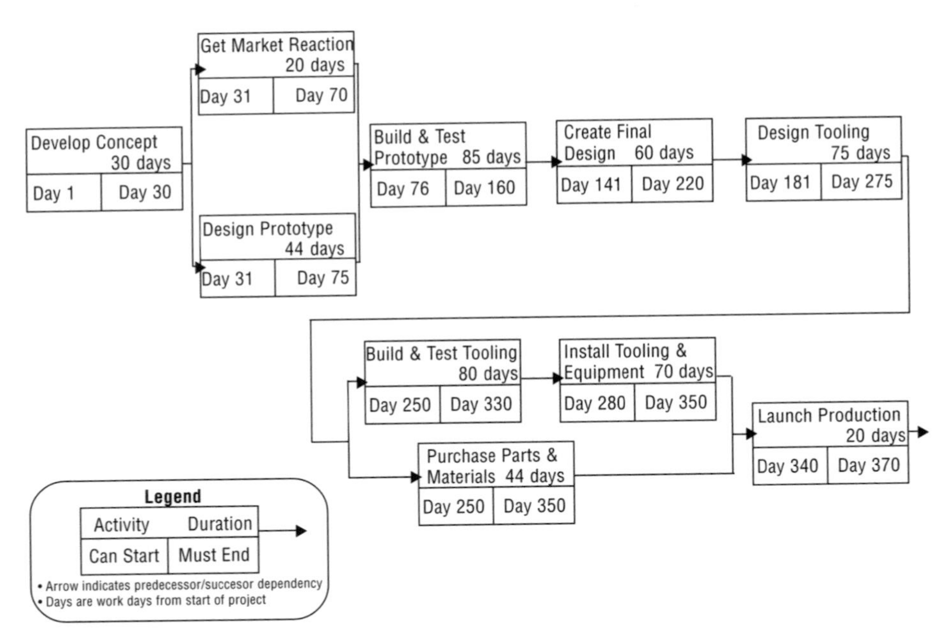

Figure 3.6 • Sample PERT Chart

[3] IDEF was originally developed under a U.S. Air Force program. See D. A. Marca and C. L. McGowan (1993) for more details.

to develop enterprise and system level architectures. IDEF has been the method of choice for many U.S. government programs,[3] as well as a large number of commercial enterprise modeling efforts worldwide. IDEF is really a set of methods. IDEF0 (pronounced IDEF zero) allows one to look at the flow of both physical and intellectual objects. IDEF1x is an approach for describing the *content* of the information flows. While there are other methods in the overall IDEF set (e.g., IDEF3 is sometimes used as an alternative to IDEF0), here we only deal with IDEF0 models.

One of the unique characteristics of IDEF0 modeling is that the model may be *nested*, i.e., the model provides its detail in *layers* that are hidden to users operating at higher levels (don't worry, we'll provide some examples shortly). This makes it easy to see the process at a high level without the distraction provided by great detail. On the other hand, it's easy to get disoriented and lost in the depths of an IDEF0 model until you become familiar with the notation and the concepts used.

An IDEF0 model starts by focusing on a single main activity to be addressed. This provides the overall context for the remaining work. Examples of main activities are:

- manufacture molds for die casting
- manufacture gas-turbine engines
- design and manufacture automotive seating systems.

This main activity is then broken down into subordinate activities which are performed to accomplish it. Examples of subordinate activities for "design and manufacture automotive seating systems" might include:

- obtain customer orders
- design parts
- create project plan
- purchase raw materials
- manufacture parts
- assemble components
- package and ship finished parts.

These activities can be further broken down into subordinates. IDEF0 shows how information, physical objects, and work flow among these various activities. IDEF0 describes a series of activities and related information. The related information is called an ICOM (Inputs, Controls, Outputs, and Mechanisms). A generic IDEF0 diagram is in Figure 3.7. It shows both the activity box and the associated ICOM.

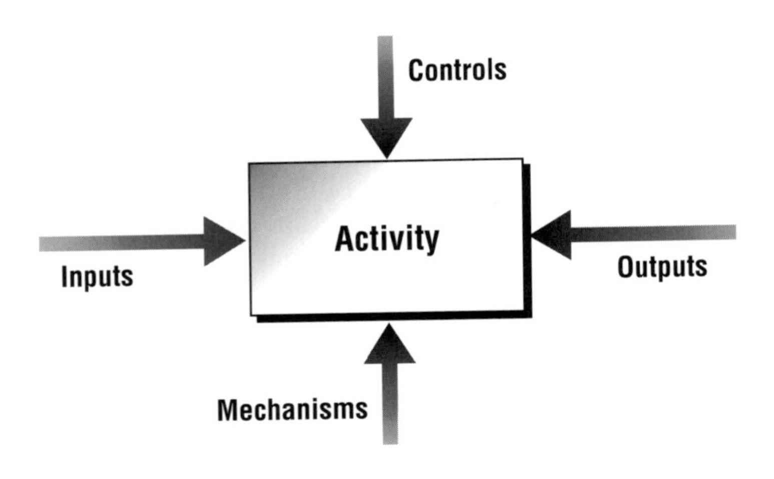

Figure 3.7 • Generic IDEF0

- ***Activity*** – what happens to the information, physical object, or work; the transformation that takes place. Activities are characterized by active verbs.
- ***Inputs*** – the information, physical object, or work that is acted upon by the activity.
- ***Controls*** – information, physical object, or work that directs, limits, or otherwise controls the activity.
- ***Outputs*** – the information, physical object, or work resulting from the activity.
- ***Mechanisms*** – the means by which the activity is accomplished. These are the tools necessary to accomplish the activity. Normally you show only those mechanisms of particular relevance, not the generic mechanisms used in most activities.

In practice, mechanisms mostly come from outside the activities, although an activity may create a mechanism used by another activity. Controls regularly come from both within and outside of a model.

The first step in developing an IDEF0 model is to develop the highest level activity and ICOM (called Level 0). An example of a high level IDEF0 is in Figure 3.8.

IDEF0 models are grouped hierarchically to show the movement of information, objects, and work in the overall activity. That flow is shown by a series of diagrams that show models linked by arrows (Figure 3.9). Figure 3.9 shows the flow within the higher level task "Design and

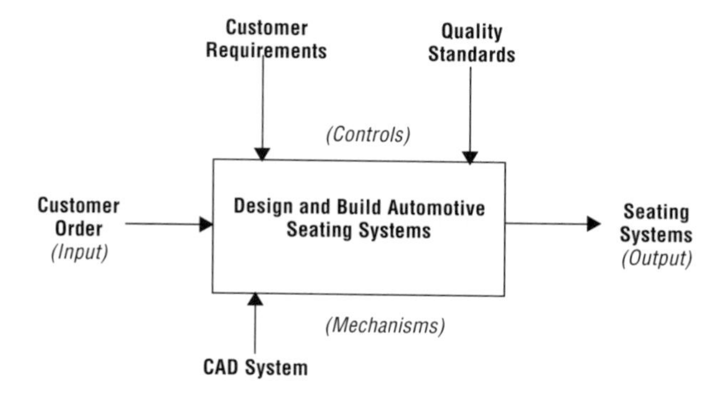

Figure 3.8 • Sample Level 0 IDEF0 Model

Build Automotive Seating Systems" which we showed at its highest level in Figure 3.8. Note that the overall input and output are the same as for the higher level model that contains all these activities.

Each box in Figure 3.9 can of course be further broken down. Under most circumstances, this second level of breakdown would be quite valuable. While you can go as deep as you need to in this breakdown of activities and information flow, the rapid growth in the number of elements present requires that some limits be placed on the process. Unless

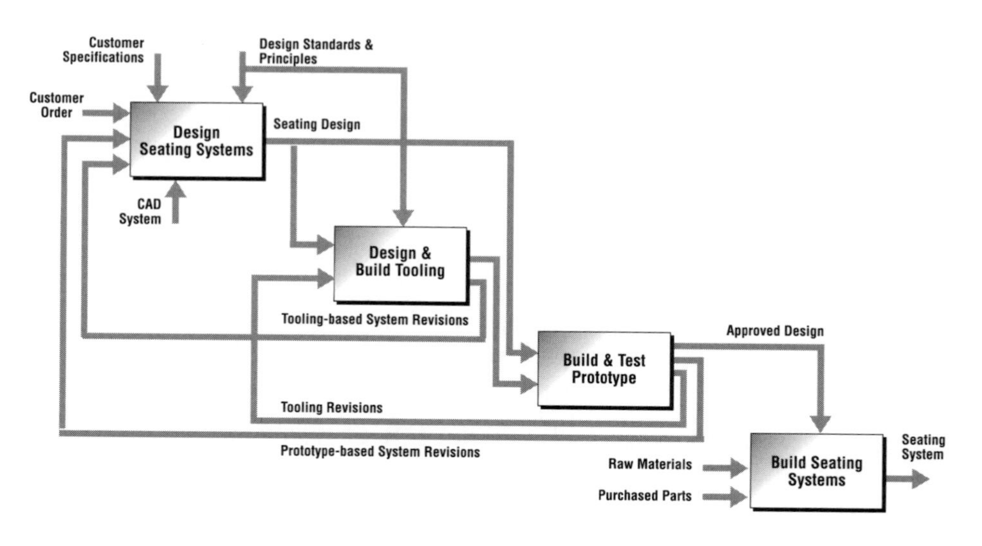

Figure 3.9 • IDEF0 Flow Diagram for "Design and Build Seating Systems"

you are actually designing a complex database (which is what IDEF was originally intended for), there may be no need to go beyond the second level of decomposition.

Each box requires a narrative. That is, each activity is described in enough detail to outline the level of breakdown that would form the next lower level IDEF0 diagram. The narratives are critical to understanding the model. Following are example narratives for two of the boxes in Figure 3.9.

- *Design Seating System.* Based on the customer's order and good seating system design practices, develop a concept for the seating system and then do the necessary detailed design to make the seating system into a manufacturable product.
- *Design and Build Tooling.* Based on the component design and good tooling design practices, design the tooling for manufacturing the component, and then build that tooling.

Once an activity is described, the inputs, controls, outputs, and mechanisms that affect that activity or come from it need further description. The purpose is to get a detailed understanding of the various flows, to better understand who supplies what to whom. Following are examples for two of the outputs in Figure 3.9.

- *Seating design* — the collection of information that describes the seating system in enough detail that it can be manufactured either as a prototype or in production. The seating design is made up of:
  - one or more mockups (physical, paper, and/or electronic)
  - a complete set of drawings (paper and/or electronic)
  - manufacturing specifications
  - bill of materials.
- *Tooling revisions* — recommended changes to the tooling design that result from using the tooling to construct the prototype. Tooling revisions are made up of:
  - zero, one, or more mockups (physical, paper, and/or electronic)
  - zero, one, or more annotated drawings (paper and/or electronic)
  - verbal or written descriptions of the recommended changes.

In many cases, the most useful part of IDEF0 modeling is the *process* of building the model. The resulting model may well be useful, but the information gained and the insights and other discoveries made in the process of building the model provide the real value. By using a structured, graphically oriented method, the model developers are forced to surface issues that might otherwise have been missed.

Strengths and Weaknesses of Flowchart Methods. All of the flowchart methods show precedence relationships.[4] This is both a strength and a weakness. The flowchart gives you a sense of the order of tasks which is useful, but also suggests that one task must be complete before another can begin. Concurrent engineering suggests that many tasks be done in parallel. In CE often one task must be under way before another can start, for example, product design must progress to some degree before enough information is available to die designers to start work. However, the first task may not have to be completed before the second begins. As Clark and Fujimoto (1991) observe, die design can begin and tooling steel can be ordered while the product design is in process, based on early product design decisions.

The simple flowchart can be very useful both in analysis and design of your product development process. For analyzing your as-is situation, you can quickly develop rather complex flowcharts of how you currently do things. It is useful to do this with a cross-functional group, as each function will understand its own activities but is probably less familiar with those of others. Our experience is that a simple flowchart like this is very enlightening to all participants. Often it becomes immediately obvious that the current design process is needlessly complex and there are many non-value added steps that add waste to the process. Convoluted flows can then be rationalized in the design phase again using a simple flowchart as a first step.

Freudenberg/NOK is an automotive supplier of seals and gaskets known for its effective use of kaizen (continuous improvement) on the shopfloor. As a first step in this process they map the flow of their current process. The approach is very simple and visual. They take photographs of each step and mount the pictures on a board. They draw arrows connecting the pictures and showing the flow. They then label each step using a color coding scheme to distinguish value-added steps, non-value-added steps that may be necessary (e.g., taking components from a rack), and non-value-added steps that are pure waste (e.g., repairing defects). They show a simple ratio of value-added and non-value-added steps. As they eliminate a non-value-added step, they remove the picture and keep track of progress over time. The flowcharts are placed in a very visible spot on the wall right near the manufacturing operations they model.

It may also be useful to develop a PERT chart for analysis in order to get a sense of the critical path and where there are bottlenecks in the current process. Without timing on the chart, one might concentrate too much effort on streamlining activities that take relatively little time and are well off the critical path. We do not believe it is worth spending

[4] Note that IDEF0 cannot show activities which take place in parallel; the other methods can do this.

a great deal of effort getting precise timing on detailed activities when analyzing current processes as many of these activities might change or go away in the design stage.

PERT can be a useful process design method. Once a simple flowchart is agreed on, try to add realistic timing numbers to it, and the critical path may suggest opportunities for further streamlining. The timing can also be used to set guidelines for a "typical" development project, though the timing needs to be kept flexible in practice as each design project is different. Developing a highly detailed PERT chart should be reserved for the management of specific design projects when the unique details of that design project need to be built into the model.

IDEF0 is of greatest value as an analysis method for very complex processes. For complex processes both simple flowcharts and PERT tend to lose their utility because everything must be shown in a single diagram. That diagram can become hopelessly convoluted and ultimately meaningless. Because IDEF0 provides a disciplined approach to layering the flowchart, it makes it relatively easy to understand very complex processes.

IDEF0 also provides some benefit as a design method. This is because it explicitly shows resources and controls not shown in any other method. Knowing these during the analysis of the current situation is of some value, but it often isn't critically important. On the other hand, since these must be designed into the new process, having an explicit way to do so is of significant benefit to you when designing the new process.

Phases and Gates Approach

The Phases and Gates approach looks very much like a flowchart, but its primary feature is the focus on decision criteria for ensuring that various sets of activities don't go past a certain point without reference to other activities. A "phase" is a set of activities which must be performed in approximately the same time frame. The "gate" is a design review point at which there is evidence provided that *all* of the tasks have been completed. Only when the gate has been passed through can

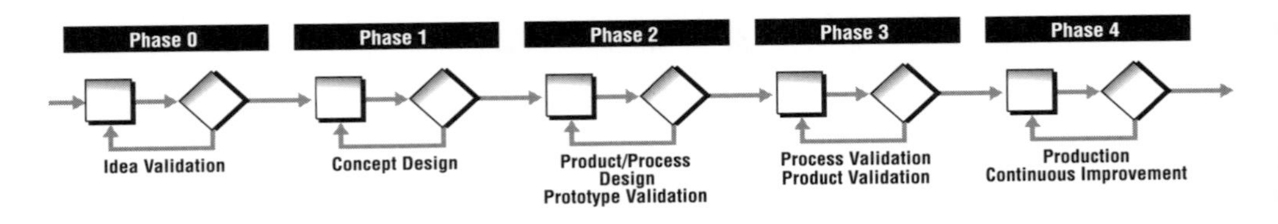

Figure 3.10 • High Level Phases and Gates

activity commence in the next phase. An example of a Gate and Phase approach can be found at the highest level in Figure 3.10.

Figure 3.11 shows some of the detail in the first phase (Phase 0 in Figure 3.10). The critical point is that everything in the phase must be done before activity can begin in the following phase. Obviously each activity listed in Figure 3.11 can be developed to a lower level of detail, and there can be "sub-phases and gates" within the larger phase.

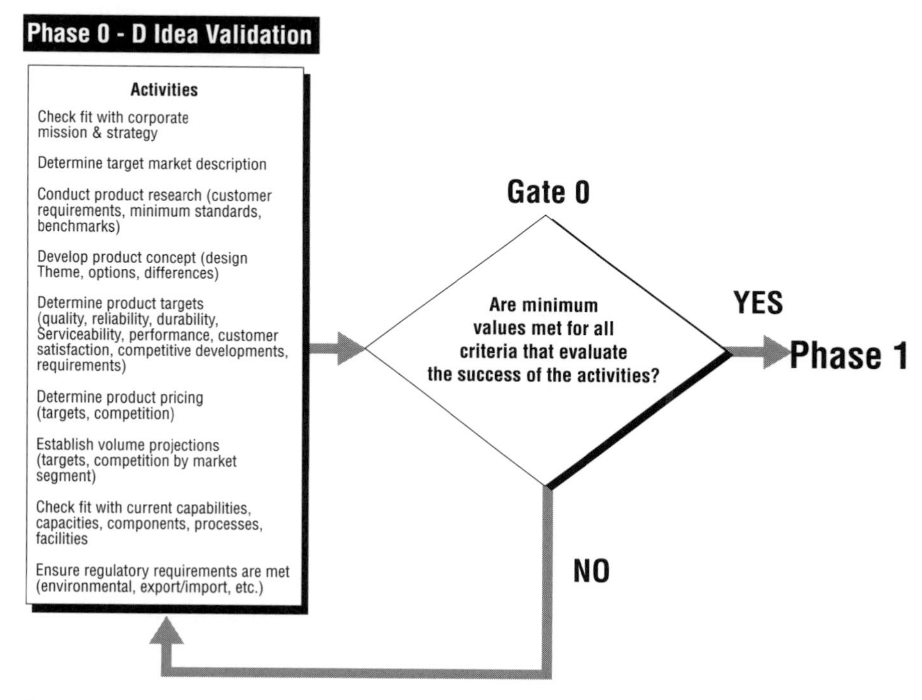

Figure 3.11 • Example Phase and Gate

In many large U.S. manufacturing companies, a gates and phases approach to modeling seems to be widely accepted as good practice. Companies from industries as diverse as automotive (GM, Ford), aerospace (Boeing), office equipment (Xerox), computers (Sun), defense electronics (Westinghouse, Motorola, TI), home appliances (Whirlpool), and power tools (Black and Decker) all use this approach to some degree. In contrast, many Japanese companies seem to use much simpler flowcharts to model their processes.

Strengths and Weaknesses of Phases and Gates Approach. The primary strength of the Phases and Gates approach is that it provides entry and exit criteria for activities to begin. No other approach explicitly provides this. It also implies a philosophy of operation which says that activities in different functions are linked in the sense that an activity in one function cannot proceed without the success of another activity in another function. Interestingly, many companies which claim to use a phases and gates approach to their product development process often ignore it in practice. They will claim to be guided by the details of the method, but when it comes time to halt all activity in a project because someone's task is not satisfactorily completed, they will allow the rest of the project to proceed. This suggests a failure to fully appreciate the links among the activities, or else a failure to build correct links.

Summary of Modeling Approaches

Table 3.1 summarizes the strengths of each of the approaches we have described (we excluded the purely descriptive approach as it is not really modeling the process). An X in a cell indicates that the approach provides coverage in that area. Each approach obviously has its strengths and weaknesses. If you want significant coverage, you will certainly have to use multiple methods. The (X) for Activity Precedence is shown to represent the inability of IDEF0 to show parallel activities.

Table 3.2 summarizes where each modeling approach seems to be most useful. The simple flowchart is most useful for assessment and the early stages of designing your process. The Gantt chart and PERT add detailed timing and are the most useful for project management, though each can also be useful in both assessment and design. The Gantt chart shows activity overlap and thus complements the PERT chart which helps analyze the critical path. PERT is somewhat more useful than the Gantt chart in our opinion for assessing the as-is work process, though in many cases you may not want to include detailed timing in this

	Gantt Chart	Simple Flowchart	PERT	IDEF0	Phases & Gates
Activity Precedence		X	X	(X)	X
Schedule	X		X		
Critical Path			X		
Entry/Exit Rules					X
Controls described				X	
Tools & Resources described				X	
Easy to use	X	X			

Table 3.1 • Summary of Modeling Approaches

phase. IDEF0 works in assessments and design when the process is too complex to be modeled with simple flowcharts or PERT. Also, IDEF0 provides information about information flows, resources, and control processes where that's important. Finally, the Phases and Gates approach is only really useful for design of the new process and for managing that process.

	Gantt Chart	Simple Flowchart	PERT	IDEF0	Phases & Gates
Assessing As-Is Work process	*	**	*	**	
Input to Organization assessment and design		*		*	
Designing a New Work process	*	**	*	*	**
Project Management tool	**		**		**

Key: blank = not useful * some use **very useful

Table 3.2 • Uses of Modeling Approaches by Tasks

Structured Product Development Process: Texas Instruments Example

One of the most common practices followed by most American companies on the leading edge of concurrent engineering is the "structured design process." While each company has its own name for the process, it is usually some version of a phases and gates approach represented by a flowchart and lots of detailed explanation. Broad phases of product development are defined with gates leading from one phase to the next. Between phases there are major design reviews, and the project must pass a set of predefined hurdles to proceed to the next phase. One possible outcome at a gate is that the project will be terminated. Obviously if project termination is going to happen, the earlier the better. The broad phases are generally broken down into subphases with more detailed tasks which can in turn be broken down even further.

The basic assumption of this approach is that, although design is a creative activity that cannot be 100 percent structured, the process can still benefit from some structure. In fact, without structure there is no opportunity for improvement, and the design process could possibly go on endlessly, or until someone just says "stop!" It is clear that a detailed map at the task level associated with standardized work sheets like we might see in repetitive manufacturing is not feasible in the case of product development. That would be too much structure. So finding the appropriate amount of detail is the challenge for a structured design approach. At the very least, a structured product development process should break up the process into a set of phases and opportunities to review the design as it progresses. Design reviews are key as they create deadlines and structured opportunities to provide feedback to designers. Some of the best Japanese companies use design reviews very effectively. Whenever possible, they prefer physical prototypes at these reviews so that there is a concrete product with test results that can be examined. The physical prototype also allows for tests of integration with related components which invariably reveals unanticipated interactions between components (Kamath and Liker, 1994).

Texas Instruments is one of the companies that has taken a leadership role in developing structured design processes. This was spearheaded mainly in TI's Defense Systems and Electronics Group which won a Malcolm Baldridge Award in 1992 in part because of their approach to integrated product-process design. Prior to that, their approach was named best in industry by the U.S. Navy's 1991 Best Manufacturing Practices Survey. The centerpiece of these awards was TI's structured design process that they call Integrated Product Development (IPD). In fact, IPD was benchmarked so frequently that TI created a separate consulting business to market the approach.

TI's process has the usual thick notebooks and half-day overview courses, but they went beyond this and created a process by which development teams can design their own customized structured product development plan for their particular project. To facilitate this customization, TI created a two-day "tailoring course" to be attended by the development team. The course is run as a workshop and the team actually begins to develop the customized plan in this session. IPD specifies approaches to forming the cross-functional teams and the identification of an IPD champion on the team who acts as the "guardian" of the process. This person becomes a link to the IPD central organization for any advice needed, and also is the point of contact for training the team and facilitating the design and execution of the process.

Like IDEF0, IPD is a nested set of flowcharts. A single box in a high-level chart expands out to a more detailed set of subtasks in a lower-level chart. An example of three levels of a generic IPD process used by TI in their training is shown in Figures 3.12 to 3.14.[5] At the highest level there are five phases, each marked by a major management decision to proceed (i.e., gate), as shown by M0, MI, MII, etc. There are many design reviews within phases shown below the phases by acronyms like SRR

[5] Source for all three figures: Texas Instruments Defense Systems and Electronics Group

(Systems Requirement Review) and TRR (Test Readiness Review). We do not provide definitions of each of these acronyms as they are specific to defense contracts or to TI, and the details are not particularly important for our purposes. Since TI is a defense contractor, most of these reviews are required by their Air Force customer. But the structured plan shows where in the flow of activities the design reviews should occur. More detailed levels of the structure show even more specifically when these reviews must occur. In Figure 3.12 we see the "engineering and manufacturing development" phase expanded out into more detailed activities which lead to the various design reviews. We then show a third level flowchart in Figure 3.13 which expands out one of the "engineering and manufacturing development" steps—detailed design—into a set of activities and reviews.

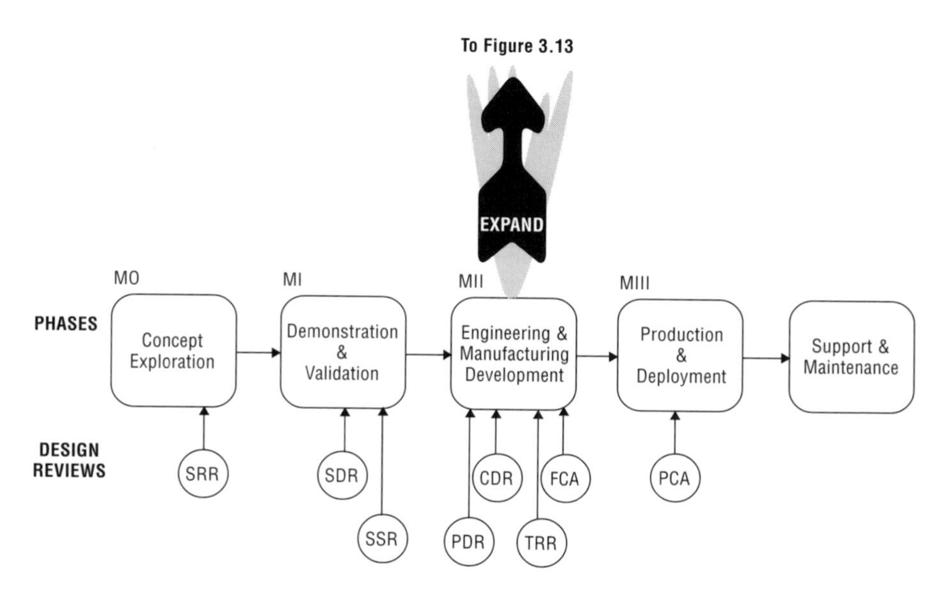

Figure 3.12 • Texas Instruments Program and Design Reviews

You can go as deep as you feel is useful in creating flowcharts. TI actually goes one step further to a fourth level which expands out each of the activities in Figure 3.14. For example, there is a very elaborate detailed flowchart for electrical design. Also, at this most detailed level, which really gets down to specific tasks a given individual might perform, they define a standardized procedure for performing the task, e.g., a standardized procedure for documenting the results of a certain test. You might feel that this level of detail is overkill. We certainly have

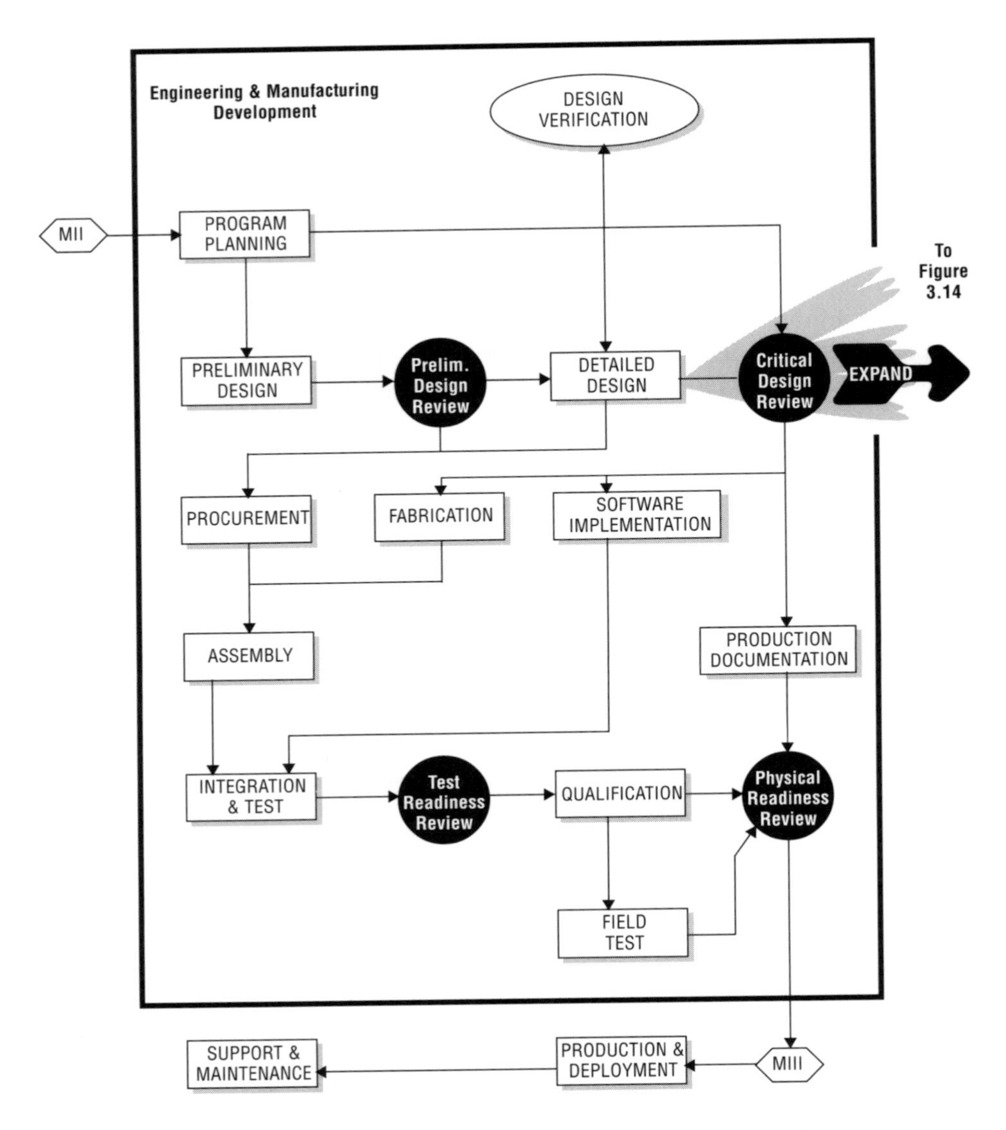

Figure 3.13 • Texas Instruments: Expanded Second Level Flowchart for Engineering & Manufacturing Development

not seen this level of detail in the excellent Japanese manufacturing companies we have visited. In part, the level of detail TI is using in this division may be the result of the need to report on progress in very standardized ways to satisfy their defense customers. What is important, however, is that the more detailed the level of flowcharting, the more likely it is that different projects will have different requirements. For example, some tests may have already been performed

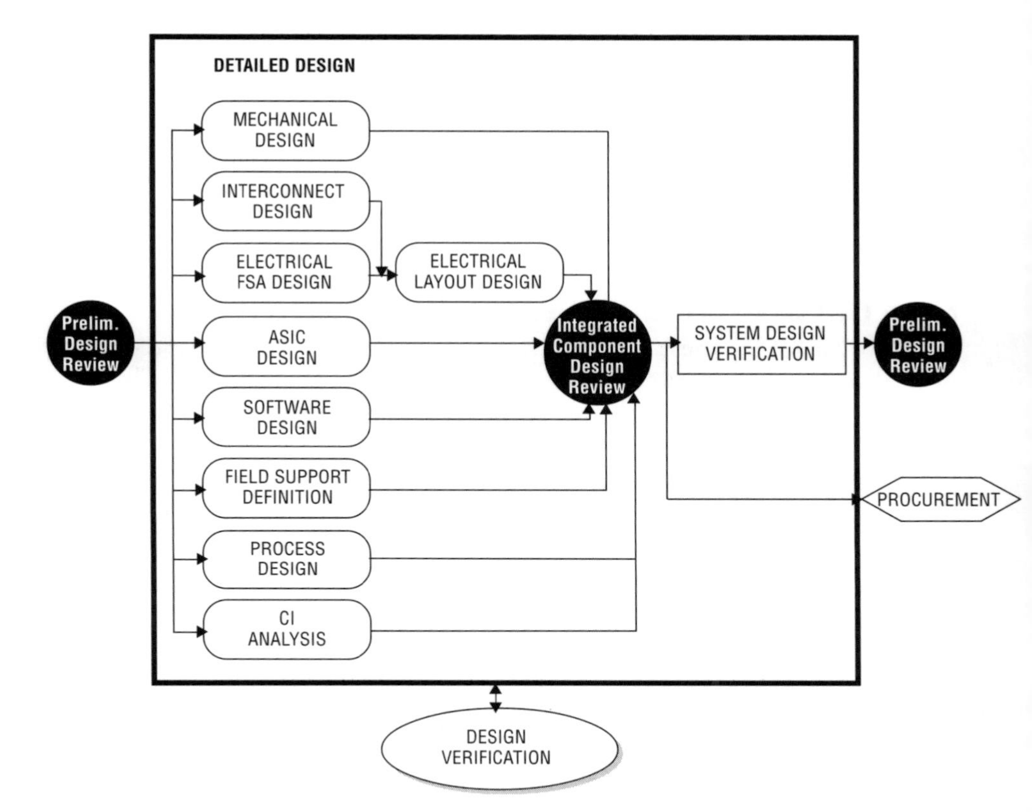

Figure 3.14 • Texas Instruments: Expanded Third Level Flowchart for "Detailed Design"

if the project is an incremental change of just a few parts of the product. Thus, it becomes important that the flowcharts be tailored to particular projects, and also that the tailoring should be done by the team members—those people who understand the project best. In this sense, TI presents an excellent model.

Using a Modeling Approach

Many companies have built elaborate models of their product development processes, only to see the models go unused. Use seems to be a function of three factors, all of which have to do with the ways in which the company approaches the modeling task.

1. *How detailed should the model be?* Should the model include every piece of information associated with the model, or should it be more of an outline in which everyone "knows" what to do?

2. *Who does the work of modeling?* This comes down to whether the task is done from the top, with a high level view dominating, or whether it should be done in a distributed manner with each level creating its own model.
3. *How standard should the model be?* Should the model change for each project, i.e., be customized, or should the model attempt to stay as-is for every project?

How Detailed Should the Model Be?

Observers of Japanese product development processes have noted the near absence of elaborate, formal processes (Ward et al., 1995). This is very much in contrast to many large U.S. companies, which tend to have highly detailed, written procedures. Curiously, the opposite situation occurs in the case of manufacturing—Toyota is famous for its highly formal Toyota Production System, in which each step of every process is written down. In comparison, many U.S. factories are running almost by the seat of their pants. In fact, many U.S. companies are now copying Toyota and are developing their own formal production systems with detailed standardized processes.

Should U.S. companies move away from their elaborate systems in the case of concurrent engineering, and copy the Japanese approach? We suggest *not* "copying" the Japanese in this case. Close study shows that the Japanese *act* as if they had the same kinds of formal product development processes that U.S. companies have (Kamath and Liker, 1994). They have a number of formal reviews which act to keep functions aligned. They "know" what tasks need to be done in which phases; they simply don't have elaborate written procedures. We believe this has a lot to do with the stability of employment of Japanese firms and the stability of supplier relations in Japan, in combination with many years of experience continuously improving their product development processes. We believe that most U.S. product development teams need the written guidelines, particularly when first introducing a new product development process. However, over time this may change. As concurrent engineering and IPPD become routine, as they are taught in design courses in engineering schools, and as U.S. firms gain more experience, the need for elaborate, written procedures may be reduced. But not yet.

However, there is elaborate, and then there is baroque. We believe extremely detailed, standardized sets of procedures are probably overkill in most cases. Developers of such processes haven't distinguished between what is critical to specify and what is "nice" to tell people. The ability of Sun Microsystems to keep almost any document down to 20 pages (as discussed later in the chapter) tells us that many others are going way too far in their level of detail. This is not to suggest all sets of procedures could be kept to 20 pages — probably not for more complex products (like an automobile engine) — but most components of most products are probably no more complex than a computer workstation.

Another issue to consider is what your purpose is in modeling. If, as we recommend in this book, you are undertaking a significant redesign of your product development process, then you may need to get relatively elaborate. If you already have a satisfactory process, and you are simply making minor changes to it, then creating an elaborate structure will be a waste of time. After all, the primary purpose for writing the process down is to make sure everyone knows what it is and is able to follow it. If they have already internalized it (as is often the case with the Japanese) and the environment is stable, then there is no need to create a large document.

Who Does the Work of Modeling?

The question of "who should do the modeling?" is in fact closely connected to the question of "how detailed should the model be?" Consider an IDEF0 model (Figure 3.8). The box titled "Design Seating System" implicitly contains another flow diagram within it, each of which has about five activities in it. Each of those has another diagram, with more boxes. The time needed for developing a "complete" model can rapidly explode, and the sheer size of the model means that no one will ever use it. Thus, as the model gets more complete, the cost of developing it rises astronomically, and the probability of it being useful goes down as fast as the cost goes up.

Many users of modeling techniques have thrown up their hands at this problem; the scenario described above seems insolvable. After all, why build a model if you're only going to stay at very high levels? — most of the real work of design is done at the lower levels of analysis. As it turns out, the solution to this problem is contained in how the modeling is done.

An all-too-common approach to modeling is for some external organization (or perhaps some staff organization) to build the complete model for the company. *This is the source of the problem.* In order to know enough to build detailed models at lower levels, the model builders spend an enormous amount of time learning about what people do at the lower levels of the organization. This not only consumes their time, but the time of the people whose work is being modeled. The solution to this is quite simple. The effort of modeling should be distributed. The people whose work is being modeled should build their own models. Thus, top level managers (or their staff or their consultants) build top level models, but not the levels below themselves. Their subordinates build models at the next level down, and so on. Each level builds the model for their own work, making sure to fit within the higher level model. *The requirement for fit with the next higher level ensures consistency and continuity of vision.* Also, no one builds a model deeper than needed for understanding their particular concerns. Top level managers may need to help coordinate the levels of those below them and work to ensure that they do indeed fit with the levels above them — but they should not be building a single huge model.

Naturally, this distributed approach requires that people be trained in how to build models. But that effort involves less time than that spent working with model builders who have to learn everything from scratch. More important, since the people who built the model will have to implement it, the model is far more likely to be both useful and used.

How Standard Should the Model Be?

Can a single model of the product development process actually be useful for every single development project undertaken by a company? Or is it necessary to build a new model for each project? In other words, are the tasks, flows, rules, and roles really the same for every project? Or do some things change each time?

In some sense, every development project is unique — the cast of characters changes, the specific details of development are at least slightly different, the market is somewhat different, etc. This suggests that a totally standard product development process (one which allows no customization at all) would not be viable.

However, there is tremendous variation in the extent to which development projects change over time in a given company. In some companies, each project is totally different from the last one. This might be the case in a company which designs custom products for each customer, or one which is in a rapidly changing market, such as computers. In other companies, each project is very much like the last (although never really identical). This might be the case when products stay very much the same over time, receiving only slight upgrades and appearance changes, such as might be found in small appliances or hand tools (or, as was the case with our example company, Attachments, Inc. over most of the past 100 years).

In the case of a rapidly changing product, it would seem necessary for each product development effort to build its own process. This does not mean they have to start from scratch each time they begin a project;

they can use templates and examples from prior projects. However, it is probably important that they not be constrained to use some single standard process whole.

We have in fact seen several examples of these different approaches. In general, attempts to define a totally "standard" product development process, with little or no flexibility, generally seem to fall into disuse. Customizable processes appear to be viable, although it appears that the extent of customizability needs to match the "newness" of the product being developed. Sun Microsystems provides an excellent example of a customizable process.

Build-Your-Own Development Process. Sun Microsystems' New Product Development Process, known as the Systems Development Model (SDM), was developed in 1992. It was the result of their experience with an earlier model which was a "big thick formal process that few people read." The old process was treated as confidential material, so many people had difficulty getting access to it. Based on the sense that this process wasn't working well because it wasn't being used, a new and simpler system was developed. The new SDM document is 16 pages long, including a list of acronyms and references. A few words from the Preface show how it is expected to be used:

> *Product development processes at Sun change and evolve faster than formal models can be revised, developed, and implemented. Any model must be capable of rapidly adapting and evolving as technology and our business demands.*
>
> *It is not the intent of this document to define and structure the procedure for individual activities of various functional organizations. That responsibility is addressed in the New Product Introductions of the individual functional groups.*
>
> *The intent is to make this model simple, meaningful, and scaleable; that is able to be customized to fit the complexity of any given new product. It is meant to be a living document that will evolve over time. As in product development, this is an iterative process - keep what works, and change what needs to be improved.*[6]

In other words, the SDM is a high level model which guides (but does not dictate) more detailed model development at lower levels. Each lower level group (and ultimately each project) is empowered to develop its own specific product development process to meet its needs. Even at the project level, the document which describes the process with key milestones is only about 20 pages long.

[6] Source: Sun Microsystems

Sun's highest level model (Figure 3.15) has only three phases.

1. *Strategic Product Planning.* This is a concept development phase. Many concepts enter this stage, but relatively few get out. This stage starts with "Project Exploration" and includes two gates — "project initiation" and "project approval." A sponsor with a good idea initiates a one-page "Project Exploration Form" which is submitted to a management group. If that group approves, a small group is empowered to explore the idea and expand on the description in the five-page "Project Initiation Form." This initiates development of a business plan for the product. Approval of the business plan is the gate into the second phase, Product Development.
2. *Product Development.* This phase includes everything from development of specifications to beta testing. The only gate at this level is the final launch review. Intermediate gates (such as monthly design reviews) within this phase are the responsibility of the New Product Team.
3. *Production and Volume.* This includes volume production as well as development of a transition plan to a new product and phaseout of existing products.

Figure 3.15 • Sun Microsystems Systems Development Model

Sun is in a particularly volatile, innovative market. A product life can often be measured in months, rather than years. Therefore, a very flexible process is needed since the exact nature of any project cannot be predicted far in advance.

How Customized Should You Be? Your level of customization should depend on characteristics of your environment and your company's competitive strategy. A higher level of environmental uncertainty probably suggests that the increased flexibility in lower levels of the company will be required. This is facilitated by greater customization of the design process, since design teams will be in a position to exercise control of their own relatively fluid circumstances.

SUMMARY

In this chapter we have described some of the methods by which you might model your product development work processes. The strengths and weaknesses of these approaches for various assessment and design tasks were discussed. We also discussed several issues involved in getting a model of the work process used in actual practice. In general, it is important to use these tools flexibly based on your particular needs, and always ask *why* you are using a particular tool.

CHAPTER 4
•
INTERNAL ORGANIZATION ISSUES

The elements of [organization] structure should be selected to achieve an internal consistency or harmony, as well as a basic consistency with the organization's situation — its size, its age, the kind of environment in which it functions, the technical systems it uses, and so on. Indeed, these situational factors are often "chosen" no less than are the elements of structure themselves. • Henry Mintzberg

THE PROBLEM OF CROSS-FUNCTIONAL INTEGRATION

In Chapter 1 we said CE included overlapping product development processes and tight links among all functions involved in the product life cycle. In our experience, most companies that wish to move to CE have relatively *loose* links between their functions. The classic case is the design engineer who barely communicates with manufacturing until the design is fully detailed. But there are many other functions that need to communicate and coordinate with design engineers, including marketing, styling, purchasing, and finance. The problem for these companies is to find ways in which links between functions can be made tighter — we will refer to this as *cross-functional integration*.

Many companies today attempt to address the problem of cross-functional integration with only a single approach — the cross-functional team. While such teams are very important, they are only one part of the solution. The rest of the solution lies in understanding (and occasionally changing) your organization structure, project management methods, office locations, communication media, standards, job design, and planning approaches.

In this chapter we will discuss the choices you can make among these approaches to cross-functional integration, and we will provide a framework for deciding among them. Much of this chapter defines the options for organization design and culture, focusing on one aspect of organizations at a time. Remember that organizations are complex systems with the parts interacting to form a dynamic whole. As you go through the options, keep in mind the broader purpose—to understand the many options available to you, and what *combination* of organizational choices will get you to your goal.

Cross-Functional Integration Mechanisms

Cross-functional integration has been the subject of a large proportion of organizational research over the years. This means that there are many frameworks for understanding it, and although most will look very different from each other, they all come to pretty much the same conclusions. Henry Mintzberg (1983), in his book *Structure in Fives: Designing Effective Organizations*, succinctly summarized the literature in the field. As Mintzberg (p. 4) puts it, "Five coordinating mechanisms seem to explain the fundamental ways in which organizations coordinate their work: mutual adjustment, direct supervision, standardization of work processes, standardization of outputs, standardization of worker skills. These should be considered the most basic elements of structure, the glue that holds organizations together." We rely heavily on Mintzberg's formulation in this chapter. Briefly, Mintzberg's five mechanisms are as follows.

1. ***Direct Supervision.*** One person takes responsibility for coordinating all tasks by telling the others what to do and keeping track of their performance. The supervisor has the big picture and subordinates need only execute their individual pieces, doing as they are told. The traditional authority structure in an organization uses direct supervision.
2. ***Standardization of Work Processes.*** Work tasks are programmed in some detail so that party A knows what to expect from party B and when to expect it, even if they have not communicated at all. Traditional assembly line work processes use this mechanism, but it is certainly not confined to the factory floor. Most design groups have a large number of standards which specify such things as how drawings will be made, what software will be used, or when engineering analyses will be performed. These standards are *coordinating* mechanisms because, if they are followed, each person knows what to expect from the others and what they are expected to do. Thus, two groups that follow a schedule (a way to standardize work processes) know what to do to coordinate their efforts

without needing to call a meeting or even make a telephone call. One can, in theory at least, prepare to receive the other's work without any added coordination effort at all.

3. ***Standardization of Outputs.*** The results of someone's work is standardized. Suppliers who provide certified parts to a customer have standardized their outputs so that there is no need to have additional coordination effort. Thus, the customer and supplier don't need to discuss what the dimensions of a part will be when an order for Part # xxx is placed; the standards for that part cover all the necessary coordination.
4. ***Standardization of Worker Skills.*** It is possible to know what someone will do on a given task based on the specific skills they bring to the task. Thus, a journeyman tool and die maker, given a task to produce a stamping die for a part, does all the work necessary to produce that die. Few (if any) further instructions are given (other than schedule, cost, and specifications). No manager, for example, tells the die maker how to produce a given feature in the finished die. While there may be some additional controls placed on the work of the die maker (program reviews, for example), most of the coordination necessary has been done through the training that the die maker has received over the years.
5. ***Mutual Adjustment.*** Coordination is achieved by continuous, two-way communication between the parties involved. This means there can be real-time adjustment of behavior and ideas based on feedback from the others involved. A vivid example is the operating room team of doctors, nurses, and technicians who need to continually communicate and adjust their actions as the situation changes second by second. Any time you call someone to discuss how you are going to go about some task, you are using the mechanism of mutual adjustment. Most teams, and indeed most meetings, take advantage of this mechanism.

Mintzberg (p. 7) notes that these mechanisms fall into a rough order (Figure 4.1): "As organizational work becomes more complicated, the favored means of coordination seems to shift from mutual adjustment to direct supervision to standardization, preferably of work processes, otherwise of outputs, or else of skills, finally reverting back to mutual adjustment." What this means is that in a very simple organization (made up of a few people, for example), informal coordination will work just fine. As the organization grows larger and more complex, someone usually "takes charge" and coordinates by giving orders. As work grows even more complex, it becomes impossible for one person to know every-

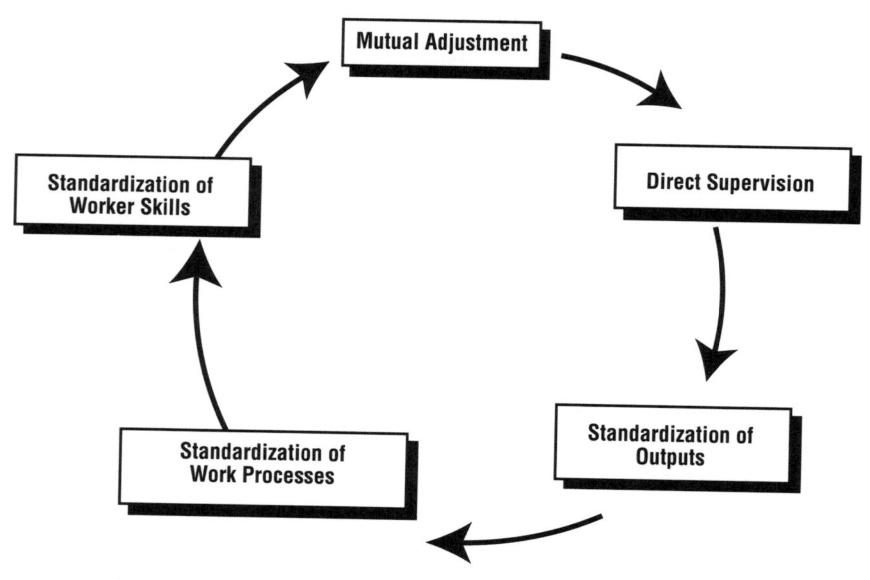

Figure 4.1 • Five Coordination Mechanisms

thing that's happening, so rules are made. Depending on how complex the work is, different kinds of rules are made. However, when the work gets extraordinarily complex, it is impossible to make rules that cover all (or most) contingencies, so the organization reverts back to mutual adjustment.

This is not to suggest that any organization (except for the most simple) would use only a single mechanism. Most organizations will use all five. All Mintzberg is saying is that the *favored means* change as the work grows more complex. Thus, we can look at product development as it takes place in almost any modern, large company. The work is done within some organizational hierarchy; there are work standards and (perhaps) a work process; outputs are usually standardized in some way (e.g., design standards or a specific form of CAD database); and worker skills are standardized in the form of college degrees required, training provided, and the like. Finally, there is a great deal of mutual adjustment that takes place in the form of meetings, telephone calls, and informal chats in the hall. Despite the fact that almost every organization will use all five mechanisms, there can be quite a bit of difference in how much of each the organization uses, and how effective the organization is at using them. Our intent is to provide you with the means to determine what combination of each of these five mechanisms makes sense in your situation.

We will discuss the details of each of Mintzberg's five mechanisms in turn (Figure 4.2), with specific attention paid to how they are used in CE. We will start with Direct Supervision since that is the easiest for most people to understand.

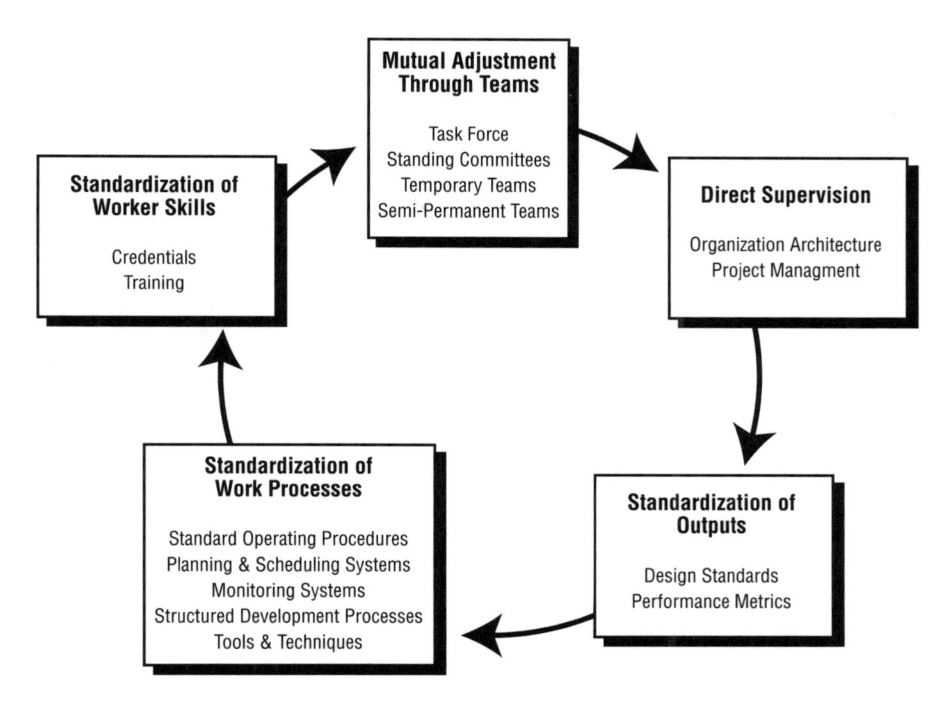

Figure 4.2 • Summary of the Five Coordination Mechanisms

DIRECT SUPERVISION

Our discussion of direct supervision is divided into two parts. The first involves what we will call the *organization architecture*, often called the organization structure or the organization chart. The second, closely related to the first, involves the management of specific projects, such as a product development effort, which we will call the *project management structure*. These two are distinguished by their permanence and their focus. The organization architecture is relatively permanent, at least as permanent as things get in the last decade of the 20th century. By this we mean that the reporting relationships in the architecture are defined as lasting until they are changed, but are usually not limited by time or product duration. In contrast, the project management structure lasts only as long as the product cycle involved.

ORGANIZATIONAL ARCHITECTURE

The most basic form of direct supervision is that provided by your "boss," the person who supervises you on a day-to-day basis. The popular notion of the boss looking over your shoulder all the time, telling you what to do is of course only a joke in most settings. Most supervision is limited to goal-setting, monitoring, assistance with exceptions, and coordination with other parts of the organization. However, goal-setting is in fact a subtle (or not so subtle in some cases) way of telling you what to do, even if it doesn't involve minute-by-minute direction. Much of coordination involves "dealing with the outside," that is, telling those who are inside what to do about the outside. Consider the following scenario.

Edgar Price is a product design engineer for Metal Desks, Inc., a manufacturer of office furniture. Office furniture is an industry in which customer preferences change suddenly, based on fashion, building trends, and the overall economy, so close contact with the market is essential. Edgar is responsible for executive office chairs and has been working on a new chair for the past few weeks. Edgar reports to an engineering manager, Harold VanDyke, who reports to the Vice-President for Engineering, Martin VanMannen. Marketing at Metal Desks, Inc. has its own Vice-President, Ted Artz. The President of Metal Desks, Marsha Ferrus, calls a meeting of her Vice-Presidents at which she complains about slow sales of executive office chairs. She says she's heard from some of her golfing buddies that Metal Desks' chairs lack features such as a heated seat cushion and lower back massage, that some competitors' chairs have. First she looks at VanMannen, then she looks at Artz, and she says "I want something done about this situation, fast!" After the meeting, Artz and VanMannen walk down the hall together. Their discussion revolves around how to get better market information to the design engineers. They agree that marketing should assign a marketing staff member to each product line. From now on, all design engineers will meet with their assigned marketing staff member at least once a week. VanMannen passes this information along to Harold VanDyke, who tells Edgar Price about it. Price then sets up a meeting for every Monday at 10:00 with his assigned marketing staff member, Belinda Stone.

One of the interesting parts of this story is the implication that the marketing staff and the engineers were in different parts of the organization, with separate VPs. In other words, the organization architecture not only determines who your boss will be, it also determines who is in which parts of the organization, dividing the organization into groups or departments.

There are two fundamental ways of breaking up the organization into these different kinds of departments: by function or by product line. In fact, any real organization of any complexity at all will be a mixture of these types.

Functional Organization. Functional organizations group people by the kind of specialized activity they perform. Generally this involves a core work process and a certain body of specialized knowledge and skill. For example, a mechanical engineer has a certain set of knowledge and skill about mechanical systems. His core work process might involve transforming a set of requirements into a detailed design of the mechanical system. A company organized functionally might put all mechanical engineers into one department reporting to a mechanical engineering manager. Another kind of functional grouping might be specific to a component of the product. For example, weapon systems packaging might be a specialized mechanical engineering function applied to developing the metal casing of a weapon system that the electronics fit in. In this case, there might be one functional department for weapon systems packaging and another for weapons electronic systems design. Another common way of separating engineering is to put those designing products into one group and manufacturing engineers into a separate group. Figure 4.3 illustrates a simple functional structure at the vice president level of a fictional company we'll call Weapon Systems, Inc. Different vice presidents are responsible for engineering, manufacturing, purchasing, and marketing. Each vice president assigns engineers to work on projects involving different products as needed.

Product Organization. In this type of organization, individuals are grouped based on their contributions to a particular type of output, such as products (e.g., washers, dryers, dishwashers) or services (e.g., home equity loans, home mortgages, savings accounts). In a *pure* product organization, once the category of output or customer grouping is defined, all of the specialists who are needed to produce that output or

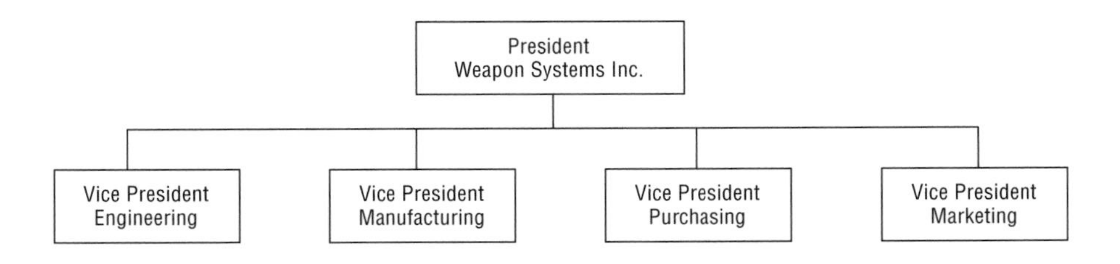

Figure 4.3 • Functional Engineering Organization at Weapon Systems, Inc.

service for that customer grouping are put together in a self-contained organization unit. For example, all of the mechanical engineers necessary for designing and manufacturing each type of product would be assigned 100 percent to that product. This produces some redundancy of functions. One would need to hire at least as many mechanical engineers as there are products that require mechanical engineering. Figure 4.4 illustrates a simple product structure for Weapon Systems, Inc. There are separate business units led by a Vice President for each of the three main product lines—airborne missiles, defense decoys, and torpedoes. Each unit has self-contained, dedicated engineering, manufacturing, purchasing, and marketing staff led by Directors. It is also possible to divide the organization by market segment, rather than products per se. In this case, all the functions are assigned to the market-based groups. Since product organizations and market organizations have much the same behaviors, we will only refer to the product organization in this book.

Matrix Organization. This form of organization, popularized when it was used effectively by NASA for getting astronauts to the moon, combines functional and product (or market) organizations. This is often used in engineering organizations where tasks are grouped into projects. Each person has at least two bosses—one functional manager and one product manager. Consider the example in Figure 4.5. The func-

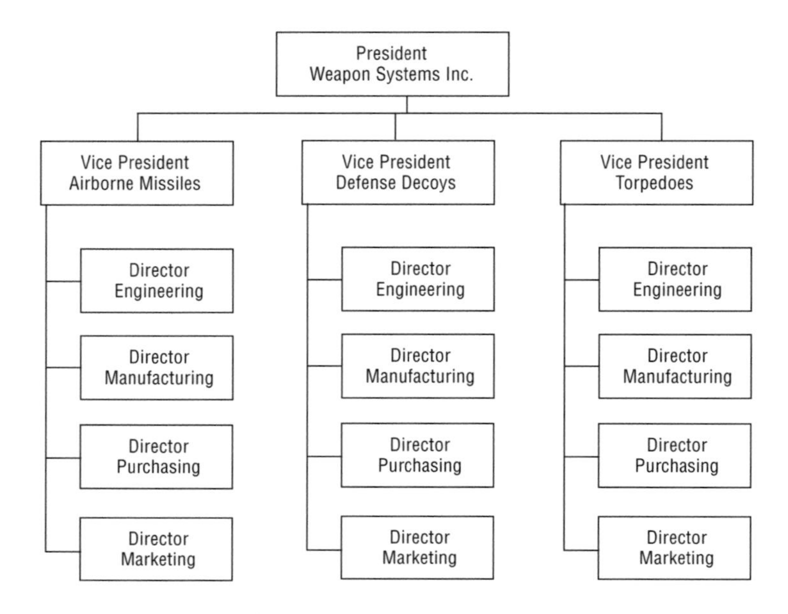

Figure 4.4 • Product Engineering Organization at Weapon Systems, Inc.

tional managers are shown as column headings and the product managers are shown as the rows. Subordinates reside in a "home" functional department reporting up to a functional manager, in this case at the Vice President level, and they are also assigned to a product group reporting to a General Manager. For example, a structural engineer might report up through a structural engineering manager to the Vice President of Engineering, but also report to the Missiles General Manager. This matrix structure fits naturally into a firm such as a weapon systems manufacturer that works on a distinct number of large government projects. But it has also been used effectively in private sector manufacturers such as Toyota. At Toyota, engineers are assigned to functional groups and to a vehicle development team for an extended period of time, reporting to the chief engineer of that development program. Day-to-day management of the engineer's project activities is usually performed by the product manager (chief engineer), while performance evaluations and career planning are typically done by the functional manager, with input from the individual's project managers.

Hybrid Organization. There are many variations on the pure types we just described. For example, functional organizations are often nested within product organizations, as illustrated by the case of AlliedSignal in Figure 4.6. Note that at both the Sector and Division level, AlliedSignal is a product line organization, but at the Division VP level it becomes functionally based. For another example, Chrysler, at the VP level, is a

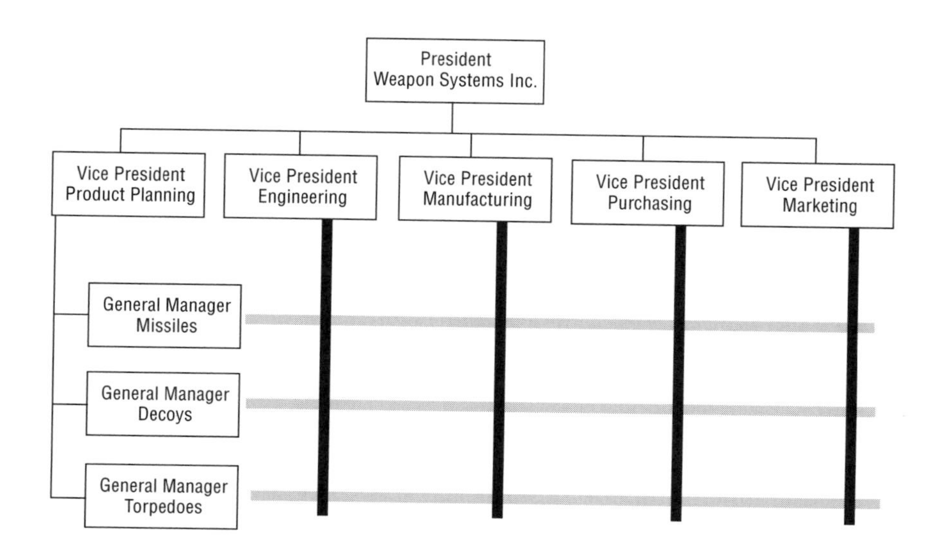

Figure 4.5 • Matrix Engineering Organization at Weapon Systems, Inc.

functional organization.[1] Within Engineering, the form switches to product lines or platform teams. Design engineers are assigned to "platform teams" which focus on a grouping of products (e.g., large cars) and some manufacturing engineers are assigned to those teams, but marketing and financial analysts are still assigned to functional groups and most of manufacturing is in a separate organizational hierarchy. Vehicle styling is even in a separate organizational unit outside Chrysler's platform teams. In fact, most organizations of any size or complexity are likely to be some hybrid—pure organizational forms are the exception, not the rule.

There is no perfect organizational architecture. Each has its benefits and disadvantages. We have summarized a set of common benefits in Table 4.1.[2] We will briefly discuss each of the categories of benefits and disadvantages.

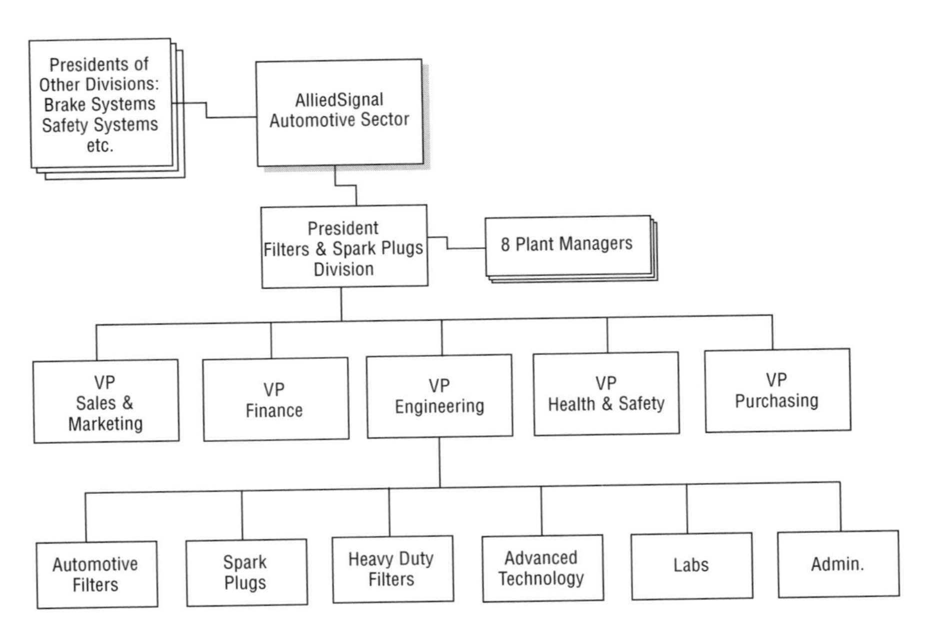

Figure 4.6 • Hybrid Organization at AlliedSignal

1. *Synergy within Functions.* Synergy is the extent to which the "whole" is greater than the sum of its parts. Synergy within functions refers to the degree to which there is interaction among those in the same discipline or function in order to stay current on technological innovations and know-how. Through this interaction, a group of en-

[1] An organization chart for Chrysler can be found in the case study in the Appendix.
[2] Many of the ideas in this table were adapted from Duncan (1979).

Benefits of Organization Architectures	Functional	Product/ Market	Matrix
Synergy within functions	High	Low	Medium
Synergy across functions	Low	High	Medium
Economical use of talent	High	Low	High
Ease of performance evaluation and control	High	Low	Low
Goals customer or end-product driven	Low	High	High
Coordination of effort toward customer or end-product	Low	High	High
Easy to manage or operate	High	High	Low

Table 4.1 • Benefits of Alternative Organizational Architectures

gineers will learn from each other, and build on each other's ideas in ways that exceed the sum of their individual capacities. For this purpose, the functional organization form is best, because those from the same functions are grouped in the same departments and have more opportunity to interact. They also will be encouraged to focus on their professional specialties by a common boss. The matrix organization is also fairly strong on synergy within functions as individuals ultimately report, for career development and performance evaluation purposes, to a functional boss.

2. *Synergy across Functions.* This is the synergy that can result from interaction among those of *different* functions and disciplines so they can learn from other disciplines and get new ideas, i.e., get out of their "disciplinary rut." The now classic image of product engineering throwing the design over the wall to manufacturing reflects a complete *lack* of synergy between design and manufacturing. Obviously a pure functional organization will be the worst at providing this kind of synergy. A product or market organization will be best. A matrix organization falls somewhere in between. If the product (or project) heads have real power, the matrix organization can be very strong on synergy between functions.
3. *Economical Use of Talent.* A pure functional organization uses talent more economically. People can be flexibly assigned to projects as needed. If they are needed for just a few hours on selected weeks, they can be assigned for just those few hours and work on as many other projects as time permits. On the other hand, in a product organization, each product group has the full array of talent dedi-

cated to it. This means there will be a certain amount of redundancy. Even if projects are winding down and a smaller number of engineers are needed in that product group, they are permanent assets of that group. This can be problematic for smaller companies where the extra cost of this redundancy can be a large proportion of payroll. Another cost of the product organization is that each product group may lack the critical mass of engineering specialists required for in-depth research, and physical facilities may have to be duplicated. But the benefits that come from synergy across functions and doing things right the first time in the product organization may well outweigh the cost of this redundancy. The matrix form, if properly managed, can lead to economical use of talent as engineers are shifted among projects based on need.

4. *Ease of Performance Evaluation/Control.* Typically, it is easier to set up uniform evaluation and control systems within functional disciplines as compared to product groups. There is a common basis for evaluating employees who share a discipline; the functional manager is likely to have a better understanding of professional expectations for that function. In the matrix organization, a major benefit over the product organization is that the primary responsibility for performance evaluation is in the hands of the functional managers. But the functional manager has less day-to-day contact with subordinates as compared to a pure functional structure. In a well-functioning matrix, the evaluation process would include input from project managers, with evaluation being a consensus decision of the functional and project managers. This need for communication and coordination increases the complexity of performance evaluation. Similarly, control can become quite complex as each employee has at least two bosses who must coordinate their supervision, oversight, and goal-setting. Employees can get conflicting direction, or even lack of direction, as each manager thinks the other is really supervising the employee.
5. *Goals are Customer- and End Product-Driven.* Those in functional departments often focus too narrowly on the local goals within their own departments, rather than focusing on the customer and/or end product as the goal. For example, a mechanical engineer may want to perform state-of-the-art structural analysis and make the structure withstand every conceivable stress and strain even if this does not lead to the most attractive or the most manufacturable design. The functional form will have the most difficulty focusing on the customer and total end product, whereas the product and market organization will be best. Again, the matrix organi-

zation has the potential of being very strong at this, if the product manager is powerful enough to influence those assigned to the team.

6. *Enhanced Coordination of Effort for Customer or End Product.* Grouping people around customers or end products as in the Product organization obviously will enhance coordination of efforts toward end goals rather than the intermediate steps taken to achieve them. This can enhance the move toward rapid product development.
7. *Ease of Management and Operation.* The more complex the organizational architecture, the more difficult it is to manage. While we would never say any manufacturing company is easy to manage, it's clear that under most circumstances, the matrix form is most difficult. This is because each individual or group has multiple "bosses." Unless this is managed very well, subordinates will find it difficult to deal with conflicting goals and priorities, and may experience a great deal of tension and uncertainty.

Choosing a Form. Given our discussion of hybrid organizations, it should be obvious that you don't get to choose a single "form" for your entire company. What you do need to decide is where the most cross-functional or cross-product line integration is needed. Grouping individuals and units under single managers provides a relatively high degree of integration. It is important to recognize, however, that the architecture only provides a small portion of the total integration needed. The other cross-functional integration mechanisms discussed in this chapter are also essential.

When choosing a form of architecture for part of your organization, you should give major consideration to the characteristics of the environment and tasks we discussed in Chapter 2: environmental uncertainty, task uncertainty, and task interdependence. Table 4.2 relates each of these three characteristics to the organization architecture forms. Product and market organizations are combined as the same pattern of supportive conditions applies. Let's consider each type.

Functional Organization. The functional organization has the advantage of strong technical strength in each of its discipline areas. On the other hand, it tends to be weak in terms of cross-functional integration and the ability to focus on customers and company-wide goals. Under conditions of high *environmental uncertainty*, and, most critically, rapid changes in what the customer wants, the functional organization will have the most difficulty learning and absorbing information about the changes going on around the company. *High task uncertainty* suggests that the company may be searching for new ways to go about its fundamental work. Having a strong concentration of knowledge about specific technologies in functions may be a useful way to find new ideas.

	Functional	Product/Market	Matrix
Environmental Uncertainty	Low	High	Medium to High
Task Uncertainty	High	Low	Medium to High
Task Interdependence	Low	High	Medium to High

Table 4.2 • Conditions Favoring Each Form of Organization Architecture

High task interdependence means that different functions need to work closely together; obviously the functional organization would be weak at this.

Product Organization. The product organization is strong in the opposite conditions of the functional organization. Product-based groupings tend to be focused on common company- or division-wide goals and are hence better able to focus on customer needs in an uncertain environment. Thus, a product or market-line organization would work better under conditions of high *environmental uncertainty*. In contrast to the functional organization, the product or market-line organization often has difficulty getting a critical mass of people in a single technical discipline together to solve a technical problem. Thus, this type of organization does not work as well under conditions of high *task uncertainty*. Finally, the product or market-line organization deliberately brings different functions together under one roof, making it easy to work under high *task interdependence*.

Matrix Organization. When it is functioning well, the matrix organization excels at flexibly adjusting to changes in the environment and there is the functional depth to keep up with changes in technology within functional domains. It also has the flexibility to deal with high *task interdependence*. The matrix seems to thrive when there are a relatively small number of distinct product lines. If there are so many different product lines that every individual in the matrix must be assigned to many different development projects, the matrix will quickly become unwieldy. It works best when individuals can be assigned to one or maybe two projects with reasonably long lives. Then the individuals assigned to those projects have time to become a team, learn how to work together, and learn how to work with the project leader.

Conclusions About Architecture. Architecture is just one of many ways to build cross-functional integration, but it is one of the most fundamental. In companies where cross-functional integration must be strongly balanced with cross-product line integration (i.e., within-function synergy), then some form of functional architecture may be highly desirable, as long as it will be combined with other, relatively strong cross-functional integration mechanisms, such as those we discuss in this chapter. In companies where cross-functional integration is paramount, then it should be built right into the architecture through some type of product organization or matrix.

Project Management Structures

Cross-functional relationships are often organized around projects. The project management structure describes the assignment and distribution of administrative and leadership responsibilities and authority for specific projects, such as the development of the next generation DRAM chip. The organization chart may reflect this structure, but it is likely to do this rather poorly. For example, the organization chart may show a product planning organization that includes a set of project managers, but it will not show their responsibilities and authority since the project manager may not formally have direct reports, or may have very few direct reports. In the case of a product organization where the heads of product groups are project managers (as in Chrysler's platform team organization), the organization chart may not be a bad picture of the project management structure.

Just as we divided the formal organizational architecture into types, we can divide Project Management Structures into the following five types: absent, liaison, lightweight, heavyweight, and autonomous. In Table 4.3, the boxes marked "X" show the project management responsibilities of each project management form.

A summary of the strengths and weaknesses of each of these five types is shown in Table 4.4, with the weakest forms of project management to the left and the strongest to the right. The primary strengths (the "highs") of each type have been highlighted with white letters and a dark background.

Absent. There is no "formal" project manager in the absent project manager type, which would be the case in a "pure" functional organization. Functional specialists work on different phases of the project, with coordination provided by a general manager (or even the president, in smaller companies). The general manager serves as the spoke of the wheel, providing very limited cross-functional coordination. As shown in Table 4.4, organizations that rely purely on the functional structure, without project managers, can have very high synergy within functions, but little synergy across functions. This works reasonably well for very small firms where informal communication across functions is very high, or on products which have functional subsystems which barely interact. This form of management is sluggish in its response to changes in the environment, and requires

Project Management Responsibilities	Absent	Liaison	Lightweight	Heavyweight	Autonomous
Distribute and share technical information among project members in different functions		X	X	X	X
Distribute reports, minutes of meetings		X	X	X	X
Set project goals		X	X	X	X
Schedule and coordinate project activity			X	X	X
Allocate funds and equipment for project			X	X	X
Select staff for project				X	X
Evaluate performance of project members on project				X	X
Evaluate overall performance of project members					X
Professional development of project members					X

Table 4.3. • Responsibilities of Project Managers by Form.

a relatively slow, sequential approach to development with long feedback loops.

Liaison. The role of maintaining the linkage between two organizations is known as a liaison role. Acting as a liaison is generally a part-time position, or in some cases a full-time but temporary assignment. The liaison usually reports to one department (e.g., engineering) and is responsible for coordinating activities between the home department and a target department (e.g., manufacturing). Sometimes this is combined with collocation when the liaison is physically located in the target department. For example, we know of several cases where a manufacturing engineer has been given an office and spends half or more of the time in the design engineering building. In other cases, the product engineer is temporarily assigned to the manufacturing plant as much as full time during the launch of a new product. The liaison is the weakest form of cross-functional project management. The liaison serves to pass information back and forth between functions, but typically lacks au-

Engineering Priorities	Absent	Liaison	Lightweight	Heavyweight	Autonomous
Synergy within functions	High	High	High	Moderate	Low
Synergy across functions	Low	Low	Moderate	High	High
Achieve customer goals	Low	Low	Moderate	High	High
Achieve company goals	Low	Low	High	High	Moderate
Responsiveness and speed of development	Low	Low	Low	Moderate	High

Table 4.4 • Strengths and Weaknesses of Project Management Structures

thority to manage the project. As shown in Table 4.4, it essentially retains a functional structure and thus synergy within functions remains strong. The liaison can be very useful for addressing short-term needs during a critical phase of the development process, e.g., the product engineer in the plant during manufacturing launch. But this often involves fixing problems that could have been anticipated earlier through a more powerful form of project management.

Lightweight. The lightweight project manager is often a design engineer or product marketing manager, mainly responsible for such coordinating activities as sharing information across functional groups, setting project goals, scheduling, updating time lines, and expediting across groups. The lightweight manager will generally control the project budget (Table 4.3), but functional or product-line managers retain authority over human resource allocation, performance evaluation, and personal goal setting. In a sense, the lightweight project manager holds authority over the project, but not the people working in it. This can be extremely stressful for the project managers as they may have responsibility without authority, and have to cajole people to do project work. Because of their limited power, lightweight project leaders are often ignored or preempted, potentially leading to costly and time consuming delays in project completion as well as compromises to quality. On the other hand, because they do control the project and its other resources (e.g., information, money), this form can be extremely effective in cases where personnel are made to be responsive to the need to acquire those resources. The *Newport News Shipbuilding Case* (in the Appendix) provides an example of extremely effective lightweight project management. In this case, since the only source of revenue for the shipbuilding firm comes from government contracts which are structured as projects, managers of those projects end up with considerable power and influence. Overall, the strengths of this project management form are in focusing project resources on company goals while retaining synergy within functions (Table 4.4). We rate this form as moderate in providing synergy across functions and focusing people on the customer. Because of the lack of power of the lightweight manager in directly influencing the people working on the project, responsiveness and speed of development is not very high.

Heavyweight. The heavyweight project manager directly supervises project members' work, and may be responsible for their hiring and evaluation for the project, although overall performance evaluation and longer-term career development usually rests with a functional manager (Table 4.3). Particularly when combined with a strong team structure (e.g., where people are dedicated and collocated for the duration of the project), heavyweight project management has many benefits, including strong employee loyalty and commitment to the project, strong coordination of different functional specialists, and a clear focus on the end goal (Table 4.4). But the heavyweight project management structure has drawbacks too. Conflicts can still occur with functional managers which must be managed from the top. Also, the heavyweight project manager may take control of resources needed elsewhere in the organization. A successful heavyweight structure depends upon a clear vision by all of the overall company direction, and a clear charter for the scope of the particular project being managed. In Japan, many companies have developed a form of heavyweight project manager they call the chief engineer.[3] The chief engineer combines the formal authority of the heavyweight project manager with deep technical expertise and experience, and the command of great personal respect within the company. A chief engineer is often able to resolve conflicts which a less respected manager would be forced to refer to a higher level. The *Toyota case* (in Appendix) provides an example of the chief engineer. Toyota only promotes engineers to this highly revered position who demonstrate *both* exceptional leadership skills and outstanding engineering skill.

Autonomous. In the autonomous project management structure, the project manager has full control over members of the project, including hiring, firing, and evaluation. Top management holds the project manager fully accountable for final project results, while allowing the project manager to create incentives and norms for employees in the group. The *Chrysler Platform Team Case* in the Appendix provides an excellent example of an autonomous project management structure. Unlike Toyota, where engineers still reside on a long-term basis in the functional orga-

[3] The "chief engineer" was referred to in Clark and Fujimoto (1991) as the "shusa." In fact, shusa is a general term in Japanese that means project manager and can include project managers for relatively minor projects. Toyota literally uses the term "chief engineer" taken from the West to refer to their heavyweight managers of product development projects.

nization and then are on loan to the chief engineer for projects, Chrysler shifted to a platform team structure and eliminated the functional organization within engineering. The head of the platform team (a general manager) has a great deal of project management autonomy. Most people who work on the project do so full time and report solely to the general manager of the platform team. This project management structure has the benefits of providing complete control in the project manager's hands, in effect creating a small, focused company which can act very quickly in a crisis. On the other hand, it reduces synergy between functions to near zero, and, if it is applied on an isolated basis to one or a small number of projects, may result in a "rogue project" which loses track of both company and customer goals. Lockheed's well-known "skunk works" approach to development of the U2 and SR71 airplanes is a good example of the successful use of this approach.

Relationship Between Project Management Structure and Organizational Architecture

When we discuss this framework, some people get confused by the difference between a project manager and a product-line manager. The difference is that the project manager is managing day-to-day work on a single project, such as a stand-alone product or a component. The product line (brand) manager is managing a whole line of products. For example, a product-line manager within Whirlpool might be managing the line of Kitchen-Aid dishwashers. A project manager in the same organization might be managing the development of model 84-97A Kitchen-Aid dishwasher. In the situation where there are relatively few products within a product line, then the project manager and the product-line manager may be one and the same. In the case of Chrysler, the platform team generally only has a single product under development at any given moment, hence the platform team general manager is also the project manager for the new vehicle. Despite this occasional overlap, the project management structures we have described are independent of the architectures. In other words, you can have any of the project management forms within any of the architectural forms.

Standardization of Work Processes

The standardization of work processes means that there are specific rules for how work gets done. In some sense, this means the work can be "programmed" to follow a specific set of steps and approaches, but not necessarily in the sense of being programmed for performance by a computer. Jobs that can be heavily programmed in this way would have *low* "task uncertainty." If we look at the work of product development, it should be obvious that the work of designers and design engineers cannot be programmed in the same way an assembly line task can; thus they have relatively high task uncertainty. On the other hand, we see a wide variety of rules and standards for how design work is to be done. Some of these rules, such as DFM rules, define the interfaces between functions (a classic coordination issue), others define how parts of the work will be done (e.g., a rule to always perform a mold flow analysis), which helps ensure that coordination issues will not arise. There are at least five approaches to standardizing work processes: standard operating procedures, planning and scheduling systems, monitoring systems, structured development processes, and tools and techniques. We will discuss each in turn.

Standard Operating Procedures

A prime example of standardized work processes is standard operating procedures (SOP), such as those for initiating and executing engineering changes, configuration management procedures to keep up the CAD database, and procedures for requesting prototype builds. They might also include procedures for forming a team, the composition of the team, the expected frequency of team meetings, the involvement of suppliers, or requirements for getting customer input throughout the design process. SOPs are particularly vulnerable to becoming rules that are enforced blindly. If the standards become rigid and inflexibly applied to all situations, they will become a barrier to effective concurrent engineering instead of an enabler. Every situation is to some degree unique, so standards need to be flexibly interpreted and continually updated with the participation of those doing the work.

Planning and Scheduling Systems

Another formal mechanism for coordinating people with different interdependent tasks is the use of a common plan and schedule. While plans shouldn't be used as a tool to browbeat managers and engineers, they must be taken seriously enough that team members are highly motivated to meet critical deadlines. For example, Toyota uses a very simple flowchart of their development process that focuses on a few key milestones such as first full vehicle prototype, second full vehicle prototype, etc. The chief engineer running the program can set the actual dates, though there is a generally understood schedule for when these milestones should occur. There is a great deal of flexibility about when different engineering groups inside the company and outside suppliers should begin their work, but there is virtually no flexibility in meeting the scheduled milestones. It is expected that all parts will be ready in time to assemble the prototype vehicles, and in Japan this almost always happens. They have experienced much lower compliance rates when dealing with American suppliers. Chrysler has used a similar approach with considerable success in the U.S.

Monitoring Systems

Once standards and targets have been set, monitoring systems are needed to provide a feedback loop on how the project is doing. Anything one can observe or measure can be monitored — quality of final or intermediate work products, amount of activity, schedule compliance, compliance with standardized practices, and so on. In the traditional Tayloristic view, monitoring systems are a means to simply control workers. Managers can look at the measures and reward or discipline workers as appropriate. In fact, the measures are often designed to be tamper proof and may not even be shown directly to the workers. By contrast, as an *enabling mechanism*, monitoring systems are an important source of feedback to workers.[4] This is the approach taken by Motorola (see the case study in the Appendix). In general, it is known that if measures are only used to externally control people and are tied to external rewards and punishment, a serious negative consequence will follow: people will work to manipulate the measures regardless of whether this is effective in reaching project objectives. This leads to inaccurate information, wasted effort beating the system, and misdirected activity. We discuss more about measurement systems and their relationship to employee motivation systems in Chapter 6, on the people side of the product development process.

Structured Development Processes

Structured development processes are standardized approaches to the flow of the design project. We also refer to these in Chapter 3. They are really a combination of work process standards, monitoring systems, and planning and scheduling systems. They provide standardized activities that should take place in the product development process, the sequence, the expected inputs and outputs, and often standard timing. The design review is a formal mechanism that provides a scheduling milestone to coordinate the activities of many different participants in the design process, and an opportunity to monitor progress and give feedback on performance relative to goals. As in all of the methods for standardized planning and control, the structured development pro-

[4] See Adler and Borys (1996) for a discussion of the distinction between "enabling bureaucracy" (use of standardized procedures to help people do their work) and "coercive bureaucracy" (use of standardized procedures to coerce people into doing things in certain ways even if it does not make sense to them).

cess can be an enabler or a coercive set of barriers to concurrent engineering, depending on how it is managed. Companies that have gone too far in detailing microscopic activities and very specific timing and sequences in voluminous notebooks find that the methods sit on a shelf and are rarely used. Broad general approaches that can be flexibly interpreted are more likely to be useful. As discussed in Chapter 3, TI and Sun effectively provide flexible approaches to structured development processes.

Tools and Techniques

Design tools and techniques are also ways of standardizing work processes. We are using the term "tools and techniques" here in a somewhat narrow way to encompass computerized or paper aids that help bring multi-functional perspectives to the design process. There are of course many, such as Taguchi methods, design for manufacturability and assembly software, expert systems, quality function deployment (QFD), and Pugh selection matrices. They all have the ability to bring data upstream into the design process to consider downstream issues, and they all represent standardized methods for doing design. Let's consider just two examples. First, Quality Function Deployment involves identifying relationships between upstream and downstream processes. For example, the first QFD matrix (often called the house of quality) provides a systematic way of linking customer requirements, often the responsibility of marketing, to engineering specifications, the responsibility of engineering. This forces engineers to systematically think about how marketing data (an upstream consideration) relates to design specifications (further downstream). The second example is Design for Manufacturability (DFM) tools which analyze the manufacturing consequences of different design alternatives. Both QFD and DFM are standardized approaches to forcing a degree of integration between upstream and downstream processes.

There is not much comparison that can be made between these different ways of standardizing work processes. All can be used in a flexible manner, adjusting for specific situations, or in an inflexible, coercive manner with little regard for individual project needs. Our preference is for greater flexibility and adjustment to specific situations. One reason standards tend to be thought of as extremely inflexible is that they often take on a life of their own. Too often, "the rules must be enforced the same for everyone" even if they do not make sense in a particular situation. By keeping the fundamental purpose of work process standards in mind, specifically the need for cross-functional integration, they can be kept flexible and useful.

Standards for work processes provide less immediate control over work than does direct supervision. However, given the realities of what is possible regarding supervision in the design task, standard work processes may actually have more impact than direct supervision on design work. For this reason, it is almost always necessary to use some or all of these methods in any product development process.

Standardization of Outputs

Outputs can be standardized in two ways: design standards and performance metrics.

Design Standards

This is a set of rules about what the final results of a design should be like. For example, a company might have rules about the use of standard fasteners, standard parts, standard part characteristics (e.g., minimum heights), and manufacturability rules (e.g., avoid curves greater than a 60° radius). These are different from work process standards, such as SOPs, since they set requirements about what the final output will include, rather than how the individual will go about producing that result.

Performance Metrics

These include rules for how the final design will perform, the cost of developing the product, and manufacturing cost targets. The most common example of this is a design specification for a product, e.g., the part will have certain dimensions and weigh so much. Another example might be a performance improvement target, e.g., each design cycle should include a 10 percent weight reduction and 5 percent fewer parts over the last cycle. Excellent companies set measurable objectives for specific projects as well as general targets to be achieved. For example, Motorola's 6-Sigma Quality is both a general quality target and a specific target for each new product. Motorola also uses very specific quantitative targets for product development projects, e.g., they will set specific targets for cost and weight.

Standardization of outputs is among the most widely used coordination methods in product development. Almost everyone uses some form of performance metric, if not in terms of improvement targets, then in terms of performance specifications. Design standards are also very widely used, but less ubiquitous. The very common nature of output standards often causes people to forget they're there or to forget why they're there.

One of the most critical purposes for these standards is to provide for ease of coordination. For example, component designers need to know a great deal about their interfacing parts in order to do their own designs. In a sequential engineering process, it might be possible for one designer to wait for the other to do their work and to then pass it along. In concurrent engineering, waiting isn't possible; coordination is critical. But, if each knows critical elements of the others' designs in advance, as a result of adhering to design standards, for example, then all can proceed as quickly as possible, without waiting.

Toyota uses extensive engineering checklists to standardize on outputs. These are used extensively for body engineering, for example. Through the years, as engineers learn what works and doesn't work, they develop standards for critical dimensions (e.g., minimum hinge size for a certain size and weight door), equations that can be used to calculate structural strength based on a given type of steel and size vehicle, minimum clearances that must be met, etc. New engineers at Toyota say that for much of body engineering they can look up information in the checklists and the design is almost complete. Manufacturing engineers also keep checklists of design features that cause trouble from a manufacturing perspective and use these to evaluate car bodies in design reviews (Ward et al., 1995).

Standardization of Worker Skills

Worker skills can be standardized in two basic ways: credentials and training.

Credentials

These are formal evidence of completion of some education or experience that provides a high level of specific skills. For example, completion of an approved apprenticeship program in die making says that the holder has a specific set of skills and experiences. In theory, if you give a person who holds this credential the task of making a die, he/she will go about it in pretty much the same way as someone else with the same credential and with about the same level of performance. All credentials are not created equal, however. A B.S. degree in some type of engineering from one university can be quite different from that received from another university, despite the efforts of professional associations to standardize such degrees. Even within a single university, the skills obtained as part of a single degree program will vary widely. Despite these differences, there are some minimum assumptions one can make about holders of an engineering degree. For example, all engineers will have been exposed to some basic grounding in physics and calculus. Mechanical engineers all take at least one course in structural mechanics. Most engineering students will likely have some CAD exposure. In this case, the credential might only serve as the minimum qualifications to receive further skill enhancement through training.

Training

We will distinguish training from education by its specificity — training focuses on some narrow, specific skill. For example, a company might train a design engineer in Quality Function Deployment (QFD), with the expectation that the engineer will be able to perform certain activities we call QFD when required, and that it will be performed in a certain way, as specified by the training program. Of course, providing training does not mean it will necessarily be used, and if it is not used the skill will not really be developed.

Much like the other forms of standardization, standardization of worker skills provides coordination by making both direct supervision and mutual adjustment unnecessary. If one person knows what the other will do, then they needn't spend time talking about it. Both credentialing and training are of course widely used in the product development arena. The typical division of labor between functions often implies that members in some functions hold different credentials (and hence different skills) than those in other functions. The engineering degree is considered an important credential by many for hiring design engineers, though it does not ensure that new engineers will have the specific skills a practicing engineer needs. Good firms have developed standard curricula to bring newly hired engineers up to speed on the specific tools used in that firm (e.g., the specific CAD system, specific design for manufacturability rules, etc.). For example, Hitachi has a two year standardized training program for new engineers (including on the job training). Until they have completed this, they cannot take an engineering assignment without close supervision. When they complete the two years, a graduation ceremony is held.

Mutual Adjustment Through Cross-Functional Teams

Cross-functional teams are the one common thread running through every case of concurrent engineering we have observed. While they are not always used as effectively as possible, in most cases they seem to help considerably in developing a more manufacturable product. In the best cases they also shorten lead time considerably.

The main advantage of a team is the richness of face-to-face communication. Properly led, a cross-functional team can help members develop common project goals and thus more easily address conflicting objectives from their different functional perspectives.

There are many different ways to characterize and form cross-functional teams. One important dimension is the degree to which the team is a relatively permanent entity versus a short term collection of individuals for a specific task. Another is the degree to which the team deals with a broad set of technical and business tasks (high scope) versus a very specific set of problems (low scope). These two dimensions are used in Table 4.5 to characterize four types of teams.

	Low Permanence	High Permanence
Low Task Scope	Task Force	Standing Committee
High Task Scope	Temporary Team	Semi-permanent Team

Table 4.5 • Types of Cross-Functional Teams

Task Forces

Task forces are temporary teams assigned to work on a specific task of limited scope. They are generally problem focused. For example, if there is a problem with paint peeling from a metal part, a cross-functional task force may be formed to identify the cause of the problem and recommend a solution or alternative solutions. Since the scope of the task is limited, task forces usually do not work on the problem full time, although it is possible that for a period of time they might do so.

Task forces are very useful in concurrent engineering to address specific problems that cut across functional boundaries. For example, the paint problem may involve the interplay between product characteristics such as shape and surface finish, and the paint processing approach. Thus, product and process experts should work together, concurrently, to address the problem. On the other hand, solving specific, local problems will not lead to the development of a complete integrated product-process system. Thus, task forces should only be used for solving specific technical problems such as investigating alternative materials, for fixing problems that occur after the fact, and for improvement projects of narrow scope.

Because task forces are temporary and typically not a full-time assignment, there is only limited opportunity for members to learn how to work together and become a highly cohesive team. Task forces should, in principle, be disbanded after the problem is solved, though often solving one problem leads to another and they become more like standing committees.

Standing Committees

These are semi-permanent teams assigned to work on a specific set of *ongoing* tasks or issues. There may be standing committees to review design standards, authorize capital expenditures, review new project proposals, and maintain or improve the computer-aided design system. Unlike a task force, they are not dealing with one specific problem but with a class of problems.

Generally, committees meet on a regular schedule for relatively short periods of time, e.g., one hour every other Tuesday. There is simply not enough time for in-depth problem solving or design in these regular meetings. Committee meetings can be effectively used as forums for information sharing, coordination, and decision making after individuals in the committee have studied the issue outside of the committee meeting. Because of the relatively brief contact periods, like task forces, committees are not likely to become highly cohesive teams or develop tightly coordinated ways of working together. Technical discussions in committee meetings are not likely to be very deep.

Temporary Teams

This is like a task force but the team has a task with a broader scope. A skunk works is an example of a temporary team. They may come together to develop a new generation product and act very autonomously to address all aspects of the product and process. Those assigned to work on the temporary team may well be dedicated full-time for the duration of the project. Then, when the product and process are designed, they will disband and go back to their functional departments or go on to join a different team.

Toyota uses temporary teams very effectively for product development. New product programs are run by a heavyweight project manager — the chief engineer. As the program ramps up, individuals are assigned by their functional bosses to work on the program, perhaps part-time at first, leading to full-time involvement during the peak period. After launch, team members return to their home function or are assigned by the functional boss to a different product program, usually under a different chief engineer.

The temporary team can be an efficient way to use people, while providing the benefits of a team for cross-functional integration. People are assigned to teams as they are needed and can be reassigned when they are no longer needed. They can even split their time among several projects if no one project needs their full attention. People can also develop a broader base of knowledge and skill by working on different products under different leaders. This rotation can bring a great deal of energy and ideas to new projects.

There are also disadvantages as compared to semi-permanent teams. Each time a team is disbanded and a new team created, the process of team formation must begin again, which is likely to lead to some loss of productivity in the start-up period. Also, it becomes more difficult to manage a project if people assigned to it have other responsibilities outside of the team. Coordinating the team, getting the time of people when they are needed, and even setting up meetings, becomes more difficult.

Semi-Permanent Teams

Semi-permanent teams are typically dedicated to one product line and are responsible for successive generations of the product, as well as improvement and maintenance of the product. This is likely to be a full-time assignment for team members who are dedicated to the one product or product line and generally have offices near each other, i.e., dedicated, collocated teams. Team members gain great depth of knowledge about the particular product and also learn ways to work together. It is easier to coordinate teams when all are dedicated to a single project. There is a great deal of ownership of the product, and the same team responsible for design will be responsible into production and next generation designs. Team members learn much about the functional perspectives of others on the team, and gain a deep personal understanding of other team members.

One major cost of the semi-permanent team is the lost assignment flexibility which can lead to inefficient staffing, such as having people on the team full time when they are not needed quite that much. Also, team members can develop dysfunctional relationships and may be stuck working together anyway.

If all or most of the staff are assigned to semi-permanent cross-functional teams, by definition this means the functional organization has been disbanded. In the auto industry, Chrysler, more than any car company, has disbanded the functions within product development and assigned individuals to semi-permanent teams by product line. The results on the first generation of new vehicles done under this team approach were very impressive in many respects. But Chrysler's within-function synergy was problematic during the development of these early vehicles (something they have actively worked to address).

You will note that we really haven't defined a "permanent" team, but rather we call the most permanent form of team "semi-permanent." In effect, a permanent team would be a department within a product-line organization. Since that is formally an element of organization structure, we have discussed such departments earlier in this chapter (see the section on Organization Architecture).

Organization Culture

So far we have talked about cross-functional integration mechanisms in the internal organization, but we have not yet brought up the subject of the organization's *culture*. For many engineers, "culture" seems to be just about anything "soft" that they can't measure using engineering methods. Even for many managers, culture seems to be anything in the organization that they just don't know how to deal with by the usual management tools. A comment we often hear when discussing the people and organizational side of CE is, "Oh, that's a culture problem." Or they say, "Yes, we've got to change the culture around here." So, for many people, culture really represents the mysterious, the unknown, the unknowable, the "dark side" of the organization. And, indeed, when we look more formally at culture, there are some aspects of this that is an "unknown" part of the organization.

Probably the best-known author on the subject of organizational culture is Edgar Schein (1992). He defines culture as follows:

> "*A pattern of shared basic assumptions that the group learned as it solved problems of external adaptation and internal integration, that has worked well enough to be considered valid and therefore, to be taught to new members as the correct way to perceive, think, and feel in relation to those problems.*" (p. 12)

As you can see from this definition, the unknown part of it has to do with assumptions — those things which affect our behavior, but which we rarely, if ever, articulate, and hence often treat as unknown. Our purpose in this section is to provide you with a framework for thinking about and analyzing your organization's culture. In doing this, we don't expect to show you how to make major changes in your culture, since that is at best extremely difficult, and some would say impossible. Rather, we are mostly concerned that you be aware of your organization's culture, and understand both the limits it places on you, and the strengths it provides.

This is not to suggest that we believe culture can't change; quite the contrary. However, we believe that culture changes slowly and indirectly. This means that the fundamental assumptions in your culture can usually only be changed through experience rather than through logical, rational thought and discussion. Thus, making a decision to change your culture, running a workshop, and leaving it at that will lead to little or no change. In contrast, changing what people do on a day to day basis will ultimately change the way they see the world, resulting in a long-term cultural shift.

As a simple example, imagine a CEO who decides that his company's culture is not participative enough — people stick to their own thing and don't ask for or provide help when needed. One of their fundamental assumptions is that people should be self-sufficient. The CEO then decides to go out and make speeches about how important it is to be participative. He also writes articles for the company newsletter and sends e-mail to employees telling them to help each other out more and to ask for help when needed. Imagine how much change this is likely to result in — probably very little! Now imagine the CEO not saying a word about participation, but rather acting participatively and rewarding others for doing so. For example, he now asks for help from others. He evaluates his direct reports on how much they involve others in decisions, and whether they evaluate their direct reports in the same way. In relatively short order, fundamental assumptions about how self-sufficient people should be would change.

The point of this example is not that the CEO has to personally initiate all change (although it certainly helps!), but that culture change most often follows behavior change, rather than the other way around. Note that Schein's definition talked about culture as being learned — and we learn best from experience.

Schein describes three "levels of culture," which are ways in which culture is represented in the organization. His three levels are as follows.

- ***Artifacts.*** These are visible organization structures and physical objects. Artifacts are relatively easy to find, but often hard to understand. For example, upon approaching an office building, if you see a small number of parking spaces with individual names on them and a large number of unmarked spaces, you might assume that the organization housed inside places a high value on individual status and providing "perks" to display that status. Where the first author of this book works, you would see just that, but the interpretation would be wrong. In fact, the spaces are assigned by lottery to people who contribute to the United Way! A significant amount of investigation is necessary in order to understand the meaning of any artifacts.
- ***Espoused Values.*** These are strategies, goals, and philosophies that are openly stated. A statement of "Corporate Values" posted on the wall, or an executive's speech about how "we value our people

above all other things," represent espoused values. Everyone does not have to agree with these espoused values, and indeed there are often conflicts present. Part of the description of the organization's culture may be to describe these conflicts.

- ***Basic Underlying Assumptions.*** These are beliefs and feelings that are unstated and taken for granted. They may be congruent with espoused values or in conflict with them. We can define some set of shared underlying assumptions among all or most of the members of the organization. Defining these assumptions is often very difficult, and in a great many cases we infer assumptions from behavior and artifacts.

If we combine these levels of culture with the elements of our CE Context Model, we get a matrix as in Table 4.6.

We have filled in the cells of the matrix in Table 4.6 with examples of how each CE element operates at each of the levels of culture.[5] In this case, there are some key contradictions between espoused values and underlying assumptions. In practice, you could use a matrix like this to make comparisons between your own culture and those of your suppliers, or between "subcultures" within your own company. A subculture is a culture which contains elements of a larger culture as well as unique elements. For example, there may be a subculture in manufacturing that is very different from the subculture in engineering. At automobile companies we have visited, we find that product engineers often live in a very different world than manufacturing engineers. The product engineers think in terms of abstract symbols on paper or a CAD tube, while the manufacturing engineers think in terms of very concrete machines and processes. These are evidenced in the form of artifacts, but also represent assumptions about what it takes to understand reality. These kinds of differences can make communication across the boundaries between these two functions very difficult. Another very common example of a cultural difference is between customers and suppliers. We sometimes find major differences in assumptions about how much one should trust someone from another company. For example, a supplier may implicitly trust a customer to give them needed information (about interfacing parts, for example), and, when they bid on work, assume they'll get that information when it's needed. If the assumptions are different on the part of the customer, the needed information may not be forthcoming, with serious consequences for the supplier.

[5] Some of these examples come from Adler (1992) and Liker and Fleischer (1992).

	Artifacts	Values	Assumptions
Work Process	process maps; design rules, observation of work, methods in use	policy statements about quality, statements by design engineers about thinking of manufacturing as their "customer"	beliefs about the value of spending more time early in the design process to prevent having to fix mistakes later
Internal Organization	organization chart, physical layout, presence of people at meetings, observation of interactions	policy statements about the importance of participation, statements of how teams should work together	beliefs that everything must be controlled from the top, providing individuals with the least amount of autonomy
Supplier Relations	physical location of supplier, location of meetings, contracts, observation of interactions	policy statements about the importance of supplier involvement in design, statements of how the company trusts its suppliers and should work to earn their trust as well	beliefs that suppliers are "lower than dirt" and can be manipulated at will by their customers
People	published reward system, job descriptions, training certificates, observation of morale	policy statements about the importance of building staff skills or rewarding people fairly	beliefs that design engineers are higher status than manufacturing engineers
Technology	location and use of specific tools and how advanced they are	statements about how "new technology will solve that problem"	beliefs that new technology will solve all problems and that "we don't need to worry about that 'soft stuff'"

Table 4.6 • CE Cultural Framework

How Much Cross-Functional Integration Do You Need?

Obviously there is no precise measure of the "amount" of cross-functional integration needed for concurrent engineering, or for anything else for that matter. However, there are ways to know *where* in your company more cross-functional integration will be needed. There are three considerations: task interdependence, uncertainty, and distance.

Task Interdependence

In Chapter 2 we defined task interdependence as the extent to which the performer of a task is dependent on another individual or department for information or material in order to perform his/her task. We also defined three types of interdependence: sequential, pooled, and reciprocal, in order of increasing interdependence. Higher levels of task interdependence between functions mean that greater levels of cross-functional coordination will be required to meet the needs of that interdependence. As a very simple example, if design engineering needs to know a lot about manufacturing processes in order to design the product, and manufacturing engineering needs to know a lot about the product to design the manufacturing system concurrently, then there is a high level of reciprocal task interdependence between them. This suggests close attention should be paid to cross-functional integration between design and manufacturing engineering.

Task and Environmental Uncertainty

In Chapter 2 we also discussed task and environmental uncertainty. Tasks that are high on uncertainty are ambiguous ones that need to be continually defined and clarified for each specific situation. These kinds of tasks will require more integration than routine tasks that can perhaps be fully described in procedure manuals. This might suggest a more functional organization or intense forms of communication. Similarly, rapidly changing environments that the firm needs to adjust to continuously will necessitate using richer forms of communication for coordination purposes than simple, less dynamic environments where standardized procedures can be used to coordinate.

Distance

Other factors that influence the need for coordination are the physical, organizational, and cultural *distances* between functional groups that need to coordinate their activities. Actually the distance does not affect the *need* for communication as much as the difficulties in making communication happen naturally. So when a lot of communication is needed, more effort must be made to ensure an appropriately high level of communication.

Physical Distance. People in different states or even different countries cannot spontaneously have a face-to-face meeting and won't naturally run across each other each day. In fact, studies have shown that even short distances can greatly reduce the likelihood of spontaneous communication. Thomas Allen's (1977) classic study of communication networks found that the probability of weekly communication is relatively high when offices are within 30 meters of each other, but then drops off very steeply. Individuals more than 30 meters apart might almost be hundreds of miles apart in terms of their likelihood of regular communication.

Organizational Distance. Organizational distance can be crudely measured by counting the number of steps in the hierarchy an individual must go through to reach another individual. Count up from the first person to a common boss or boss's boss and then count down to the second person. Organizational distance is almost as important as physical distance. Thus, two engineers who are located in adjacent offices may find it hard to collaborate because of their "location" in the organization structure. If each reports up to different bosses who have conflicting objectives, the engineers may find it difficult to establish common ground to work together cooperatively. For example, if the product designer reports to a product head who is judged based on product performance and warranty, manufacturing cost may not be an important objective. But that may be the most important objective to the manufacturing engineer who is judged on productivity and achieving cost reduction targets. Organizational distance is certainly an issue when individuals from different companies must work together. Consider an engineer in a manufacturing firm who is trying to purchase parts at the lowest cost and is working on a design with a supplier engineer whose objectives have been defined as maximizing short-term profits.

The examples above suggest that, at its root, organizational distance is a result of conflicting objectives which place individuals in an adversarial position; and this is particularly a problem when they are working on interdependent tasks. Of course, just because two engineers report up to different parts of the hierarchy does not automatically mean they will see themselves as having competing objectives. For example, Toyota has a very strong functional organization that could well have conflicting objectives with the product organization led by chief engineers. Yet all of the evidence we have suggests there is an extraordinary degree of cooperation and a strong sense of common objectives. This could be influenced by a third factor – organizational culture.

Cultural Distance. Cultural distance arises when several distinct cultures have conflicting beliefs or values. This can occur between different firms, e.g., between manufacturers and suppliers, or within a firm. There are many different ways to slice up and combine organizational members to identify distinct cultural groupings. One can look at different occupational groupings, different levels in the hierarchy, different genders, different educational levels, and different friendship cliques. One classic example in design is the product engineering/manufacturing cultural distance. This was described by Liker and Fleischer (1992) in a case study of a U.S. automaker. Cultural distance was created between product engineers and manufacturing engineers because of differences in education level (product engineers had more years of college from more prestigious institutions) and different status levels within the firm (product engineers were more valued, paid more, and had nicer offices). Product engineers valued technological complexity of products whereas manufacturing engineers valued simple, efficient systems that work. Product engineers thought and talked in more abstract conceptual terms while manufacturing engineers thought and talked in concrete, nuts and bolts terms. These and other differences made it difficult for product and manufacturing engineers to communicate effectively and jointly make decisions about product and process design. It fed the adversarial relationship that was also reinforced by physical and organizational distance.

Choosing Cross-Functional Integration Methods

As we noted in the beginning of this chapter, Mintzberg suggests that the five basic coordination mechanisms fall into a rough order of mutual adjustment, direct supervision, standardization of work processes, standardization of outputs, standardization of skills, and finally back around to mutual adjustment. You move up the scale in terms of overall preference as work gets more complex. By complex, we mean the difficulty of coordination and the extent of task interdependence.

Almost all of the work in product development is relatively complex and requires relatively high levels of task interdependence, and hence cross-functional coordination. So, there is really no chance that you might cover all or most of the coordination needs with one or two mechanisms. In other words, you almost always have to do some of everything. The basic approach we suggest is to think of your problem as having to empty a huge "bucket" of task interdependence. If you can offload some of it through coordination by one mechanism, that's great — you move on to the next and see how much you can offload there.

The best starting point is with direct supervision. How much coordination is filled through the use of the organization architecture and project management? Didn't get it all? Move on to standardized work processes. Still didn't get it all (of course not!); move on. Finally, you will reach mutual adjustment. This is your residual — all uncovered coordination needs have to be taken in by mutual adjustment, i.e., the use of teams and unstructured communication mechanisms.

Chapter 5

•

Supply Chain Involvement

Historically, relations between the firms arrayed along a value stream have been rather like the behavior of the United States and the Soviet Union during the cold war. Some minimum level of cooperation was necessary in order to keep from blowing up the world . . . but the operative assumption was that both sides would take advantage of each other in any way they could short of mutual annihilation. • James Womack and Daniel Jones

For most manufacturing companies, a very large percentage of their product is actually made by their suppliers, who produce the individual parts and components which make up the product. These "supplied components" are often in critical areas of the product, yet just as often suppliers are almost an afterthought in the product development process. Typically, a manufacturer will design a product and all of its noncommodity parts, and then go look for a supplier to make those parts which can't be made efficiently in-house. Just given what we know today about the need to design for manufacture and assembly, it's quite clear that such an "arm's length" relationship is no longer viable. If it makes sense for manufacturing to be involved up front with engineering in the early stages of product development, then it also makes sense for suppliers to be just as involved; otherwise the benefits of early involvement by manufacturing will only be obtained for that small proportion of parts made in-house.

Early and strong involvement of suppliers in product development is a hallmark of best practice in CE. Unfortunately, many U.S. companies treat their suppliers like they were the enemy — to be taken advantage of as much as possible, and then forgotten until the next time they're needed. In this chapter we will first look at some of the reasons for involving suppliers in product development. We'll move from there to a discussion of the Japanese approach to supplier involvement, since they are widely acknowledged to be the best in the world at this. Finally, we will provide you with a framework for thinking about different ways to enhance supplier involvement.

Why Involve Suppliers In Design?

The traditional model of component supplier management views suppliers as additional manufacturing capability for items that are either not profitable to make inside, could be made more cheaply outside, or for additional capacity during peak demand periods. In this model, one gives as little responsibility as possible to outside parts suppliers because they would exploit any opportunity to make additional money at the buyer's expense (Williamson, 1975). All purchasing agents quickly learn that they should always get multiple bids and pit suppliers against each other to drive down price. In the case of a commodity (e.g., fuel, plastics resins, lug nuts), the market used in this way does indeed serve to keep suppliers honest, and the purchasing agent is wise to use natural market forces for supplier selection. A supplier who provides poor quality products or services or who exploits opportunities to raise costs is quickly and easily replaced by an alternative. But what about non-commodity parts and components? Consider the following fictional scenario.

Lonnie Frazier, a lead design engineer for Weekend Farmer, Inc., was given responsibility for a project to enhance the company's eight horsepower engine for a new line of rider mowers. A critical component was a new powdered metal connecting rod which would be lighter and could be machined to tighter tolerances than forged connecting rods. Lonnie was very interested in what he had read about progressive Japanese approaches to supply chain management and decided to bring a new supplier, Powdered Inc., a leader in this technology, to participate in design early in the concept phase. Powdered, Inc. was enthusiastic about participating so early in the process. They saw this as an opportunity to get their foot in the door early in the new engine program and to make sure the product design fit their manufacturing capabilities. Their best engineer joined Weekend Farmer's design team, and Powdered invested $75,000 of their own funds for time and materials to bench test early prototypes. Lonnie was very excited about the test results and very pleased with Powdered's performance. He enthusiastically recommended that Weekend Farmer's purchasing department give a sole source contract to Powdered, Inc. Unfortunately, purchasing came back with an alternate source that *claimed they could make the connecting rods for 10% less than Powdered, Inc. Purchasing awarded the business to the competitor. The competitor's process turned out to be different from the one Powdered, Inc. had developed. Despite their promises, the competitor's parts did not have the promised strength at the specified weight. The added weight in the connecting rods forced an expensive redesign of other engine components to reduce noise and vibration. As a result, the engine as finally produced was heavier and less powerful than Weekend Farmer's closest competitor.*

Powdered, Inc. wasn't providing a commodity, something that could easily be replaced on the open market. They had worked closely with their customer to develop new technology and designs that were unavailable to their competitors. The experience of working closely with Weekend Farmer's design team gave Powdered, Inc. knowledge about their product that simply could not be contained in a Request for Quote that might be distributed through the mail. If we look at this scenario we can see several advantages that were obtained when the supplier worked closely with their customer.

- ***The supplier designed parts which best fit the end product.*** By working with Weekend Farmer's design team, the supplier learned details about the new engine which enabled them to design a better part for their customer.
- ***The supplier developed new technology for the product.*** Based on their deeper understanding of the end product, Powdered, Inc. was able to apply its expertise to develop new technology to benefit their customer. Powdered's competitor may have been just as competent, but they had to start from scratch after the design was already done.
- ***The customer designed an end product to take advantage of the supplier's technology.*** Based on the super-light rod which Powdered, Inc. showed they could make, Weekend Farmer designed an engine to take advantage of those characteristics.
- ***The supplier invested capital in the product.*** A substantial investment by the supplier meant that the customer needed to invest that much less.
- ***The supplier was committed to the new product.*** Because of its level of investment, Powdered, Inc. was highly committed to doing whatever was necessary to make the end product work right. As problems arise in a development project, they get solved smoothly and quickly because of this close working relationship.

The story of Weekend Farmer, Inc. and Powdered, Inc. has been played out hundreds of times in the last decade or so since the West discovered Japanese-style supply chain management. Engineers may follow the *form* of early supplier involvement in design, but their companies often lack the business infrastructure to guarantee the supplier the actual manufacturing business at the end of the day.

New product designs are highly customized efforts. Once a supplier is permitted to design a custom component, become part of the customer's design team, and develop a unique manufacturing capability for that component, it is more difficult to fire that supplier and hire a replacement. This gives the supplier a certain degree of power in the relationship. The supplier is also privy to confidential, proprietary data and can potentially use that information with other customers to get additional business, perhaps negating a competitive advantage of the original customer. For these reasons, many firms have been reluctant to involve suppliers in the early stages of product development and have preferred to keep an arms-length relationship with suppliers until the product and its components are designed and suppliers can competitively bid on the detailed design. But as we have noted, this can result in the loss of major opportunities for product improvement.

Japanese supply chain management is possibly one of the most benchmarked aspects of the Japanese manufacturing system. Much of the benchmarking at first focused on mechanics of the business relationship (e.g., target pricing). But as they investigated further, these Western companies discovered that a critical factor in the success of the best Japanese firms is the way they work together with suppliers on new product development.

The Japanese Approach to Supply Chain Management

Over the past decade, Western companies have been attempting to emulate the practices of successful Japanese manufacturers which use what some call a "partnership" model of supplier relationships. Japanese firms outsource a much larger portion of product content, mainly to a small, close-knit group of suppliers. For example, Asanuma (1985) found that while Japanese automakers produce certain functionally and strategically important parts in-house, the large majority (as much as 70 percent) of the cost of the vehicle was purchased from outside suppliers. They will work closely with strategic suppliers based on long-term contracts, intense communication, and often they will hold a small equity stake in the supplier's company. These suppliers are viewed as part of the extended enterprise.

It is clear that the success of many of the best Japanese firms depends on their ability to gain competitive advantage from their supplier relationships (Richardson, 1993; McMillan, 1990). A key part of this is collaboration in product development. Intensive communication during these long-term relationships allows the parties to fully understand each other's products and processes. This is not to say that all Japanese suppliers are given the autonomy to design their own parts. This is mostly reserved for large suppliers in the "first tier." The tier structure is much like an organizational hierarchy within a firm (Figure 5.1). First-tier suppliers are the first link in the supply chain which deal directly and intensely with the end-product manufacturer. They often are responsible for complex subsystems. First-tier suppliers in turn purchase components and materials from other suppliers and manage the "second-tier" of suppliers, who may in turn manage a "third-tier," and so on.

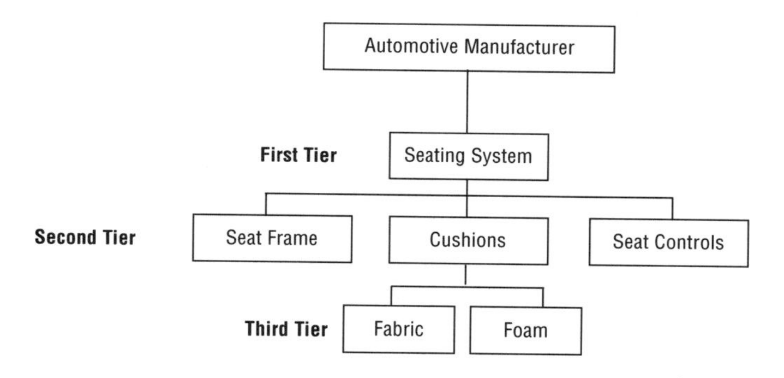

Figure 5.1 • Example Tier Structure in Automotive Industry

Here is one example of how this works in the auto industry, although the principle applies to any industry which makes complex products. For a complex component, a Japanese automaker will not design the parts needed for a new model. Instead, they will specify exterior dimensions and performance characteristics and let the supplier design the part to best match its process.

Generally speaking, except for parts that are vital to the customer's view of their internal core competencies (e.g., engines, transmissions, major body panels), first-tier Japanese suppliers have full responsibility for designing and making component subsystems of the car (Womack, Jones, and Roos, 1990). The first-tier supplier may or may not involve the second-tier suppliers in the design process. According to Clark and Fujimoto (1989, 1991), the level of joint product development accounts for a significant portion of the Japanese lead time advantage. They also state that supplier involvement and strong supplier relations account for about one-third of the person-hours advantage the Japanese automakers have over other producers.

The reliance of Japanese manufacturers on suppliers for high value-added component systems evolved gradually after World War II. A good example of this is Fuji Electric (Nishiguchi, 1994). One of their small subcontractors in 1954 made only stamped parts. At the request of Fuji, and with Fuji's support, this subcontractor expanded its range of technical capability, and by 1966 was making the complete subassembly of printed circuit boards. It continued to diversify at Fuji's request and ultimately was assembling protective relay systems, peripherals for electronic control systems, and telecommunication products. From 1974 to 1990, Fuji Electric's main plant in Tokyo went from subcontracting 49 percent of product content to 65 percent, and in that same period reduced the number of first-tier suppliers from 93 to 60. Similar stories can be found at Toshiba, Mitsubishi Electric, and many other Japanese electronics firms. The first-tier suppliers of complex subsystems not only have manufacturing responsibility, but also a great deal of responsibility for design, testing, and procurement of parts for their subsystems.

So why do Japanese firms depend so heavily on outside suppliers? In two surveys done to investigate why Japanese firms outsource (McMillan, 1990), important reasons included "to use the unique technology of vendor" and "subcontractors have expert skills parent company does not have." Over time, the Japanese OEMs selected a number of reliable and technically strong suppliers and had these selected suppliers provide parts in increasingly more assembled forms. In this process, OEMs sometimes came to entrust a substantial portion of the development stage of such assembled parts to the suppliers concerned, saving engineering person-hours of their own.

Developing technologically sophisticated suppliers can have large payoffs in product innovation. One excellent example is the case of wire harnesses in the automotive industry. As the number of electronic parts in cars have increased, so have the number of electrical wires. At the high end, the Lexus SC300/400 had 1500 to 1600 wires. As Nishigushi (1994) explains (p. 129):

> *The conventional one-big-unit design of wire harness assemblies was thus impractical. Seeing the problem (and opportunities), several Japanese wire harness suppliers proposed to divide the big unit into blocks and connect them with small "junction boxes" in plastic housings. They further suggested incorporating microchips and microprocessors into these junction boxes, with the result that a new "intelligent" wire harness system came to replace the conventional, bulky wire assemblies. This new modular design radically reduced cost, weight, bulk, and manufacturing complexity while substantially increasing reliability. More important, it helped accommodate frequent design changes throughout the model cycle of the car.*

Japanese companies take the selection of which suppliers will be given this level of responsibility very seriously. In order to be qualified as a superior supplier of customized parts, a company has to satisfy rigorous requirements in terms of technological capability, quality, delivery reliability, and price reduction capabilities. Japanese firms typically use rigorous methods for grading suppliers in these areas. Normally it is only the top suppliers that are given subsystem responsibility. Ultimately when a supplier consistently reaches the top level, they are self-certifying and do not need to go through customer audits.

The Japanese approach to effective supplier management can lead to higher quality, lower cost, faster time-to-market, and more innovative products for the manufacturer. Of course, this depends on properly selecting the suppliers, assigning them appropriate roles in product development, integrating them into the product development process, and managing them so their goals are aligned with those of the buyer. We take up these issues in the remainder of this chapter.

SUPPLIER ROLES IN PRODUCT DEVELOPMENT

It is easy to get the impression from the literature on supplier management in Japan, or indeed from visiting a small number of Japanese companies, that there is only one model of supplier management: that all relationships are full partnerships. Upon closer examination of the practices of successful Japanese firms, it is clear that there are many different roles in which the nature of the relationship varies substantially. In Japan, Rajan Kamath and Jeff Liker (1994) observed four roles suppliers take on in which they *are still involved in product development with their customers.* We call these roles: partner, mature, consultative, and contractual. The roles are defined based on the following characteristics of the design process, the product, and the supplier's capabilities.

1. ***Design responsibility.*** Who is responsible for designing the part or component?
2. ***Product complexity.*** How complex is the "product" being supplied?
3. ***Form of specifications provided.*** How are specifications provided to the supplier?
4. ***Supplier influence on specifications.*** How much influence does the supplier have on the specifications?
5. ***Timing of supplier involvement.*** When in the product development process does the supplier get involved?
6. ***Component testing responsibility.*** How much responsibility does the supplier assume for testing the part or component?

7. ***Supplier product development capability.*** What resources does the supplier have to develop new products in-house?

Figure 5.2 summarizes the four roles in terms of the seven elements of supplier involvement.

	Contractual	Consultative	Mature	Partnership
Design Responsibility	Customer	Joint	Supplier	Supplier
Product Complexity	Simple Parts	Simple assembly	Complex assembly	Complete subsystem
Specifications Provided	Complete design	Detailed specifications	Critical specifications	Concept
Supplier Influence on Specs	None	Present capabilities	Negotiate	Collaborate
Timing of Supplier Involvement	Prototyping	Post-concept	Concept	Pre-concept
Component Testing Responsibility	Minor	Moderate	Major	Complete
Supplier Development Capabilities	None	Little	Moderate	High

Figure 5.2 • Supplier Roles in Product Development

THE CONTRACTUAL ROLE

The Contractual Role is not very different from a traditional "parts supplier" role. The customer designs the component and only asks the supplier to provide simple parts. A complete design is provided to the supplier who has virtually no influence on the specifications. The supplier does get involved in the prototyping phase and may actually build the prototypes or have some input into the design at this point. They also may have some responsibility for testing the prototypes. Such a supplier would have virtually no product development capabilities, but still can represent the voice of manufacturing in their customer's product development process. Suppliers serving in this role will be increasingly vulnerable to competition in the future since they provide little else beyond manufacturing competence. On the other hand, if these suppliers have unique manufacturing capability and can deliver high quality, low cost products on time, they may carve out a strong continuing niche, particularly if there are no rapid changes in the technology of their product.

The Consultative Role

The Consultative Role brings the supplier more fully into the product development process. The supplier may provide a simple assembly and have joint responsibility for product development with its customer. The actual split of responsibility is likely to be weighted toward the customer making the major design decisions. The customer will tend to provide detailed specifications, but not a complete design. The supplier can influence the specifications by making sure the customer knows what capabilities the supplier has. Intense involvement in product design begins typically after the concept has been developed by the customer, but before detailed design work takes place. The supplier takes more responsibility for testing than in the contractual role, and has significant product development capability in-house. Product development resources are likely to focus almost completely on specific development activities for their customer, with little devoted to more pioneering R&D. This is a good transition role for suppliers moving on to greater capability. It is also a viable long-term role for suppliers with deep process expertise which is not easily duplicated. Many small suppliers will find this is the best they can do if they are to stay small, since the additional capability required to move to the Mature role may require substantial additional staff.

The Mature Role

The Mature Role shifts much more responsibility to the supplier. Typically they are producing a complex assembly. The customer only provides critical specifications and the supplier does all the design work. The supplier gets involved right at the concept stage and negotiates specifications. This negotiation process can enable new capabilities which the supplier has developed to be brought into the early product design, or it may simply allow the specifications to be set more realistically. Substantial product development resources are devoted to R&D to stay ahead of their customers and competitors on the technology for their particular component. Moving to this stage will require substantial investments in product development resources and related internal organizational changes to meet these requirements.

The Partner Role

The Partner Role is the ultimate "full-service" supplier. The supplier is involved from the very beginning in their customer's product development process, and is responsible for a complete system or subsystem. The customer provides a concept for the overall product (e.g., automobile or computer system), while the concept for the supplier's particular subsystem is likely to be based on a collaboration between the customer and the supplier. The supplier is apt to have extensive long-term R&D capabilities, and to be leading the industry in that subsystem technologically. In effect, they act as a specialty subsidiary of the customer and are given the trust and responsibility of an internal organization. Very few firms at the time of this writing, inside or outside Japan, can act in this role. Even those firms which are able do this often do not get the opportunity because their customers may be unwilling to accept the idea of a supplier taking this much responsibility. One American firm that has clearly grown into a partnership role in product development is Johnson Controls.

As one of the leading manufacturers of automotive seats, Johnson Controls has a major R&D center which develops complete seating systems, from styling to final detailed design and testing, for a large number of major automakers. There was a time in U.S. industry when auto companies designed their own seats, purchased components, and assembled the seats in feeder lines in assembly plants. Now Johnson Controls performs the entire engineering process, builds the entire seat, and sends it in sequence to the assembly plant ready for installation. In fact, there is a movement among seating suppliers like Johnson Controls and Lear Seating to move into other interior trim systems that complement the seating. In order to develop seats from the concept stage, Johnson Controls must be privy to early styling concepts of the automaker—among the most privileged information in the industry. To maintain confidentiality across customers' designs, Johnson Controls is organized into separate divisions, each catering to a different automotive customer. Each division in the R&D center has autonomous capability to design, build, analyze, and test prototype seating systems.

The full range and variety of actual supplier relationships is of course more complex than can be captured in four simple role descriptions. This means that it would be difficult to map any single Japanese supplier completely into one of the roles. They may look like a partner in testing responsibility, yet supply only a complex assembly rather than a complete subsystem. And, of course, their role might vary somewhat depending on the nature of any individual product being made.

The point of this classification is to remind us that there is no one ideal role for all suppliers. Selecting appropriate roles is a matter of matching product needs, customer capabilities, supplier capabilities, and the strategic goals of the customer. For example, even if the product includes several relatively autonomous subsystems, this does not mean that each should be outsourced to a "partner." There may not be a partner currently in existence for that subsystem; the customer may already have invested substantially in R&D capability for that subsystem; or the subsystem may be so strategically important to the product that the customer may want to maintain control. On the other hand, just because a current supplier for a product is only operating at the consultative role does not mean that the supplier is incapable of evolving further to a more mature role. The customer may want to make a concerted effort to develop the supplier's capabilities.

In the U.S. there is a general trend as part of CE toward promoting "full service suppliers" which would provide a component or system rather than just parts, take major responsibility for the design of their subsystem from the beginning of development, manage lower tier suppliers, and in general work in a team or "partner" role with the customer. The message to suppliers seems to be: either get heavily involved in your customer's product development and develop autonomous product development capability, or get out of the business. This blanket policy does not allow for productive variations in suppliers' roles and thus may be leading some industries down the wrong path. Assigning your key suppliers to appropriate roles using the classification in Figure 5.2 will be central to both the assessment and design activities in Part II.

Integrating Suppliers into Product Development

Ideally, the product development process of your suppliers should be seamlessly integrated into your company's process. Even in the contractual role, where suppliers are building parts to customer specifications or supplying parts off the shelf, the supplier may play an important role in some stages, such as in the prototyping process. As suppliers move up the ladder of responsibility toward the partnership role, they will play a more integral part in your process.

As separate corporations, your suppliers are likely to have their own internal design philosophy and process. After all, you are not their only customer. The challenge is to develop an *alignment* between your process and that of your suppliers. There are three ways to do this.

1. ***Require that they adopt your process.*** If you are a big enough customer to the supplier, you may have the leverage to insist they adopt your process. General Motors, for example, expects key suppliers to adopt GM's four-phase process. Some suppliers will

establish divisions to serve different suppliers, and these divisions will develop processes that fit their particular customer's approach. For example, Johnson Controls, Inc. supplies seats to General Motors, Ford, Toyota, Chrysler, and other major automakers. They have separate divisions for each major customer. The GM division uses GM's four-phase process. The Chrysler division works within the platform team structure of Chrysler and uses a CAD system compatible with Chrysler's. The Toyota division has learned to adapt to Toyota's approach, including adopting Toyota timing and sending guest engineers to reside in Toyota's Technical Center. Toyota also taught Johnson Control engineers to use Toyota's internally developed CAD system.

2. ***Find a fit between the customer and supplier's design process.*** You may find your design process is not all that incompatible with that of your suppliers. It may be that, with some minor modifications, the processes can fit together. For example, Delco Electronics (a division of GM) supplies climate control systems to Chrysler. Although Delco uses a version of GM's four-phase process internally, they have amended the process for their Chrysler work in order to meet Chrysler's timing for prototypes.
3. ***Use simple timing milestones to coordinate the design process.*** One way successful Japanese automakers have done this effectively is by focusing on physical prototypes. The supplier is expected to deliver a fully functional prototype that can be made on production tooling by certain planned dates. *Lateness is not tolerated.* The total vehicle tests are critical for determining system performance, and data from these tests are fed back to suppliers. As long as suppliers deliver prototypes on time and then revise the designs appropriately based on the feedback they receive, the customer does not particularly care what design process they use.

No one approach is best for all circumstances. Excellent companies will expect some close suppliers to adopt their design process, but other suppliers may adopt only some aspects such as the timing. If your company is going through a major reengineering effort, it is probably not realistic to expect your suppliers to immediately adopt your design process, and your design process may not even make sense for the supplier's particular circumstances. What is important is that you consider what you need from each supplier to effectively coordinate your development activities with those delegated to that supplier.

Can We Develop Supplier Partnerships in the West?

The ability of Western firms (those in North America and Europe) to emulate the success of the Japanese is often called into question because of cultural differences which, it is said, keep us from engaging in the same kinds of practices as the Japanese. While there is some truth to this (there are some culturally important differences we either cannot or do not want to change), there is often more myth than fact in these arguments. Two myths in particular pertain to Western firms' ability to emulate Japanese supplier management practices. These concern the keiretsu and sole sourcing.

The Myth of the Keiretsu

Relationships in Western firms between customers and suppliers are often characterized as arms-length, adversarial negotiations in which suppliers try to bid up prices and customers try to knock them down with the threat of switching to cheaper sources. In contrast, the Japanese are often characterized by cooperative relationships as epitomized by the famous "keiretsu."

The *keiretsu* is a group of companies that buy from and supply to each other, exchange personnel, and have complex equity relationships. A firm may own an equity stake in its suppliers within the keiretsu. The keiretsu's bank may own equity in all the companies. There may be unwritten agreements that the customer will ensure a steady flow of business to the supplier, and that the supplier will not do business with the customer's competitor without first asking permission.

The close-knit relationship between customers and suppliers in these keiretsu relationships has been referred to as "partnership sourcing." The partnership is characterized by mutual interest and even a sense of mutual obligation. Characteristics of the partnership relationship include the following.

1. ***Slow and thorough screening of new suppliers.*** This can take years, beginning with careful investigation of the supplier's products and reputation, followed by small orders. Based on positive experiences, larger orders and greater responsibility may be given to the suppliers.
2. ***Presourcing during the design stage***. There is an informal understanding that suppliers will get the business if they participate effectively in the design stage. In some U.S. firms (e.g., Whirlpool), a formal written contract is actually signed at this time.

3. *Use of target pricing.* Since the supplier is presourced, often before the product is even designed, the competitive market cannot set prices. So the manufacturer will determine what the end customer is willing to pay for the end product, subtract their profit, and break down the remaining cost into targets for each component or subsystem. The supplier is offered that target price and must develop a product and process that can make the product at a profit.
4. ***Early involvement and responsibility in product design and development.*** As our model in Figure 5.2 suggests, the actual roles and responsibilities vary across suppliers, but some level of involvement is critical.
5. ***Expected cost reductions and continuous improvement over the life of the product.*** In some cases, this is spelled out as an expected percent cost reduction each year over the life of the contract. The rationale is that, through continuous improvement, the supplier should be able to drive down cost and the customer should share in the benefits of those cost improvements.

Companies outside of Japan often believe that they cannot emulate this "partnership" model because they do not have a keiretsu structure. They believe that independent suppliers which are not partially owned by the customer and which supply competitors cannot be trusted with proprietary information and will exploit opportunities to make the customer dependent on their technology by raising prices. However, the Japanese situation is far more diverse than outsiders often realize. For example, consider the following.

1. ***Equity holdings are smaller than is often recognized.*** Japanese automakers have an average 18 percent equity stake in their most important first-tier suppliers of major subsystems, and less than 9 percent ownership of lower-tier suppliers. Even Nippondenso, the largest member of Toyota's group of suppliers, is only 24 percent owned by Toyota (Sako, 1995).
2. ***There is more business across keiretsu lines than is often realized.*** Over time, there has been an increase in business transacted outside the keiretsu. For example, Nissan suppliers historically were not permitted to do business with Toyota and vice versa, though they could supply other smaller automakers. Yet, there have been many publicized cases of selling across boundaries. For example, Nissan buys electric fuel pumps from Nippondenso; and Tachi-S, a Nissan-affiliated seat manufacturer, supplies seats for a compact car model made by Toyota.

3. ***There is cross-membership in the supplier associations of rivals.*** Large manufacturers have supplier associations, an elite club of major suppliers which serves as an opportunity for information exchange, business transactions, and socialization into the philosophy of the "parent" company. For a company the size of Toyota, only about 200 suppliers are in the association in Japan. Yet, 50 of these are also in Nissan's supplier association (Sako, 1995).

Our conclusion is that developing cooperative relationships is not dependent on buying shares in a supplier's corporate assets or giving them exclusive license to all the business of your company. Cooperation results from a sense of mutual dependence and trust built up over a series of transactions. The important thing is that the dependence should be two-way so that one party does not dominate and hold the other hostage. Trust is earned when the supplier repeatedly delivers quality products on time or when the customer and supplier share in the pain during a market downturn (e.g., they split the pain of a cost reduction needed to maintain business in a down market). What this means is that Japanese style partnership relations are not dependent on the presence of the economic creature we know as a keiretsu. They are not easy to build, since they require changing practices and a high level of trust, but that need not be a stumbling block.

The Myth of Sole Sourcing

Another myth is that Japanese companies rely heavily on "sole sourcing;" that is, they find one reliable supplier for a commodity and then depend on that supplier exclusively. In fact, Japanese purchasing departments generally work very hard to avoid dependence on one source for any major component. However, they also do not want a large number of sources because that is difficult to manage, and it is also hard to develop the kind of close-knit relationships that enable early supplier sourcing and sharing of cost reductions. Japanese purchasing departments will generally prefer two suppliers. Typically a supplier is the exclusive source for a particular *model* (e.g., a particular car line), but another supplier will provide that same part for another model, and each knows the other is capable of taking away future business. The Japanese approach has been referred to as a "parallel sourcing" strategy (Richardon, 1993). That is, there are a small number of suppliers for a particular component (e.g., compressors) who *in parallel* have the exclusive business for one particular product line. This approach provides the benefits of long-term, exclusive relationships with suppliers in quality, timing, and technology sharing while maintaining the motivation to reduce costs by free market competition.

Parallel sourcing is slightly risky in that a problem at a given supplier for a critical component can completely shut down your assembly operation for a model. For example, if there is a strike at the one supplier plant which supplies a given assembly operation, that operation will rapidly close down. This has happened several times recently to General Motors, for example. However, this risk of a long-term shutdown (e.g., due to bankruptcy or natural disaster) is usually not too large, since there is an alternative (the parallel source) which could begin to supply the same part on relatively short notice. Most companies which have gone to parallel sourcing (including an increasing number of Western firms) agree that the benefits outweigh the risks.

Recently there has been a convergence of supplier partnership practices between Japanese and Western companies. For example, target pricing and value engineering are increasingly being used outside of Japan, and suppliers in the U.S. report more frequent communications with their customers about product development as well as a greater commitment to longer term relationships. The evidence from these firms suggests that Western firms can indeed use supplier partnership practices successfully. Firms as diverse as an aerospace first-tier supplier like Rockwell-Collins, and automotive OEMs like Chrysler and GM, report major gains from these practices.

Coordinating Supplier Involvement in Product Development

Assuming you have decided to involve suppliers in some role in product development, then it is necessary to coordinate with them during the product development process. For the most part, the same fundamental methods are available for dealing with suppliers as when trying to coordinate functional groups internally. For example, the task of coordinating supplier and customer engineers is similar to the task of coordinating the contributions of different functional groups, such as product engineers and manufacturing engineers, within a single firm. Hence, the coordination mechanisms are also the same as those we discussed in Chapter 4. We will consider each of the major cross-functional integration mechanisms (Figure 5.3) as they relate to supplier involvement in product development.

Mutual Adjustment

Two-way communication is important in every case when suppliers have even a small role in product development. But the frequency of communication, and the extent to which rich communication mechanisms such as face-to-face meetings are used, will depend on the nature

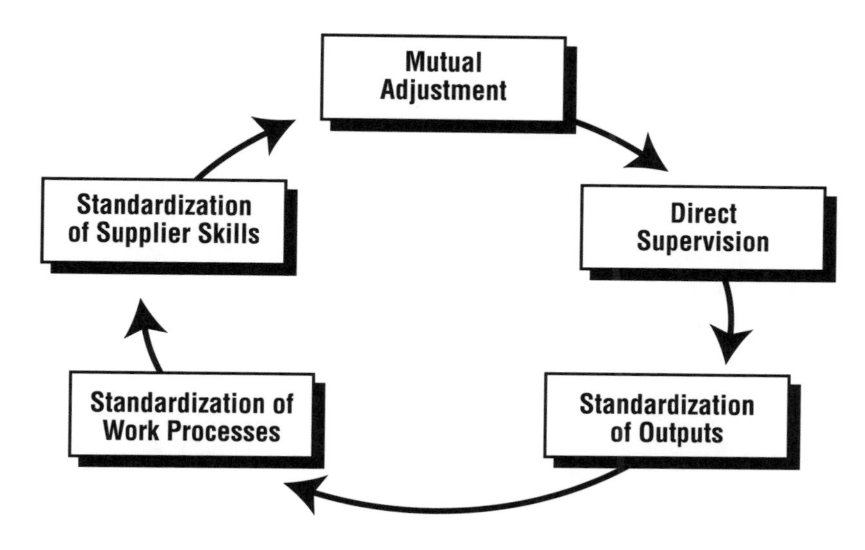

Figure 5.3 • Supplier Coordination Mechanisms

of the relationship and characteristics of the product. As you move toward the partnership model, and as the product becomes more complex and uncertain, more two-way communication is needed. As in the case of coordinating functional groups within a firm, there are several different organizational approaches to facilitate effective two-way communication between firms.

- ***Teams and Task Forces***. The buyer organization can make supplier representatives into members of any type of cross-functional teams which they might use for product development, including task forces, temporary teams, standing committees, and semi-permanent teams. See Chapter 4 for a detailed discussion of the different types of teams. An increasingly common form of team for suppliers to participate on is called a "Commodity Team." A Commodity Team may go so far as to help plan the technology roadmap for the customer along with the suppliers.
- ***Collocation.*** The customer organization can locate engineers at the supplier location or bring supplier engineers to sit with their in-house product development groups. Japanese companies often use the liaison role of guest engineer in which suppliers send product engineers to reside full-time in their customer's engineering offices for up to three years. This generally applies to complex subsystem designers and suppliers who are acting in mature or partnership roles.

Collocation can be used in conjunction with any of the design team approaches to strengthen the supplier's integration into that team. Or any of the team approaches can be used without collocating the suppliers. However, be aware that the supplier's engineers will have to travel more often and keep involved, or there is a danger they will become a second-class citizen on the periphery of the team.

Direct Supervision

Even though engineers from the supplier firm do not formally report up their customer's hierarchy, direct supervision can be an important means of coordination and control. This supervisory role can be handled by purchasing, product engineering, or a combination of the two. The form and intensity of supervision will vary with the supplier role. If the supplier is in a contractual role, then the customer engineer (or buyer) typically supervises any contributions of the supplier and decides what advice to take seriously and what to reject. In a partnership situation, the supplier is given a great deal of autonomy, and the customer engineer is more likely to play a facilitator role, coordinating the activities of different suppliers and providing access to needed resources and information.

The organizational architecture and project management structure of the customer organization will have implications for supplier management. If the customer uses a purely functional organization and the supplier provides a product that fits cleanly into one of the functional engineering groups, there may not be much coordination difficulty. On the other hand, if the product crosses functional lines it may be more of a problem. For example, if the customer has separate mechanical and electrical engineering departments and the supplier's component is electromechanical, the supplier must take direction from and coordinate with two engineering departments. And if they also have to work through purchasing, it is likely that there will be conflicting directions and critical information will fall through the cracks. This problem is not much different from that faced by the internal project engineer, who also must deal with multiple functions. But all of these issues get magnified significantly when the person dealing with it is from outside the company. The solution to this problem may be to revise the organization architecture, or it may be to make greater use of other coordination mechanisms.

Standardization of Outputs

To some degree, output standardization is automatic in many supplier relationships when there is a formal contract that includes deliverables and specifications. However, this is only a powerful coordination mechanism if the specifications are clearly communicated to the supplier and they do not change very much over time. U.S. suppliers are much more likely than their Japanese counterparts to report that product specifications from their automotive customers arrived too late, were unclear, or changed in midstream. For example, in a survey of U.S. and Japanese suppliers, 46 percent of the U.S. suppliers reported that their largest customer changes requirements in midstream, as compared to 23 percent of Japanese suppliers (Liker et al., 1996).

An in-depth study of Toyota's product development process found that they are unique even among Japanese automakers in the way they set requirements. They follow a logic that Allen Ward and associates (1995) call "set-based design." The easiest way to explain this is to contrast it with the more usual approach to design which Ward calls "point-based design." In point-based design, the customer selects a particular basic design for the component and says, "design me one of these that fits into this exact space and meets these exact requirements." This means the customer has already made many decisions about the component and its specifications prior to dealing with the supplier. The supplier in turn does the design and hands the design to the customer for *just that one alternative*. In set-based design, the customer gives a more fuzzy specification, e.g., "it should fall within this range of specs," or "it should be about this weight." The supplier then tests some alternatives that fit that range of specifications. Through testing and negotiation, the customer and supplier converge on a solution. In some cases, the customer presents more vague specifications early, before the complete design has taken place, and then gives more specificity over time, partly as a result of input from the supplier. This type of approach maximizes reciprocal interdependence. Ward argues it is more likely to lead to a global optimum whereas point-based design tends to lead to suboptimum solutions.

In a variation of this theme, Nippondenso, Toyota's largest supplier, uses an approach they call "standardized variety" (Ward et al., 1995). Nippondenso develops a product and manufacturing process simultaneously that represent a *portfolio* of components. For example, they designed a single-row radiator with the cooling capacity of a double-row radiator which weighs less and is less expensive. They can make a large variety of sizes and shapes on the same production line — although not *every* possible size and shape. They take this set of alternative radiators to their customers like a catalog, and the customers can choose from the large number of standard alternatives. This only applies to the radiator core; the outside packaging still must be customized to the customer's engine and compartment. Whirlpool, as discussed in the case in the Appendix, used a similar approach for wire harnesses. They realized large cost and time savings by standardizing the lengths and colors of wires in the wire harness, and constrained design engineers to pick from the menu of standard offerings.

Another increasingly common way to standardize outputs is called "target costing." In target costing, a company develops a price up front for what it is willing to pay for a component. This is based on data about what the end customer is willing to pay for the end product, as well as information about manufacturing costs at the supplier. The supplier agrees to design and make the component for the target price, which includes the supplier's profit. Target costs can be adjusted if conditions change, but there is usually a strong emphasis on sticking to it. This has been used successfully by Honda, Chrysler, and Motorola, among others.

Standardization of Work Processes

As discussed in the section "Integrating Suppliers into Product Development," it is necessary that your supplier's design process to some degree fits with your design process. This is particularly important as companies shorten development lead times. Supplier inputs must come in on time, and that can only happen if there are clear deliverables and schedules. As noted earlier, many Japanese companies tend to focus on physical prototypes as deliverables and have very rigid schedules for delivery of those prototypes.

The full array of standardized work processes can apply to supplier development activities. Suppliers may be expected to follow your standard operating procedures (e.g., prototype release systems, control procedures of the CAD database, standard contractual language, etc.), collect measures that are part of your monitoring systems (e.g., testing procedures, meeting measurable quality standards), and adopt your structured development processes. Some firms expect suppliers to adopt tools and techniques used internally, such as QFD or a particular design for manufacturability program; and provide training and software to suppliers.

Of course, it is not reasonable to expect a supplier to learn new design methodologies and change their internal design process if they are only a short-term supplier of a part that represents a small percent of their business. This only makes sense for long-term suppliers who expect to get a steady and substantial stream of future business.

Standardization of Supplier Skills

While the mechanism "Standardization of Worker Skills" may not seem to apply very well to relations between customers and suppliers, if we reinterpret it slightly it makes perfect sense. Think about "skills" at the organizational level and you will quickly realize that they are simply the company's capabilities. A very common way to standardize on capabilities or skills is to *rate* a supplier on its capabilities. For example, it is quite common to assess a supplier on its ability to do statistical process control (SPC) in their plant; or to assess their ability to conduct certain kinds of engineering analyses within their engineering function. Most large companies today have some system for rating their suppliers on a comprehensive list of factors.

Another approach to this is to provide suppliers with assistance in achieving certain skill levels. Often called *supplier development* programs, these efforts usually involve customer staff working with the supplier to develop some specific capability. Toyota is well known for doing this, as are many other companies. In the U.S., Toyota has established the Toyota Supplier Center to help develop both current and potential suppliers to institute the Toyota Production System. Honda of America Manufacturing has a similar program that they call "BP," which stands for Best Position, Best Productivity, Best Product, Best Price, and Best Partners. Honda's BP program has a dedicated staff of over 100 people who work exclusively with suppliers to help improve their processes.

To summarize this, in Figure 5.4 we provide a list of some common supplier management mechanisms, with their definitions.[1] Each definition will also suggest how it fits with Mintzberg's five coordination mechanisms. Figure 5.5 shows the relationship between the supplier management mechanisms and the coordination mechanisms more clearly.

Exchanging Product Data with Suppliers

As suppliers get more involved in the product development process, more data about the product has to flow between the customer and supplier. When the supplier simply made parts to the customer's design, it was usually sufficient to send the supplier a two-dimensional drawing of exactly what was to be made. Now data must flow both ways in order to make sure engineers on both ends know what the others are doing and to ensure that everyone's datasets are consistent. Not only must data flow both ways, but the exchange of data must take place more frequently as work is done more or less concurrently.

This exchange of data about the product (product data) is made more difficult by the incompatibility of computer-aided design (CAD) and

[1] These definitions are from Fleischer, White, and Carson (1996).

Early Supplier Involvement. A practice that involves one or more selected suppliers with a buyer's product design team early in the specification development process. The objective is to utilize the supplier's expertise and experience in developing a product specification that is designed for effective and efficient manufacturability. *Mutual Adjustment*

Commodity Team. A group of experts that represent various functions of the suppliers and customer to reduce the total cost and enabling other gains in productivity and benefits for all organizations involved. *Mutual Adjustment*

Collocation. Placing staff from different companies together in a single location. This may mean placing a "guest engineer" at a customer's plant, or having the customer's guest engineer at the supplier's plant. *Mutual Adjustment*

Supplier Data Integration. The ability to easily exchange information between customer and supplier. This includes the ability to exchange information in electronic form, such as product data between CAD systems or business data between information systems. *Mutual Adjustment*

Long-Term Contracting. A decision to contract with a particular supplier over an extended period of time. *Direct Supervision*

Preferred Suppliers. Those suppliers chosen for enterprise partnership based on performance which has or is anticipated to satisfy or exceed customer demands. The Preferred Supplier performance criteria exceed traditional metrics of product quality, schedule, and price. *Standardization of Supplier Skills, Work Processes, or Outputs*

Target Cost. A modified standard cost of a product for a given quality level which is derived from customer requirements and competitive constraints. As this cost relates to the entire life cycle of the product, the setting of Target Costs has to take into account the estimated future technology and process developments inside the company (company related dynamics) as well as anticipated market movements (market related dynamics). *Standardization of Outputs*

Supplier Development. A systematic enterprise effort to create and maintain a network of competent suppliers, and to improve various supplier capabilities that are necessary to meet its increasing competitive challenges. *Standardization of Supplier Skills*

Supplier Qualification. The assessment of suppliers, identified through supplier selection, to determine which supplier can best meet customer requirements. It may require evaluation of supplier capabilities against key requirement criteria through SSCM/QCAE Evaluation, on-site audits of the manufacturing or service provisioning systems and quality systems, and/or review of supplier provided data. *Standardization of Supplier Skills*

Supplier Rating. An indicator of supplier performance based on criteria such as quality, price, responsiveness to customer inquiries, problem resolution and support, product improvements, etc. Supplier ratings can be used for feedback to suppliers to encourage improvement, as input to supplier selection on subsequent negotiations, and for supplier classification. *Standardization of Supplier Skills*

Figure 5.4 • Supplier Management Mechanisms

	Mutual Adjustment	Direct Supervision	Standardization of Skills	Standardization of Outputs	Standardization of Work Processes
Early Supplier Involvement	X				
Commodity Team	X				
Collocation	X				
Supplier Data Integration	X				
Long-term Contracting		X			
Preferred Suppliers			X	X	X
Target Cost				X	
Supplier Development			X		
Supplier Qualification			X		
Supplier Rating			X		

Figure 5.5 • Supplier Management Mechanisms and Coordination Mechanisms

other computer-aided engineering systems. While translators are available, the exchange of data between systems is often difficult or impossible, and even when done may result in hard to find errors. Poor data exchange can result in large increases in cost and substantial delays. In a survey of tool and die shops, cost increases of about 25 percent and delays of over 20 percent in total delivery time were found to result from poor quality exchanges of CAD data between the tool and die shops and their customers (Fleischer et al., 1991). In Chapter 7 we discuss some of the technical issues involved in translation. Here we will briefly discuss some of the business practice issues that can make product data exchange easier.

One attempted solution to these problems is to require suppliers to use the same CAD system as their customers. For example, most major automakers require their immediate suppliers to use the same CAD system as they do. Unfortunately, they almost all use different systems! This means they are adding huge extra costs to their suppliers. In the aerospace industry, Boeing has made the same requirement for its suppliers. For very large companies, working with very large suppliers, the added costs are probably worthwhile, at least until an adequate translation method such as STEP[2] becomes mature. But for smaller companies such a requirement is impossible. As a customer, the smaller company does not have sufficient influence on its suppliers to make such a request. As a supplier, the smaller company cannot afford to purchase and manage multiple systems.

The problem can be managed, but it must be done so very carefully. We recommend the following steps be taken when exchanging product data in a heterogeneous environment.

- ***Know what systems will be exchanging up front.*** If you know what systems will be exchanging at the beginning of the project, you can begin planning for translation early on, when things are less pressing. Our experience is that good translation (regardless of method) takes a lot of learning in terms of what works and what doesn't. If you do some practice exchanges before the workload gets too heavy, you may prevent a crisis later on. Also, keep in mind that data exchange between different versions of the (nominally) same system may also be problematic.
- ***Design for exchange and downstream data use.*** Depending on the systems involved, some kinds of CAD entities and shapes will translate, more or less. Part of your preparation for exchange should be to find out what translates well and what doesn't. For example, you may have no trouble translating a shape made one way but not another — as in a circle based on a radius as opposed to a circumference. When you learn these "tricks" they need to become design rules for designers — they have to use only those entities which will translate.
- ***Minimize exchange in the design process.*** You can minimize data exchange by specifying when product data will need to be exchanged in the process. For other times, use video conferencing, personal visits, and sketches rather than exchanged CAD files. Some forms of groupware will allow you to see a CAD image without having to exchange a file. Also keep in mind that the site of a visit should depend on which is the lowest cost site to visit. Don't require all meetings to take place at the customer's site if that requires all sorts of preparation in advance that doesn't add value to the process, like exchanging and translating large CAD files.

Task/Environmental Contingencies and Supplier Relations

Just as the organization to some extent depends on task and environ-

[2] See Chapter 7 for a discussion of STEP.

mental contingencies, so do your supplier relations. We will consider the environmental contingencies first.

Environmental Uncertainty

To the extent that a firm deals with many different suppliers, and critical aspects of the supply system are changing rapidly (e.g., suppliers going out of business, new suppliers being added, or process technology changing rapidly), the supply base will be a source of greater environmental uncertainty. Customers can manage their supply base to reduce this uncertainty. For example, by reducing the number of suppliers and developing long-term relationships, environmental uncertainty is reduced. Also, by selectively changing the technology of the product and process, one can reduce uncertainty. For example, Toyota makes a point of controlling the amount of technological change in new car models. If they are trying major new technologies for the dashboard, for example, they will probably not simultaneously change the rest of the interior trim very much. In general, when environmental uncertainty related to the supply base is higher, it is particularly important that you have strong communication and coordination with your suppliers. For example, when technologies are changing rapidly, effective coordination and communication with suppliers is very important.

Technological Complexity

Technological complexity has been described in terms of degree of uncertainty and degree of interdependence. Technological complexity acts in a similar way as environmental complexity. When technology is less well understood and less predictable, then it is particularly important that you have strong coordination and communication with suppliers. The use of a guest engineer, for example, is particularly important when technological complexity is high.

There are also great advantages for having stable, long-term relationships when environmental and technological uncertainty is high. Long-term relationships are more conducive to effective coordination and communication because engineers in each organization grow to know each other, and each partner gets used to the other's operating style. The exception to that is if the technological changes are coming from firms outside the customer's supply chain, and the customer needs to bring in a new supplier in order to exploit the new technology. For example, air bags were brought into automotive firms by hiring new suppliers that had this technological capability. To the extent that the new technology replaces an old technology, work for a long-term supplier may be reduced as new work goes to a new supplier.

Task Interdependence

Task interdependence between the supplier's design task and the customer's design task also requires closer communication and coordination. In the extreme, if the supplier's component was a commodity that simply plugged into the customer's product, the only communication that might be necessary is for the customer to give the supplier specifications and then wait for the completed design. For example, if an automaker wants a mostly standard radio that fits into a standard space, but also wanted to add higher power output, they might simply specify this and the date they need a prototype to their radio supplier. On the other hand, if there are complex interactions between the supplier's part and the overall system, a great deal of coordination and communication is probably needed. For example, Mazda uses Hirotech to design and build complete doors. The door must fit with the entire body, and there are many wires routed into the door, as well as components installed in it. In this case, close communication with Hirotech is facilitated by their location next to the corporate design offices of Mazda. Communication is both frequent and intense.

Nippondenso's standardized variety approach described earlier has the benefit of reducing task interdependence. The customer can choose radiators from a standard set of options, for example, after which it is not necessary to communicate a great deal about the design.

In general, by sourcing entire subsystems rather than simpler components, task complexity and interdependence can be reduced. If the supplier is responsible for the design of a complete subsystem, the supplier can worry about the interaction of parts in that subsystem. For example, a supplier of exhaust systems can worry about the integration of the pipes, catalytic converter, and muffler, but if the customer buys each of these items separately from different suppliers, the customer must act as the system integrator. On average, Japanese automakers communicate *less* frequently with their suppliers during the product development process than do American automakers when working with their suppliers (Liker et al., 1996). They attribute this to the prevalent Japanese method of purchasing complete, autonomous subsystems from first-tier suppliers, providing critical interface specifications, and then expecting the suppliers to deliver complete prototypes that meet these specifications by a scheduled due date.

Implications

From the perspective of concurrent engineering, outside suppliers can be viewed as another division of your company. They must be an integrated part of product and process design. Most of the communication and coordination issues we discussed under internal organization also apply to outside suppliers. However, the challenges are amplified by the fact that they are not a part of your company, but rather are separate companies with their own cultures, structures, and profit-seeking goals. In designing your new concurrent engineering approach, you may not have the power and influence to get your suppliers to completely reengineer their processes to match yours. But you can design the roles of suppliers in product development and encourage and support them to develop appropriate human and technological resources to be assigned to work for your company. You can also rethink your purchasing policies as they concern supplier involvement in design.

In the methodology provided in Part II, you will go through a parallel process of analysis and design for your suppliers. You will consider how their product development process fits with your to-be product development process, coordination and communication mechanisms, the human resources required, and the technologies they will need to work with you on product development. For this to be more than an exercise it will be critical for purchasing and engineering to work together as a team. This in itself may be a major challenge, but given how central suppliers are to your design and production process, you cannot afford to ignore their roles.

Chapter 6
•
People Systems

In every sector, old and new, I hear a renewed recognition of the importance of people, and of the talents and contributions of individuals, to a company's success. People seem to matter in direct proportion to an awareness of corporate crisis. • Rosabeth Moss Kanter

So far we've talked about the work that needs to be done and the organizations in which it will take place. But organizations don't design things, *people* do. People systems are those parts of the organization which help bring the right people with the right skills together and motivate them to work toward common organizational goals.

The people systems in any company take up a very large portion of its effort, from performance appraisals to health and safety programs to benefit plans to hiring and firing processes. While all of these systems are needed to run a company, they don't all have a significant impact on cross-functional integration. Our focus on cross-functional integration allows us to ignore many of the "maintenance" aspects of the human resources function in a company. For example, the computer systems by which people get paid are vitally important to any company's health, but it has little to do with helping integrate across functions. On the other hand, as we will see, the ways in which jobs are designed and the metrics used to appraise behavior make all the difference if cross-functional integration is to succeed.

There are three primary people systems which have the greatest impact on CE and which often need to be changed in order to do CE effectively (Figure 6.1). These are *job designs, skill acquisition systems, and motivation systems*. Job designs divide the work process into discrete "chunks" which an individual can perform. Systems for skill acquisition provide an individual with the ability to perform the required work. Systems for motivation increase the probability that the individual will apply those skills to the work as needed.

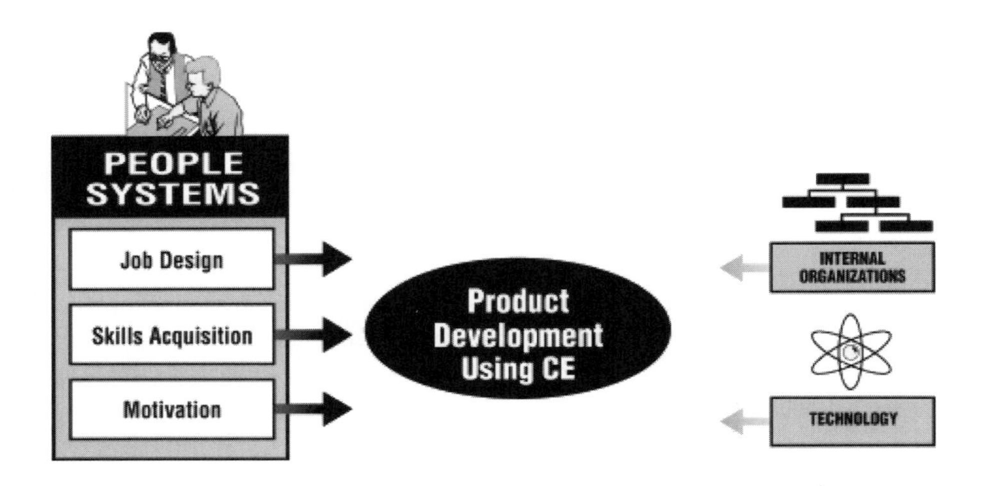

Figure 6.1 • Impact of People Systems on the Work of CE

JOB DESIGN ISSUES IN CE

Job design is the process of taking a set of work tasks and combining them into a "job" that a person can perform. This does not imply that a "job" is something a person does all alone. Work (especially product development work) obviously involves a lot of interaction with others, on both a one-on-one and a larger team basis. Nonetheless, each individual person has a job, and should know what tasks he/she is to perform, even if those tasks include working with others.

There are two approaches you can take to job design.

1. You can start from a set of tasks which need to be performed and then proceed to "clump" those tasks into jobs.
2. You can start with your existing jobs and try to decide how they need to be changed to meet some objective, e.g., improving cross-functional integration.

In this book we will recommend you generally follow the second approach. We do this for three reasons. First, it is much easier — unless your current job designs are awful, in most cases tinkering with the system will take much less time and energy than starting from scratch. Second, starting from scratch is probably out of scope for most people involved in an effort to introduce CE to their company. Finally, it is probably unnecessary — most of the time, job designs in product development and related areas follow fairly typical industry-wide patterns that usually need only minor modification to make them more effective.

This is not to suggest you should never design new jobs wholly from scratch. There will obviously be cases in which that will become necessary. For example, if you are creating a wholly new product development capability, you will have no jobs to modify. If that is your situation, you will need to refer to one of the many good guides on job design.[1]

A Framework for Modifying Job Designs

While there is a large literature on job design, much of it is concerned with how to make boring, deskilled jobs more interesting and motivating. Most of those concerns are not very relevant to jobs involved with the development of new products. Despite substantial differences in pay, status, and content, the jobs of manufacturing engineers, design engineers, designers, draftsmen, and the like are really quite varied and meaningful. Almost all jobs in the product development arena would fall into a category that Mintzberg (1983) would call "professional." Nonetheless, there are several approaches to job design that can be used, in conjunction with the other methods in this book, to improve the effectiveness with which work is done in CE.

Based on the work of Hackman and Oldham (1980), there are five approaches to modifying job designs (Figure 6.2). These are:

1. combine tasks (horizontal enlargement)
2. form natural work units
3. empower (vertical enlargement)
4. establish relations with the environment[2]
5. give feedback on work.

These five approaches provide sufficient coverage to be used as an organizing framework for the modification of job designs in CE. We will first discuss the five approaches and then show how you can choose among them based on different characteristics of your work process.

Combine Tasks (Horizontal Enlargement). This means to increase the scope of a job by assigning it more activities. For example, if we take

[1] For example, see Hackman and Oldham (1980).
[2] Hackman and Oldham refer to "Establishing Client Relations." We have expanded this approach to include all relevent aspects of the environment.

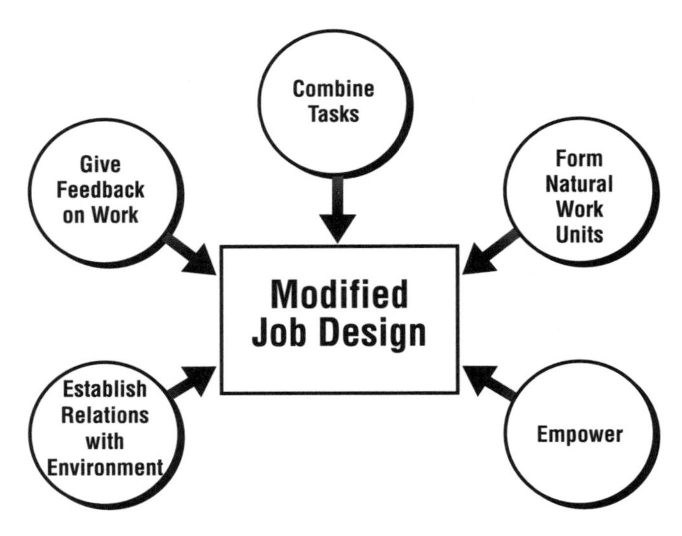

Figure 6.2 • Approaches to Job Design

a traditional designer's job and assign it additional tasks by combining it with what used to be the draftsman's job, then we have enlarged it *horizontally* by combining tasks. Alternatively, we might take just one or two of the more routine design engineering tasks and assign it to the designer.

Form Natural Work Units. This means to form a team that will work on a particular task or set of tasks. These should be regular design-related tasks, not just special temporary projects. In this case, the work is assigned to the team as a whole, rather than to the individuals in it. Each individual also has a role within the team. The typical cross-functional team within a CE product development environment probably falls into this approach. However, note that this type of natural work unit might be a cross-functional team, but it doesn't have to be. It's also important to realize that every group that is called a team isn't necessarily a "natural work unit." In other words, we often form teams for other purposes. For example, a task force formed to brainstorm problem solving ideas when a crisis arises is a temporary team, whereas a natural work unit is a more stable group around which we have designed our work.

Empower (Vertical Enlargement). This means to increase the scope of a job by assigning more decision-making and control. For example, if we took the traditional designer job, which involves following the directions of a design engineer, and gave designers more responsibility for making their own design decisions, such as interpreting engineering analysis results or taking responsibility for the work of a component supplier, then we would say the job had been empowered by having it *vertically* enlarged.

Establish Relations with the Environment. This involves changing an individual job so that it includes contact with relevant elements of the environment. In many cases, this will be with the customer of the work being done. For example, if we take a traditional designer's job and assign the additional responsibility of getting feedback from the customer about the adequacy of the design, then we have added environmental relations to the job. Another common example of this is when an engineer from an R&D laboratory is assigned to spend time with end user customers to find out how they use the product being developed. You can also apply this concept to relationships with an internal customer; for example, bringing a product designer into contact with a manufacturing engineer. In some cases, relations will need to be established with non-customers, such as technology vendors or R&D providers. This would be the case if the technology was changing quite quickly.

Give Feedback on Work. This means to provide feedback about job performance to someone who traditionally hasn't gotten very much. The most common application of this is on assembly line tasks where the individual worker may know little or nothing about product quality being produced. It can sometimes apply to designers and draftsmen in product development in situations where the design engineer and/or an account manager is the sole outside contact and fails to communicate with other people working on a project. Actually, when we establish environmental relationships, such as going to the assembly plant to see how the designed part is installed into a larger system, we are also providing a form of feedback on work. So there is some overlap in these categories.

Job designs can be modified using any and all of these approaches. One very common example is the combination of tasks from the traditional roles of draftsman and designer in durable goods manufacturing industries such as automotive and aerospace. Traditionally, in those industries, the designer was responsible for building a detailed design on a drawing board, and the draftsman was responsible for cleaning up the designer's work and inserting a variety of details into the drawing. A very common career path was for someone to come out of school and get a job as a draftsman. As he/she learned about the parts and components being designed, a major promotion would be to the position of designer. With the increasing use of Computer-Aided Design (CAD) tools, draftsman positions have all but disappeared in many companies as the responsibilities of that position have been combined with that of the designer.

Another example that is talked about a great deal is the merger of the traditional designer and design engineer roles—or at least vertically expanding the designer role to include more design decision making (Fleischer and Liker, 1992). In the traditional model, design engineers are responsible for understanding the customer's requirements and developing the conceptual design based on them. The conceptual design is passed along to the designer, who builds a detailed design. This traditional division of labor leaves the designer as the primary CAD user. Indeed, in most companies using these job designs, the design engineer will rarely, if ever, use the CAD system. Some companies, such as Toyota, have combined the design engineer and designer jobs into a single "super design engineer" position. Thus, the design engineer is responsible for both concept design and detailed design using CAD, as well as coordinating with suppliers and other design engineers, and communicating with manufacturing to make sure the component gets made correctly. By doing this, the company has combined tasks, established environmental relations, and empowered the design engineers. This model is in widespread use in the electronics industry, but is relatively rare in durable goods with a high mechanical engineering content, such as autos or aerospace. We suspect at least some of the difference has to do with the high level of design automation available in electronics as compared to mechanical design. Nonetheless, some durable goods companies, such as Toyota, have succeeded at using this model.

Another example of horizontal enlargement is the attempt by some companies to combine the jobs of design engineer and manufacturing engineer. In this position, the engineer is responsible for the design of the component and the design of the system for manufacturing it. Typically, this is only successful when one job or the other is not very complex. For example, we have seen this done in a company which produced wire fasteners — product changes were minor and mostly involved modification to the manufacturing equipment. In this case, the combined role was really filled by a manufacturing engineer who did product development as needed. There has been some discussion of doing this in more complex products, such as autos, but we have not actually seen it done successfully. We suspect that in most cases the complexity of the two positions makes it likely that only a few extraordinary individuals are capable of handling such an enlarged job (Liker and Fleischer, 1992).

Job Designs and Task Characteristics

So far we've discussed job design approaches in the abstract, but how do you decide whether to enlarge or specialize any given job? One clue to this is how you characterize the task to be performed and the environment in which it is performed. There is a strong relationship between skills required to perform the task, task and environmental uncertainty and task interdependence,[3] on the one hand, and the job design approaches, on the other (Figure 6.3).

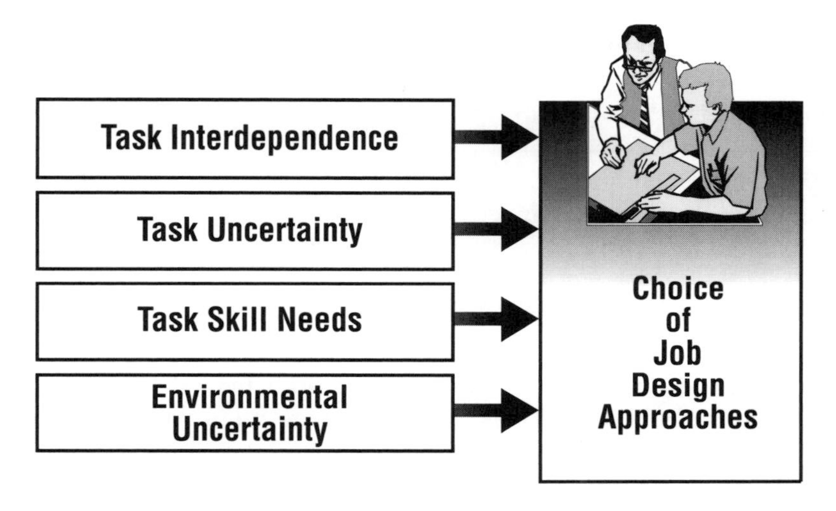

Figure 6.3 • Determining Job Design Approaches

Table 6.1 shows what levels of the task or environmental characteristics would be associated with the use of a given job design approach. Each approach is best suited to a different combination of conditions. In the following sections, we describe these relationships in detail. Note that the first three approaches are pretty much mutually exclusive, since they have significantly different contingencies. The last two approaches could be used separately or in combination with any of the first three, as appropriate.

Combine Tasks (Horizontal Enlargement). Combining tasks is the approach to use if there is high task interdependence, low task uncertainty, and the tasks require relatively low skills to perform. Skill needs should be low because the overall skill level required for the combined job will become too high if the skills needed for the initial positions is high. Equally, task uncertainty should be low because the overall level of uncertainty would become unmanageable if the starting points were also high. Finally, task interdependence should be high or there would be no benefit from combining tasks into a single position. As an example, consider the combination of designer and draftsman positions we described above. Task interdependence was at least sequential and

[3] See Chapter 2 for a discussion of these topics.

Job Design Approach	Circumstances in Which They Apply
Combine Tasks	High Task Interdependence *and* Low Task Uncertainty *and* Low Task Skill Needs
Form Natural Work Units	High Task Interdependence *and* High Task Uncertainty; *OR* High Task Interdependence *and* High Task Skill Needs
Empower Individuals	Low Task Interdependence *and* High Task Uncertainty *and* High Task Skill Needs
Establish Relations with Environment	High Environmental Uncertainty
Give Feedback on Work	High Task Uncertainty

Table 6.1 • Contingencies for Job Design Approaches

often reciprocal between the designer and draftsman. Task uncertainty for the draftsman position was low — although it is a skilled task, drafting is well understood. Finally, skill needs for the draftsman position were *relatively* low, as far as a designer is concerned, since almost all designers at one time had performed draftsman tasks.

Form Natural Work Units. The formation of natural work units is the approach to use if there is high task interdependence and either (or both) high task uncertainty and/or high skill needs. In a way, the formation of natural work units is very similar to combining tasks, except that the combination involves multiple jobs in a team, rather than combining them into a single position. Thus, it only makes sense if task interdependence is high. However, you would only form a team if task uncertainty were too high for a single person to cope with, or if the skill needs were so high that no single person could be that highly skilled. The most common example of this is the cross-functional CE team which might combine people from marketing, design engineering, manufacturing, and purchasing.

Empower Individuals. Empowerment of individuals is the approach to use if there is low task interdependence, high skill needs, and high task uncertainty. If task interdependence is low, then there is little sense in combining tasks into a single position. However, if task uncertainty is high, then the performer of the task needs to be able to make decisions without having to ask a supervisor for permission all the time. This is only possible if the holder of the position has a high level of skill. As an example of this, consider the Chrysler case study (in the Appendix) in which the design engineer has assumed some of the decision-making authority that once resided with purchasing. In this situation, once purchasing has made the determination that a supplier is the one who will provide a given component for a vehicle, then the design engineer can work out all details with the supplier. For example, if purchasing has chosen a seat supplier, the design engineer has the power to make decisions about the cloth, even though there are price differences between materials — as long as the engineer stays within budget. In the past, purchasing would have been the final decision maker about any price change of this nature. Purchasing retains some oversight, but the clear decision maker is the design engineer.

Establish Relations with the Environment. Making environmental relations part of the job is an approach to use if environmental uncertainty is high. The reason for bringing immediate first-hand information about customers into someone's work is that they are uncertain about what the customer wants or about the customer's reaction to something being done, usually because the customer needs change rapidly. If uncertainty is high, you need to make sure the people doing the work have the most immediate contact with customers so they can learn what is needed as soon as possible. As an example of this, consider an automotive parts manufacturer we know. They have made it part of their design engineers' jobs to visit customer plants to learn how well their parts are fitting into the automobile. This approach could be combined with any of the first three approaches: combining tasks, empowering, or forming natural work units.

Give Feedback on Work. Providing feedback on the work is an approach to use when task uncertainty is high. More immediate feedback about performance, preferably on the spot as work is being performed, is most needed when the work itself is not so well understood. An example of this in a product development environment is when we provide design engineers with access to simulation and rapid prototyping technology, such as stereolithography. The rapid feedback on a design provided by the simulation or the prototype is most crucial when developing new kinds of products, where task uncertainty is high. This approach could be combined with any of the other approaches that involve high task uncertainty, such as forming natural work units or empowerment.

Job Designs and the Organization Design

As you have probably already figured out, there is a link between job design and organization design. The primary link is that both are ways of dividing up work among people and groups. The organization design does this at a fairly high level, deciding which very large "chunks" of work will be assigned to certain types of large groups (e.g., departments). The job design does it at a much lower level; within those large chunks, how will work be divided up among individuals and smaller groups? For this reason, you usually want to make your organization design decisions prior to starting work on job designs.

Job Design and Culture

The way work gets done is highly dependent on culture. In some organizations, the restrictive nature of some of the formal job designs is just ignored. For example, it might be that the formal job description for a designer's position has him/her sitting at a computer all day doing nothing but developing CAD models. In a reasonably flexible organization, this person might be asked by a design engineer to go along on a visit to a customer site to better understand certain problems encountered by the customer. Everyone in the organization knows that they need to do what's necessary to meet the company's goals. In another, less flexible organization, the designer's supervisor might refuse to allow such a visit, since it's "not his job." This inflexibility is part of the organization's culture. In the flexible organization, job designs don't matter too much, since everyone knows how to work around the rules as needed. In the inflexible organization, much more attention needs to be paid to keeping the job designs up to date and accurate. We don't mean to suggest that total flexibility is inherently better. It can result in all kinds of problems, from things not getting done, to everyone stepping on everyone else's toes as they all try to do the same thing. The point is to know what your culture is like and to adapt your rules and procedures accordingly.

Providing Skill

Job designs pretty much determine what skills are needed for each position — if you know what you want an individual to do, you're then in a position to know what skills he/she will need to do those things. Our focus here will be on the kinds of skills necessary to improve cross-functional integration. We'll divide our discussion of skills into two parts: technical and social skills.

Technical Skills

We can divide the technical skills that an individual has into two types: core skills and cross-functional skills. The core skills are those required to do the core work of their job. For example, the core technical skills of a designer might include the following.

Drafting

- Ability to physically manipulate 3-D structures
- Ability to manipulate objects and surfaces
- Ability to draw with a computerized tool
- Knowledge of the elements and capabilities of CAD packages

Design

- Knowledge of design standards
- Knowledge of product features and functionalities
- Ability to interface designs and produce parts specifications
- Knowledge of product-data-related Electronic Data Interchange
- Understanding of STEP Application Protocols
- Ability to troubleshoot design function

The cross-functional skills are those required to work together with those from other functions. These are needed to enable the individual to understand what the other function does, to communicate needs to the other function, and to understand needs expressed by the other function. For example, cross-functional skills for a designer might include the following.

Manufacturing

- Understanding of the manufacturing system in sufficient depth to consider manufacturability and assemble-ability in the design
- Knowledge of material properties

Engineering

- Knowledge of procedures for doing design tests (e.g., instrumentation, feasibility testing)
- Data interpretation and modeling (e.g., interpreting engineering analysis results)

The designer's knowledge in these areas may not be very deep, but it must be deep enough to be able to communicate with the functional specialist, to ask appropriate questions, and to know when an answer seems appropriate. Each job will have a set of core technical skills and a set of cross-functional technical skills that are required to perform the job.

Social Skills

Social skills are required in order to be able to work with other people. It is a common stereotype of engineers that they lack social skills. As with many stereotypes, this one holds a germ of truth — many engineers do tend to focus more on technical details and less on the social situation in which those technical details will be used. This may result in the engineer having the right answer, but not being able to get anyone to listen to it. Or it may result in the engineer having the right answer to the wrong question, since he/she didn't spend enough time listening to what was wanted. Of course, we should not let stereotypes rule our thinking — many engineers have excellent interpersonal skills and go on to make effective managers.

We have divided social skills into four types, all of which should enable people to work more effectively with each other.

Communication. There are two aspects to communication: expressing yourself and listening; both are essential for cross-functional integration. Expression involves the ability to convey facts and ideas in ways that others will easily understand. We all know people who have great technical ideas, but are unable to help others to understand them. Unfortunately, unless others can be made to understand, those great ideas may be wasted. Equally important is listening — taking the time and effort to try and understand what others are saying. Effective listening is a critical component of effective expression. What this means is that if you are trying to convey a difficult idea to someone else, you need to listen to their feedback about how they understand what you are saying. If you listen carefully, you may get cues which suggest what they don't understand. This provides the opportunity to reformulate what you are saying so that it is easier to grasp. Everyone who is working in a cross-functional situation needs good communication skills.

Conflict Resolution. This is the ability to work with others to solve problems and resolve conflicts. Few people enjoy conflict, but it is a fact of organizational life; and, indeed some level of conflict can actually energize an organization and make it challenge the status quo. Conflict can emerge from many sources, including differences in personality, differences in values, and differences in interests of people in separate specialties. A classic case is the conflict between the design engineer who cares about product performance and the manufacturing engineer who is concerned with ease of manufacturing at low cost. There are a number of ways to deal with conflict, some of which are better than others.

1. ***Avoidance***. At one extreme is avoidance—you pretend it does not exist and hope it goes away. Unfortunately, serious conflicts of interest rarely go away on their own, and if the problem is not discussed it is unlikely you will arrive at the highest quality decision.
2. ***Accommodation***. A second approach is for one party to give in to the desires of the others, which is called accommodation. One party might say: "Have it your way. It's not worth arguing about." This can reduce the immediate tension but also may not lead to the best

solution. For example, if the manufacturing engineer were to give in to the design engineer, it may lead to a suboptimal design that cannot be efficiently manufactured.

3. ***Competing***. A third way is competing. You can attempt to win at the expense of the other department. Design might go to top management and get a ruling in favor of tighter tolerances despite the fact that those same tolerances will be difficult for manufacturing to achieve. Design will have won, and manufacturing will have lost. Unfortunately, the company may also have lost if this leads to endless rework while the tight tolerances were not all that critical to begin with.
4. ***Compromise***. You may bargain leading to a compromise—you give up something in exchange for the other party giving up something. Imagine the product engineer agreeing to loosen the tolerances in one area so it is easier to manufacture, in exchange for tightening up tolerances in another area. This may lead to a solution that stops the conflict, but is it a good one for the customer?
5. ***Collaborate***. The final approach to conflict is to openly share information and ideas and work collaboratively on a creative solution where everyone wins.

These five options are shown in Figure 6.4. In this diagram, department A attempts to satisfy its own concerns, or does not consider its own interests and is thus either assertive or unassertive, as shown along the vertical axis. Department A can either cooperate with the other department or not along the horizontal axis. If department A is assertive and uncooperative, it is competing; if it is unassertive and uncooperative, it is avoiding, etc. Note that as you move along the line from competing to accommodating, the sum of what is being split up between departments is a constant. Some call this the distributive dimension, and the issue is how we distribute the spoils. If I win, you lose, and vice versa. On the other hand, along the line from avoiding to collaborating, the pie actually grows and there is more for everyone. Thus, collaborating can actually lead to everyone winning.

One of the key characteristics of creating an environment supporting collaboration is a "superordinate goal" that all parties believe in and strive for. If the goal in the dispute over tolerances becomes satisfying the customer at low cost, this may lead design and manufacturing to solve problems and collaborate on a solution which is good for both parties. For example, it may lead to a new material that can achieve the same functionality as the old material but allow for looser tolerances. The purpose of bringing functional specialists together in a cross-functional

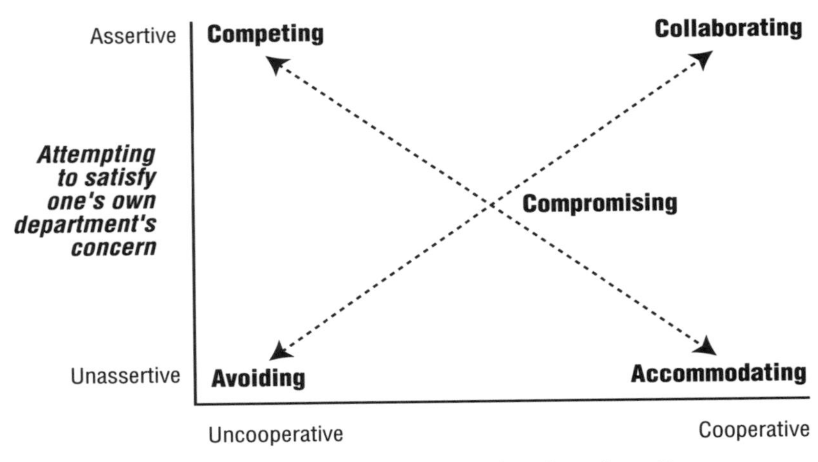

Figure 6.4 • Approaches to Conflict Resolution

team with a heavyweight project manager is to foster collaboration and the search for win-win solutions from multiple perspectives.

Group Facilitation. This is the skill of helping to move a group through a defined process. It may be as simple as reminding the group that they have an agenda and a time limit, or it may involve standing up in front of the group, writing notes on a flip chart, and acting as the chairperson. Group facilitation skills usually include having a "bag of tricks" for helping a group achieve its purpose. For example, most good facilitators will use methods such as affinity diagrams, fishbone analysis, and flowcharts to help a group work toward its goal. A good facilitator will also work to include everyone in discussions, often by asking quiet individuals to speak, even if they haven't volunteered. Someone serving in a pure facilitation role (as opposed to a leadership role) usually wants to be seen as very neutral in any discussions, and must be able to maintain that neutrality even if he/she holds strong positions on the issues being discussed. Group facilitation is particularly important for cross-functional integration because so much work takes place in group settings. Not everyone needs to be able to formally facilitate a group, but it helps a great deal if most members of a cross-functional group can do so.

Leadership. This is the ability to motivate others to move in a direction you want them to go. Leadership takes many forms, but what is most essential about it is that people *want* to follow a leader. We say this to distinguish leadership from management. Management is the formal exercise of control through the rules and hierarchy of the organization. Any individual *can be* both a leader and a manager (a happy combination!). What we generally find in effective cross-functional groups is that different people take on leadership roles at different times — depending on the topic at hand and the emotional needs of the group. Leadership can be taught, although some people are inherently unsuited to be leaders. Fortunately, not everyone needs to be a leader, and indeed a group with too many leaders will probably fail to get anything done.

Getting the Right Skills

There are two things you can do to make sure the people in your organization have the right skills to enable them to perform their jobs. You can pick people who have the right skills to begin with, or you can provide training and education to give them skills they didn't bring with them when you brought them into the organization. Rather than assume everyone in a given job needs the same training, we recommend an assessment to tailor the training to individual needs.

Assessment of Skills

If you're going to make sure your people have the right skills, you need to find a way to *assess* those skills. This is as true of new people whom you hire into your company as it is for people currently on staff. This is not to suggest that you find some formal testing service to test each individual on each skill we've described. In fact, most of these skills, especially the social skills, cannot easily be assessed using formal testing methods. You are probably already assessing some of these skills on a regular basis, but instead of thinking of it as an assessment, you think of it as a performance appraisal. Many performance appraisals in fact ask the manager (or the appraisal team) to assess skill level as well as performance — in other words, you may already have processes in place for assessing many skills.

The Skill Profile. You should have a skill profile for each job. This is the specific set of skills required to perform the job and work with other functions effectively. You then need to match the skill profile with the actual skills of the people who currently hold the jobs, and whom you are considering for placing into the jobs. These profiles need to be kept up to date in the face of constantly changing work and technology. In Chapter 11 we'll provide you with some suggestions for how to assemble such a profile.

Conducting a Skill Assessment. There are two approaches you can take to the assessment of skills needed for cross-functional integration.

1. ***Take data from performance appraisals.*** Find out which skills are assessed as part of your appraisal process. You should try to find out how well they are assessed. Often, a well-informed manager who observes the individuals involved may in fact be making the appraisal. In other cases, the appraisal may be the result of a polling of peers or customers of the individual being appraised. In either case, the assessment can be done effectively based on the appraisal. In other cases, the appraisal may be clearly uninformed or biased, so you wouldn't want to use it for this purpose.
2. ***Conduct a new assessment.*** For those skills which are not assessed in the appraisal process, you should go to the trouble of conducting an assessment. This need not be an enormously elaborate task. For our purposes, ratings of the individuals involved by a well-informed manager will probably be sufficient. If the current appraisal is done only by a functional manager (e.g., an electrical design engineering department head assesses all electrical design engineers), the assessment is likely to focus mainly on core functional skills and should be expanded to include cross-functional skills (e.g., by polling other functions affected by the engineer's designs).

SELECTION FOR SKILLS

There are most likely three occasions when you will have the opportunity to select people to work in a cross-functional environment.

1. ***Creating a new team to develop a product in an existing organization.*** You may have the opportunity to select from existing staff for a new team you are creating to develop a new product. In this case, you will be working with people who are pretty much known to you or at least known to others in your company. In addition to the usual assessment you might do through interviews and demonstrations, you are also in a good position to discuss skills with previous supervisors and peers.
2. ***Hiring new staff for replacement or growth.*** For individuals who are unknown to you or your organization, you must pay special attention to assessment during the interview. This is particularly true for social skills, since you cannot look at someone's credentials and hope to get much sense of their social skills. We recommend placing the individual in one or more situations which provide the opportunity to demonstrate social skill development. For example, you might have the individual make a presentation or lead a group as part of the interview.

3. ***Creating a new product development organization.*** You may have the challenge of creating a whole new product development organization from scratch. Good luck! This is the most difficult situation, since you not only won't know much about the people you're hiring, you also don't know much about the work they're going to do, making it exceedingly difficult to specify skill needs. This puts you very much in the same situation as if you were hiring new staff for replacement or growth, in that you need to formally assess the skills of the people you're hiring. However, you also need to make sure you've carefully done your job designs beforehand, so you'll have a good basis for the assessment.

TRAINING AND EDUCATION

Regardless of how you got them, once you have a staff you will find some or all lack some of the skills you want them to have. That being the case, you will need to bring the skills to them. If you have done an assessment, training needs are usually pretty obvious. We will only make a few comments about training providers and the pace of training.

Education versus Training. Training usually involves providing someone with a specific skill, while education usually involves providing more fundamental background. For example, sending someone to college to get a B.S. in Engineering would usually be called education, while sending someone to a course on how to use a specific CAD system would be called training. Almost everyone is hired into a job with a certain level of education, and many will receive little or no further education during their career. On the other hand, almost everyone will receive significant amounts of training during their career. Most employers would prefer to avoid providing education to their employees, since it is relatively expensive (both in dollars and time) as compared to training.[4] Hence, selection is often based on education credentials, since training credentials can be obtained more easily.

Training Providers. Providers come in all forms, shapes, and sizes. Since large organizations already have access to most of these, we will confine our advice to training providers for smaller organizations which have a limited budget. Low-cost training is usually available through at least three sources.

- *Local community colleges.* Community colleges in most locations are very adept at providing both semester-long courses and short courses on industry-relevant topics. Since most community colleges are subsidized by both local and state government, costs are

[4] Of course, many employers pay for tuition as a benefit, but this is almost always done on the employee's time.

very low. Even if your company is in a rural location, there is likely a branch of your local community college within easy driving distance. In many cases, even a relatively small company can arrange for a custom developed course or even for a course to be taught on site. In some states even custom course development is subsidized. In Michigan for example, the state spends about $100 million per year for such development.

- *Government funded economic development organizations.* Both the federal government and many state governments have funded economic development organizations to provide assistance to small manufacturing firms. As of this writing, the federal government's efforts in this area are known as the Manufacturing Extension Partnership (MEP) program. MEP programs, some of which are known as Manufacturing Technology Centers or Manufacturing Outreach Centers, are funded by the National Institute of Standards and Technology (NIST) to provide services at low cost (sometimes no cost) to small manufacturing firms in their region. The services they provide are often limited in scope or targeted to certain industries. Check with your local economic development agency for information on MEP programs in your area. Some state governments have also developed a network of assistance programs. Pennsylvania's Ben Franklin Partnership and Ohio's Thomas Edison's programs are examples of these.
- *Government funded training assistance.* As of this writing, the federal government seems to be mostly getting out of the business of funding training for individual companies. However, many states are using this kind of assistance as a competitive weapon to attract and retain manufacturing companies. Some of these efforts are specifically targeted to upgrading workforce skills. For example, Michigan recently created the Michigan Jobs Commission which funds training programs with the intent of creating a more highly skilled workforce that will attract and retain companies in the state.

On-the-Job Training. In our experience, excellent Japanese companies are masters at on-the-job training. A typical Japanese university's engineering curriculum tends to focus on general mathematics and science and less on practical skills and tools. Companies assume they need to train the "freshman" engineer hired directly out of school through some short courses, but mostly through on-the-job training. For the first year or two, it is assumed the new engineer will contribute little of value to the company and is mainly in training. This period can include broad exposure to the business. For example, at Toyota, first year engineers will spend a month or two working in a plant as an operator, another month or two selling cars door to door, and another couple of months in the advanced engineering lab. Once assigned to a particular engineering department (e.g., body engineering), Toyota assumes it takes about 10 years of experience in a given specialty for most engineers to really be excellent at their job and ready to manage other engineers. In the U.S., Ford provides a similar two-year training program for its new engineers. Hitachi has a similar program in Japan. Of course, this level of investment in the early training of engineers only makes sense if the engineer is likely to stay around for a while—a good assumption in Japan but sometimes a bad one in the United States. The point is that mentoring combined with on-the-job training is an important part of training (perhaps the most important). It is a mistake to view training and education as something that happens only in a classroom with someone lecturing in the front of the room.

Motivation Toward Common Goals

The final People System element that affects how cross-functional integration works is the system to motivate people to work toward common goals. As discussed earlier in our discussion of conflict resolution, one way to resolve conflicts that arise from parochial functional interests is by getting everyone focused on higher goals that will help the company be competitive and profitable. In our experience, most professionals in companies are quite highly motivated. The question is whether they are motivated to do the right things. In the case of concurrent engineering, this means the right things for the customer, which often means stepping outside narrow functional interests and considering the total design and manufacturing process. There are two related methods used by most organizations to enhance and maintain motivation: measurement and rewards. Both have traditionally caused people to focus primarily on local, functional interests, but both can be used effectively to motivate people toward cross-functional teamwork.

Measurement

Motivated people are people who want to succeed at something. *Goals* provide an objective target to define what success will look like. *Metrics* are the ways you measure the extent to which you are achieving those goals.

It's hard to overestimate the importance of goals and metrics as a factor in individual and group motivation. People will almost inevitably use the goals and metrics established for them as fundamental guides to their behavior at work, yet many organizations fail to use them to their full motivating potential. There are several issues which must be addressed in developing goals and metrics for CE at the individual and small group level. In the sections below, we discuss our observations of good and bad practice in this area, and provide some rules of thumb for moving toward doing better.

Consistency of Goals. Goals are what direct individuals and groups in a particular direction. Without common goals, individuals may all be motivated but pulled in different and even conflicting directions. The classic problem of design engineers throwing unbuildable blueprints over the wall to manufacturing is a problem of conflicting goals. Design engineers in this scenario have goals related to product performance (e.g., minimum weight, maximum power, maximum durability, low noise), and as long as they can develop a prototype with these performance characteristics, they assume manufacturing can build it. For their part, manufacturing is concerned with ease of assembly, ease of machining, manufacturing cost, utilization of existing equipment, etc., even if it means building something the customer does not want. The solution to this problem of *goal conflict* is to develop "superordinate" goals that both design and manufacturing are linked into. A process for doing this is called "policy deployment." There are many ways to do policy deployment, but what it comes down to is rather straightforward. Goals for a period of time (e.g., a year) must start at the top of the firm, with the next level down defining goals consistent with the level above. The next level below that then defines goals consistent with the level above them, cascading down through the organization. If design engineers and manufacturing engineers have defined goals that are consistent with the higher level goals, they will generally be complementary with each other rather than in conflict.

Goals versus Metrics. One of the most serious flaws we have observed are goals which are in conflict with their metrics. Although we earlier defined metrics as being inextricably tied to and in support of goals, in practice the link between them is often hard to find. For example, we once saw a design department in which stated goals all had to do with designing quality products that meet customer needs and that are producible in the manufacturing plant. However, *metrics* for designers included number of drawings produced, time to produce a drawing, and mechanical errors. It's not that these "productivity" metrics are inherently "bad," it's just that they're not linked to the stated goals of the department. In this company, there were no metrics linked to the goals, except the obvious "Don't make any errors!" Does this mean that designers ignored the department's goals and just rushed around all the time rapidly producing garbage? Of course not. The designers (and their supervisors) mostly recognized the need to do the right thing, and they did so. But, every now and then, someone, usually at higher levels of management, would focus on the productivity metrics ("Why is that designer talking on the telephone, rather than sitting at the CAD terminal working?"). At times like that, everyone (designers and supervisors) would get mad and cynical, and feel very unmotivated. Furthermore, some of the designers and supervisors would occasionally take the productivity metrics to heart and produce relatively junky designs that still enabled them to look good on the metrics.

The problem with many metrics is that they are often chosen because they are easy to measure, rather than because they measure anything important. In other words, it's easy to measure how many drawings a designer produces, but it's much harder to measure the quality of the design. In general, you're better off without a metric for a goal than you are with an irrelevant metric, since the irrelevant metric will pull some significant portion of effort away from the real goal. As the example above shows, most people will focus on the goal anyway — why demotivate them with irrelevant and possibly harmful metrics?

Goals, Metrics, and Rewards. How closely should goals and metrics be tied to rewards? Should people be explicitly rewarded for accomplishing goals? This is the point of programs like "Management by Objectives" as well as many recognition programs, such as those that provide restaurant coupons or certificates for special performance. Our experience suggests these don't work very well. One common problem is that the metrics they're tied to often can be manipulated, as when someone focuses all their energies on maximizing their measure on a single metric, to the detriment of others. Another common problem is that the metrics may be viewed as being irrelevant, such as the metric of producing lots of drawings, when the real goal is something else. Finally, they are often perceived as being unfair, as when a manager or a committee makes subjective judgments about who should receive an award.

This is not to suggest that people shouldn't be recognized for special accomplishments or that there should be no link between pay and performance. The problem comes when we try to use the reward system to "fine-tune" behavior. Our levers are so gross and the behavior we need to manipulate so fine that we almost always fail when we make too close a link. In general, it makes sense that people get raises, bonuses, and promotions based on good overall performance, but a tight linking between specific metrics and specific rewards all too often backfires.

Group versus Individual Goals and Metrics. Should goals be set for individuals as well as for groups? How about metrics? In our experience, it's pretty much universal for formal groups such as departments to have plenty of goals and metrics. Most less formal groups, such as cross-functional teams, will also have goals and metrics. In most companies, individuals will have some kind of goal-setting process, but it may be pretty well divorced from any group level goals. As an example, individuals may have goals related to their professional development or even for their social development, but there is often no attempt to develop individual goals that link to their group's. Surprisingly, this may well be the right thing to do, especially when task interdependence is high, as it often is in CE. There are relatively few circumstances in CE in which individual goals and metrics can be easily linked back to group goals. For example, a salesman's sales goals can appropriately be linked back to some group level sales goals. But does it make sense to try to create a timing goal, for example, for a specific design project for a specific designer? Probably not — what's needed is for the designer to accept ownership of the group goal. Individual goals are often in subtle conflict with group or team goals. For example, if an individual designer has a goal to take a certain number of continuing education credits, that may conflict with taking a workshop for a skill needed by the group, but which does not offer continuing education credit.

The Goal Setting Process. While people are usually highly motivated to achieve some goal that is set before them, that goal must be perceived as being *legitimate*. By legitimate, we mean the goal makes sense and it has been put in place by people who have the authority and the wisdom to do so. An example of an illegitimate goal might be one established by a manager who wields little power and who is respected by no one in the group, i.e., people in the group don't think the manager knows anything and he is unable to enforce his decisions effectively. In such a case, the manager might decide to establish a goal (e.g., reduce engineering changes by 50 percent) and everyone would probably ignore it. In fact, it's not so much the legitimacy of the manager that's the problem here, but the legitimacy of the *goal setting process*. For example, if this same manager set up a team within the group to establish goals, then the same goal might have high legitimacy — since members of the group were involved in setting it, it would probably be perceived as being very sensible. Equally, a goal established by the well-respected president of the company would also be perceived as being highly legitimate, since the process of receiving orders from this well-respected individual is perceived as quite appropriate.

Appraisal Systems and Rewards for CE

Appraisal Systems

Everyone needs feedback on their performance. In particular, they need feedback about how well they're doing toward meeting their individual goals (if they have them) and how well they are contributing toward their group goals. This feedback is the purpose of an appraisal system. Too often, "appraisal time" is looked at as getting a "grade" or being placed at "risk." The appraisal should be an opportunity to learn rather than something to be feared.

The only essential features of a good appraisal system are that it be fair and accurate. Unfortunately, if nothing is changed, the introduction of CE may make these characteristics difficult to achieve. The most serious problem likely to be encountered when a company changes to CE is that the system is based purely on individual performance and is the result of evaluation by a single individual. The cross-functional integration inherent in CE usually suggests a greater reliance on team-based appraisal and evaluation by multiple individuals. It may also suggest more frequent appraisal. Let's consider why this is the case.

Task Interdependence. As we have already suggested, CE typically involves a higher degree of task interdependence and task uncertainty. High task interdependence suggests that team feedback should be given high priority, since the interdependence means that it is the coordinated efforts of a group that matter more than individual efforts. High task interdependence also suggests that the appraisal should be based on input from multiple individuals, rather than a single evaluator.

Task Uncertainty. High task uncertainty suggests that more frequent, more intensive feedback may be necessary. This is not to suggest conducting an appraisal monthly or even quarterly, but it might suggest twice yearly, rather than annual appraisals. Equally, the appraisal might involve more careful investigation by the appraiser — e.g., rather than soliciting feedback about performance using a simple survey, it might be necessary to discuss performance personally with those peers or managers in other departments who have been asked to provide input.

Who Contributes to the Appraisal? Appraisals seem to work best when they are based on a variety of sources. An analysis of task interdependence can provide suggestions for who participates in any given review. Obviously fellow team members should participate. It's important to note that everyone with high task interdependence need not participate in every appraisal — only a representative sample of such people need do so. GE Aircraft Engine has a process it calls a "360 degree appraisal process" in which every individual is reviewed by people below them and at their same level, as well as by their superiors in the organization. Each person is asked to nominate about 10 people who might serve in a review capacity for them. Their manager then selects four or five to actually participate in the review. The manager consolidates input from all reviewers. The major benefit is that people are now formally reviewed by both their peers and their direct reports.

Reward Systems

We haven't seen much evidence that the reward system within a company is a particularly useful lever to use when introducing CE. We have seen most of the possible alternatives in place, and it doesn't appear to us that any of it matters very much. In fact, we have seen systems which seemed terribly unfair and inappropriate, and it still didn't hurt things nearly as much as we thought it would! We have a few ideas about why this is the case.

Why the Reward System is a Weak Lever: Intrinsic versus Extrinsic Rewards. Intrinsic rewards are those which derive from the work itself. An example is the sense of well-being you get when you know you've done a good job by your own standards, or when you've achieved a significant personal goal. Quite simply, you feel good about yourself. Extrinsic rewards are those provided from the outside in the form of pay, bonus, praise, or a coffee cup provided in recognition of good performance. For most professionals, intrinsic rewards are more important than extrinsic. There are several reasons for this. Perhaps the most important is that, as a professional, your sense of personal identity is strongly tied to the work you do, and a sense of having done well at work is closely tied to feeling good about yourself. Another reason is that most professionals are reasonably well paid — any likely bonus adds only a small increment to total compensation. What this means is that people involved in product development are usually already highly motivated to do a good job. You can alter this slightly by manipulating the reward system, but you have to be careful, since it's a two-edged sword.

Rewards as Disincentives. There are many characteristics of a reward system that can be *demotivating*. This means that people feel less motivated to behave the way you want them to because of something wrong with the reward system. There are two likely reasons for this.

1. ***Using the Wrong Contingencies***. This means that you are rewarding the wrong behavior — either you're rewarding things you don't want people to do, or you are at least missing out on rewarding specific behavior you want to happen. For example, a common mistake is to reward highly individually assertive behavior — such

as being very well known. This may be the result of someone making sure they get all the credit for something that happens, when many others had a lot to do with the success. This not only will make the others unhappy, but they will not work as well with the individual who took all the credit. Normally this is bad enough, but in a cross-functional team based situation, like CE, the workings of a major team might be disrupted.

2. ***The System is Unfair***. The system may be perceived as being administered unfairly. At the worst, this can be the result of apparent favoritism. But it can be more subtle, when people simply don't know what the criteria are for distributing rewards. In such an arbitrary system, people will often believe the worst.

SUMMARY

The three People System topics discussed in this chapter — job design, skills, and motivation — all have powerful effects on your ability to do CE well. A better distribution of work through good job design will mean that people have well-rounded meaningful jobs to work on that make more sense and are easier to perform. Making sure the proper sets of skills are available to accomplish the jobs you've designed will help to ensure those jobs are done correctly. Finally, establishing suitable goals and metrics will help to ensure that individuals and teams know what they *ought* to be doing. In Part II we will discuss methods for assessing your current systems in these areas and designing improvements in them, if necessary, so they can better support concurrent engineering.

Chapter 7

•

Technology[1]

The future offers very little hope for those who expect that our new mechanical slaves will offer us a world in which we may rest from thinking. Help us they may, but at the cost of supreme demands on our honesty and our intelligence. The world of the future will be an ever more demanding struggle against the limitations of our intelligence, not a comfortable hammock in which we can lie down to be waited upon by our robot slaves. • Norbert Weiner

The final piece in our model of the context for concurrent engineering is technology. Technology includes all of the tools and methods used by people in the product development process. Our purpose is not to tell you all about the latest in CE tools and methods; there are plenty of places you can learn about that. We will focus on how those tools and methods *fit* with the work process, organization, and human systems in your enterprise.

Technology is vitally important in CE because technology mediates many of the work processes performed during product development. If we can think of the fundamental work being done during product development as a series of decisions, with the results being stored and managed in some way, then it becomes easy to see why technology plays such a central role. Designs are created using computer-aided design (CAD) systems, analyzed using computer-aided engineering (CAE) systems, stored in databases, accessed and managed with product data management (PDM) systems, and communicated to others using computer networks. In a sense, the work of product development and its technol-

[1] Dr. Thomas Phelps from the Industrial Technology Institute assisted with writing this chapter. Dr. Phelps holds a Ph.D. in Mechanical Engineering from the University of Wisconsin.

ogy is inextricably linked. The very nature of the work changes depending on the technology being used. Work processes which once required a series of manual steps being done by different people and which were difficult to coordinate are now tightly integrated, to be performed by one or two people, and instantly communicated to anywhere in the world.

This tight integration of technology with the other CE elements leads to our decision to reserve the discussion of the technology for last. You will note in Part II that technology is the last thing assessed and the last thing designed in our methodology. This is not simply a result of our biases as organizational specialists. We know the technology is extremely important, but all too often we see what has been called "technology-push" in the companies we visit. An executive decides that only the hottest, latest technology will do, and so the company spends large amounts of money on tools it later finds it can't use very well. Or, a customer demands that a supplier should do all its CAD work in the customer's native system, and so the supplier is forced to adopt a system that is expensive and may be ill-suited to its design tasks. This may be good for the customer in the short run, but it may be very bad for the whole enterprise in which the customer operates, and ultimately bad for the customer in the long run.

In our view, technology should generally be planned in such a way that it is subservient to the work process and the people doing that work. It should *support* design work; the design work should not be changed just to fit into the capabilities of the technology. For example, designing in solid models places some limitations on the way a designer is able to do his/her work, and it can be very cumbersome. Unless you need to take advantage of the benefits of solid modeling, it can lead to higher costs and significant delays. Yet some companies blindly charge into this and other technologies without fully understanding their implications and requirements. Naturally, new technology can often provide opportunities to improve the work process; we are not suggesting that these opportunities should be ignored.

We often hear dreams of a purely electronic design process with one of the central themes being the elimination of physical prototypes. The Boeing 777 is one famous case of doing this, and certainly a passenger jet is as complicated as any product ever made. The Seawolf submarine, described in the Newport News Shipyard case study in the Appendix, is another example. In neither case were *all* physical prototypes eliminated — only the full system prototype was eliminated. Thus, no full airplane or submarine prototype was built in those cases, as had been the practice in prior programs. This certainly resulted in a huge savings. Still, we cannot help but wonder what it will take for analysis and simulation to replace all of the functions of the prototype. The Toyota case in the Appendix is one example of a company that structures its entire design process around physical prototypes. They serve to test for interactions between components that Toyota believes cannot be completely anticipated with analysis. They also serve to integrate across functions as people stand around and watch the prototype go together and make notes on how their particular systems fare. Thus, some of the social and technical functions of prototypes in integrating mating systems may be difficult to completely replace with computers. This is not to say it will never happen, but full attention will need to paid to *all* of the functions of the physical prototype before they will disappear forever.

Our primary emphasis in discussing technology is its ability to help integrate across various boundaries, be they functional boundaries inside a company, boundaries between companies, or operational boundaries between steps in a work process. Technology affects and is affected by three primary CE elements: work process, internal organization, and supplier relations (Figure 7.1). The internal organization and supplier relations provide the cross-functional integration requirements for the technology. The work process provides core work requirements for the technology. The technology also "pushes back" on the work process, in that the commercial availability of technology makes certain kinds of work possible and other kinds impossible (hence the double arrow). Finally, the technology influences skill, motivation, and job design requirements of the people systems which must be considered when designing the technical system. We will start our discussion with core

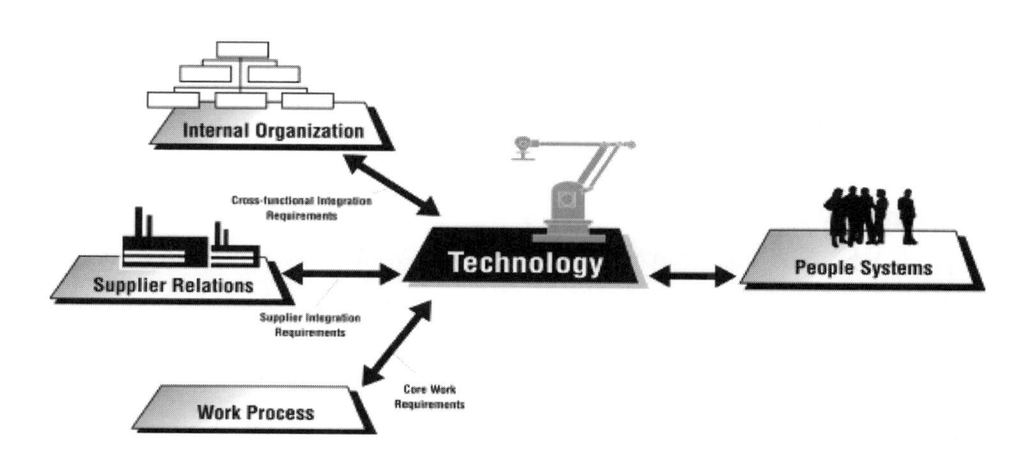

Figure 7.1 • Technology and Other CE Elements

work, then move on to cross-functional integration requirements, and finally discuss the people system issues. The chapter will conclude with a discussion of the problems of technology-push in product development organizations.

Core Work Requirements and Technology

The task breakdown inherent in the work process suggests what core design technologies are needed at each step in the process. Note that these technologies may or may not be focused on integration. For example, technology used for building prototypes may be irrelevant from a cross-functional integration perspective. On the other hand, a technology may *appear* unrelated to integration but in fact may be very related. As an example, a CAD system may appear to be simply a tool for creating a design, but it usually has great cross-functional integration implications due to the need to exchange data with other systems.

There are many possible core work requirements for technology in the product development process. We will discuss some common requirements below. Not all companies have these requirements, although most have at least some, and some companies will have additional requirements.

Provide for Rapid Visualization

Visualization is a fundamental activity in the design of any mechanical object. Rather than having to imagine what an object would look like based on 2-D drawings or having to build a physical model, technologies like CAD use surfaces and solid models to make it possible to see what the finished product would look like right on the screen. Visualization is not limited to a single object. Objects can be combined into assemblies, checking for fit and interference between the objects. The ability to visualize individual objects and assemblies means that decisions about the product can be made much more quickly than if a physical model has to be built. When on-screen visualization is not enough, the mathematical data can be used to drive a rapid prototyping system which builds a non-functional physical model, usually from some soft material like plastic, wax, or paper.

Represent and Build Complex Shapes

Because of the power of math to represent complex shapes, electronic models can represent shapes not documentable in 2-D form on paper. This is especially true of complex curves and surfaces. The swooping, compound curves found on automobile body panels in the 1990s are a direct result of this capability to represent complex shapes. Without the link between CAD and Computer-Aided Manufacturing (CAM) systems, the dies required to manufacture such body panels would be very difficult to make.

Support Complex Calculations

Mathematical models are more than just representations of what something looks like. Because of the richness of the underlying representations, more advanced forms of electronic models, such as solid models, can support the calculation of physical properties of the modeled object like volume, surface area, moments of inertia, and dynamic properties. On a complex shape, this kind of information can otherwise only be determined using costly experimental techniques.

Encourage Reuse of Old Designs

Reuse of electronic models can take a number of forms. For companies that make products which are similar to each other, the ability to reuse product data can save enormous amounts of time. But this is only true if the datasets from older products are stored in a way that permits designers to know they exist and to find them when they need them. An electronic product model can be used for other applications as well, such as CAE or CAM programs. These applications require the same kind of electronic models. Without an electronic model from a CAD system, users of the other applications have to build what are essentially the same models from scratch.

Facilitate Communication

Sharing of product information is essential between customers and suppliers, and between different functions in the same firm. Electronic data are often easier to send to others than either physical models or large drawings. Electronic data can also be much more usable. A particular example of this is in tooling design. Since the tooling design is based on the product shape, designing the tooling by starting from an electronic model of the part saves time and helps ensure accuracy in the result.

These requirements can be met by a wide variety of different technologies. The following list describes some of the technologies which help to meet them.

- ***Computer-Aided Design (CAD).*** CAD systems enable designers and engineers to create a design in electronic form. Where once a design could only be modeled as either a two-dimensional (2-D) drawing or as a physical model, it can now be electronically represented as a mathematically defined 3-D model, such as a wireframe or solid. A wireframe model is, as the name implies, a set of points, lines, curves, and (sometimes) surfaces that, taken together, describe the geometry of the object. A solid model is one in which the mathematical model properly represents a true solid, with an explicit representation of where the object's material is and where it is not. The use of CAD is now very widespread, although quite a bit of design work is still done using manual drawing, particularly in smaller firms. CAD not only aids in rapidly building a product model, it also aids in visualization.

- ***Computer-Aided Engineering (CAE)***. CAE refers to the use of computers to conduct engineering analyses, including simulation. For example, there are many programs to conduct stress, thermal, mold-flow, crash, and other forms of analysis. All of these analyses are ways to test some form of product model. These electronic "tests" augment or replace physical tests which are expensive and much more time consuming, if not actually impossible to carry out. In some cases, the information derived from an analysis package may not be obtainable any other way. CAE systems often can make direct use of the CAD product data model, which greatly speeds up the ability to use them. There is often a more direct electronic link between design and analysis in electrical design. CAE typically is used to facilitate rapid visualization and to support complex calculations.
- ***Computer-Aided Manufacturing (CAM)***. CAM programs generate programming used by computer-numerical control (CNC) machine tools in manufacturing to make the product. A common use of CAM is to generate cutter paths used by cutting machines, such as a mill or lathe. CAM systems can also check CNC programs by simulating the cutter process, allowing the operator to see just how the cuts will be taken. This allows checks for errors in the programming or for interference between the cutting tool and the workpiece. CAM is used to represent and build complex shapes.
- ***CAD/CAM Systems***. CAD/CAM refers to a system that integrates both CAD and CAM capabilities, so that CNC programming is based directly on the CAD data. This does not mean there is no human intervention in generating the CNC program, since no CAM system is currently that fully capable. But direct use of the common data saves time and reduces the possibility for error.
- ***Rapid Prototyping Systems***. Rapid prototyping systems are tools that quickly produce physical models of electronically modeled objects. Such physical models are generally not fully functional, but they can serve many purposes that require someone to physically work with a model that may not require the strength of the

actual product. They can help visualize the product in a way that many people prefer to 3-D displays on a screen. The best-known rapid prototyping technology is stereo lithography, but many others are now available. All rapid prototyping systems require electronic mathematical representations of the object as the starting point of the process. Rapid prototyping systems are used to support rapid visualization.

The specific technology used by any individual on their job will be a function of their job design and their skill level. We will discuss that later in the chapter when we get to the link with people systems.

Cross-Functional Integration Requirements for Technology

The cross-functional integration requirements for technology depend on your cross-organizational design internally and with suppliers (see Figure 7.1) We will divide our discussion by the relevant subelements of cross-organizational design.

Project Management Structure

Depending on your project management structure, you may have a need for highly sophisticated technological support or very modest support. For example, a liaison will have lighter needs than a heavyweight project manager. The heavyweight and, to a lesser degree, the lightweight project manager will generally have a staff and be guiding and monitoring the project at every step of the way. They certainly need project management software and basic clerical tools to generate documents. They may also need communication technologies, tools, and techniques used to standardize work processes and perhaps access to data transfer. In contrast, the liaison is mostly acting as an intermediary between others who are the real managers of the work processes, hence he/she will probably not need elaborate project management tools. Which other technologies they might need should come from further analysis of functional involvement, communication patterns, and coordination needs. Outside suppliers will not need access to project management software if they have a minor role in design; but if they serve in a partner role, you may want them to use the same project management structure as your firm so they can plan their activities and then plug them into your larger work process model.

Functional and Supplier Responsibilities

Each function and each supplier is involved to a different degree in major tasks in your product development process. This has implications for the technology needed by the members of each function and each supplier. For example, tooling design might best be done on a particular tooling CAD system. If you want toolmakers to provide significant support to the tool engineers in the design of tooling, you would probably want to give *both* engineers and toolmakers access to the tooling CAD system. This would not be necessary if the tool engineers did the entire tool design and then handed off the final database in the form of NC tapes to the toolmakers. The involvement of outside suppliers in different design tasks is of equal concern. For example, if they will be responsible for prototype testing, they will need appropriate testing facilities.

COMMUNICATION

Ultimately, cross-functional coordination requires communication. But the form, frequency, and intensity of communication will vary depending on what needs to be coordinated and the coordination method used. For example, standards may be effectively communicated through written guidelines or a computer database. On the other hand, joint problem solving between different functions to resolve a serious design problem almost certainly requires a face-to-face meeting. Different functions and different individuals within them will require access to different communication technologies. For example, design engineers might require convenient access to the Internet because they interact with both customers and suppliers so much. In contrast, toolmakers may only need access to an intranet for e-mail and file exchange with internal staff.

Table 7.1 summarizes communication technologies roughly rank-ordered from weakest to strongest. The relative strength of the technologies, called "richness" in Table 7.1, refers to how much information can pass between parties per unit of time. For example, face-to-face communication is higher in richness than written communication. When face-to-face, the receiver not only gets the content of the uttered words, but can also hear voice intonations and see body language. Moreover, the receiver can immediately ask questions of the sender to clarify the meaning of the message or add additional information to build on what was just said.

While the "richness" levels are subjective, they are useful as rough guidelines. The richness of each medium is related to another concept displayed in Table 7.1 — synchronicity. In asynchronous communication, there is some time delay between when a message is first sent and when it is received, and a time lag in the response from the receiver. Mail (hard copy, e-mail, or voice mail) is a form of asynchronous

Communication Technologies	Description	Richness	Best for
Formal written messages (paper or electronic mail)	One-way communication of words only, without intonation, body language, or immediate feedback.	Low	← One-way, low frequency
Shared databases	Rated higher than other formal written messages because of the quantity of information and because changes to the database are rapidly accessible to all users.	Medium	←→ Two-way asynchronous, low frequency
Computer conferences	Provides delayed feedback; not as strong as face-to-face meetings since body language and verbal and visual cues are removed.	Medium	←→ Two-way asynchronous, low frequency
Personal written messages (paper or electronic mail)	Two-way communication without audio cues, but personalized and directed to the individual receiver; can be direct feedback to a message from the receiver.	Medium	←→ Two-way asynchronous, high frequency
Voice mail	Can be personalized and provide audio cues; asynchronous so feedback from the receiver is not immediate.	Medium	←→ Two-way asynchronous, high frequency
Telephone	Provides audio cues and immediate feedback; not as strong as face-to-face as body language and visual cues are removed.	High	⟷ Two-way synchronous, high frequency
Video conferences	Allows reading some body language and visual cues; less complete and vivid than face-to-face meetings.	High	⟷ Two-way synchronous, low frequency
Face-to-face meetings (coming together from distant places)	Immediate feedback allows understanding to be checked, interpretations corrected and ideas to build on each other. Allows reading of body language and visual cues.	Very High	⟷ Two-way synchronous, low frequency
Face-to-face meetings (collocation)	Same as above, but can meet with greater frequency.	Very High	⟷ Two-way synchronous, high frequency

Table 7.1 • Communication Technologies from Low to High Richness

communication. In synchronous communication, you get the message immediately and can respond then and there, as in phone calls, face-to-face meetings, or video conferences. The richest communication technologies are synchronous.

It is easy to see in Table 7.1 that there is a tradeoff between the richness of each communication medium and the cost of using it. Formal written messages are the fastest and cheapest, and can go out to many people simultaneously, whereas face-to-face meetings are encounters that are costly in terms of the time spent by the people in the meeting, and may involve travel costs. Thus, it makes sense to ration judiciously the use of richer media.

Cross-Functional Coordination

Design Tools and Techniques. There are many design tools and techniques that help coordinate across functions. We are using the term "tools and techniques" here in a somewhat narrow way to encompass any computerized or paper aids that help bring cross-functional perspectives to the design process. Some prominent examples include the following.

- ***Quality Function Deployment***. QFD is a series of matrices that link dimensions from one functional perspective to another, and in this way is a tool for cross-functional coordination. For example, the first matrix in QFD (called the House of Quality by Hauser and Clausing, 1988) provides a systematic way of linking customer requirements (typically a responsibility of the marketing function) to engineering specifications (a responsibility of the engineering function). While the method for doing this is a subjective judgment about the relationship, even this is a great improvement over typical practice. Often this linkage takes place by a blind hand-off from marketing to engineering, and it is not clear to anyone how a vague market research finding (e.g., "the product should last") translates into hard specifications. QFD forces the design team to make such transitions explicit. Thus, QFD can be an enabler for integration of marketing and engineering. Other QFD matrices can help integrate downstream processes, e.g., the translation of engineering specifications to manufacturing process specifications. Some QFD users have found it to be awkward and time consuming. Others have found ways to use the method flexibly in a wide variety of situations.
- ***Design for "X."*** Design for Manufacture and Assembly (DFMA) is a methodology with supporting software for making sure design

engineering is aware of the effect their decisions have on cost and ease of manufacture. A DFMA system has rules built into it that score a part on how easy it is to make or assemble. Other systems have been developed which provide rules and scoring for Design for Maintainability, Design for Disassembly, Design for the Environment, and others. These "design for . . ." systems are commonly referred to by the term "Design for X." One of the best known of these tools is the Boothroyd and Dewhurst (1989; Boothroyd, Dewhurst, and Knight, 1994) Design for Manufacture and Assembly system. It asks the designer to input information about the design, labor costs, etc., and then evaluates the cost implications of the design. The designer can play "what if" games and test the consequences of design changes. The algorithms used greatly simplify the reality of manufacturing and are generic rather than specific to a company's own processes. But the program can often lead the designer to question assumptions about the design, and generally leads to desirable manufacturability features like reduced part counts and modular designs. As one engineer we know put it, the software acts as a "neutral arbitrator." He put it very well when he said, "We can sit around and argue over the best design in endless meetings but the program provides dollar and cents numbers that the team cannot argue with." These systems are not a substitute for actual dialog between designers and manufacturing engineers who understand the subtle complexities of their manufacturing processes. Yet they can provide a starting point, raise questions worth addressing, and be seen as having a legitimacy that human judgment may not have.

- ***Product Data Management (PDM)***. PDM systems are packages that control the access to and versions of documents and electronic representations of the end product or its manufacturing system. The sophistication and richness of the coverage of such systems varies widely, but the basic concept is the same: provide a managed central location for data users to store and retrieve information. While there are a number of commercial PDM products available — some free-standing, some associated with a specific CAD system — many companies have built their own PDM systems. "Homemade" systems, and even many commercial systems, are often unable to integrate well with other systems used by the company. A major issue with PDM systems is making sure they are used consistently. Substantial discipline is required to maintain the information stored in PDM systems. Another major issue is that PDM systems are focused on *internal* data management. As product data increasingly move across supply chain boundaries, the ability to manage data that flow across these boundaries will become increasingly important.

- ***Workflow Management***. Workflow management systems attempt to organize and control the flow of work through the organization, typically using some software system to do so. A typical workflow manager will prompt an individual that a certain task needs to be done, and will monitor its completion. When that task is completed, other tasks which are dependent on it will then be prompted.

These examples illustrate that CE tools and methods can be useful enablers of cross-functional coordination by facilitating standardization of work processes. Yet, there are also important limitations to these tools. They are simply tools, not substitutes for direct coordination between experienced professionals. For example, in QFD the usefulness of the exercise depends on the quality of the customer requirements. Simply getting marketing to send over tables and computer output summarizing customer preferences is not enough for engineers to generate a useful list of requirements. Considerable work must be done by marketing to translate the raw data into a useful list of requirements, and further work is needed to clarify what those requirements mean as the engineering requirements are being generated. This depends on effective two-way communication between marketing and engineering. DFMA also requires accurate data as inputs, and the conclusions that come out of these exercises need to be checked against reality through discussions between product engineers and manufacturing experts. This may be most effective if product engineers and manufacturing experts work through the DFMA process as a cross-functional team.

Computer Networks and Coordination. Computer networks connect individual computers to each other so they can communicate. This can include all types of computers, not just personal computers or workstations. A Local Area Network (LAN) is confined within a local area, such as a building or corporate campus. By itself, the LAN is a closed system, often situated behind a "firewall,"[2] and hence is relatively secure, although any dial-in capacity through a modem compromises that security to some extent. The Internet is a worldwide "network of networks" that link together using the Internet Protocol (IP) to form one large network in which any computer on the network can access any other computer. The Internet Protocol forms the basis for most

[2] A firewall is a hardware and software mechanism that isolates the local network from the outside world. The barrier provided by a firewall can vary greatly, from a simple mechanism such as a password to more secure techniques.

electronic mail and for the World Wide Web. When the Internet Protocol is used on a LAN, the result is often called an "intranet." The computer network makes it quick and easy to share product data and other information with anyone else on the network. As we will discuss below, the ability to share data requires much more than a simple network connection. It requires that the data be compatible with the system using it.

World Wide Web. These days anyone who watches television or reads is inundated with news about the Web and how it will transform the very fabric of social life in the global community. The Web is a way to use the Internet by linking documents located on different computers. Information put on the Web (e.g., in a home page) can be accessed by people anywhere in the world. Obviously companies do not want to give millions of people access to their sensitive product data. There are security measures that can be taken to restrict access to information placed on the Web, though clever hackers can sometimes break through these. As security on the Web improves, it should offer a relatively inexpensive way of storing design information that can be shared by remotely located designers and suppliers throughout the world. It also provides a means of communication for those individuals to communicate using e-mail, computer conferences, and shared applications. This technology is moving very fast right now, so it's important to keep up to date on what's happening with it.

Coordination and Product Data Exchange. Coordination is often highly dependent on the ability to *share* information, especially product data. Often two organizations could make use of the same dataset. For example, a company that makes a product which is an assembly of parts, but has suppliers which design some of those parts, will need to bring in product data from those suppliers so it can have a complete model of its assembly. Even a company that is sourcing a "black box" component[3] will need to have packaging information so enough space can be provided and interfacing components designed. Unfortunately, sharing data from CAD and other downstream computer systems can be very difficult to do from a technical point of view. Each vendor of these systems uses different formats to represent their data; indeed, some use different math to represent a simple shape, such as a circle. This means that the direct exchange of data between different vendors' CAD systems, or

[3] A black box component is provided by the supplier without the customer needing to know anything about what goes "inside the box." The customer simply has to know the performance specifications and interfaces to the component. Some customers are not satisfied to have the insides of such components be completely invisible, so this is sometimes called a "gray box" component to signify that there is some visibility provided.

between CAD and other applications, is almost always impossible; some means of translation is required.[4] There are three basic approaches to this issue of translation.

- ***Direct Translators***. A direct translator is one written specifically for the two systems exchanging data. These are expensive to build and go out of date as quickly as vendors ship new versions of their systems. Some direct translators are provided directly by the vendors of CAD systems. For example, we know that some CAD systems provide a translator to translate the CAD file into STL, the language used to drive stereolithography equipment. These translators usually only apply to standard languages, not to other vendors' proprietary equipment. Other translators must be purchased from third party vendors. This means there is usually some lag between the arrival of a new version of a given CAD system and the translator for it. The quality of such translators is also uneven and highly dependent on the given vendor.
- ***Initial Graphics Exchange Standard (IGES)***. IGES is a standard for translation. IGES has never been a very effective standard, although it is widely used and provides considerable savings over direct translators in many applications. For a variety of reasons, such as the standard being implemented differently by different vendors, data exchange using IGES often results in many errors and incomplete exchange. This means the data need to be corrected by the receiver. This is particularly problematic if the data will be used to drive numerically controlled equipment since the data must be near perfect for this type of application. The number of errors may be so large that in many cases a model will have to be completely reentered by the receiving party, at substantial cost and with some probability of additional error. Even when the model is received intact, IGES only handles the geometry and not all the other data that can be critical to design (e.g., it does not handle configuration management data and numerical controlled process plans).
- ***STandard for the Exchange of Product Model Data (STEP)***. The STandard for the Exchange of Product model data, better known as STEP, is a new International Organization for Standards (ISO) standard, formally known as ISO 10303. STEP is different from other product data exchange standards such as IGES in two important ways. First, STEP is intended to support all product data,

[4] For the real novice at this, the problem is actually very much the same as moving word processing files between a Windows based computer and a Macintosh, or between two different word processors on the same computer. The word processing file exchange problem is now pretty much solved, but CAD files are much more complex.

> not just geometry. Second, STEP data exchange is based on formal models of the product data within specific application areas. In other words, you don't just exchange product data using STEP, you exchange complete product models. This allows for substantially greater coherence of the data and better chances for complete and accurate data exchange. STEP moves substantially beyond the concept of graphical product data to encompass all product data. The kinds of data already being added includes configuration management data, product life-cycle data, and numerical controlled process plans. Ultimately, all of the information needed in the product development process should be available for exchange using STEP.

This issue of product data exchange makes clear the need for compatible tools for work processes that have interfaces. Until recently it was common practice for each user of information to have to create their own version of it. For example, if you needed interfacing part information as input to your own design work, you simply entered that information yourself off of a blueprint. If you needed to program a Coordinate Measuring Machine (CMM) to verify part geometry on a prototype, you programmed the CMM from scratch. With advances in product data exchange and other methods for interoperability, the same product dataset can be used to help program the CMM, start a CAM program, and provide interface data for other designers. As better standards become available to ensure interoperability, greater use can be made of the single, master product dataset, and less redundant data will be required.

Technology and Skill, Motivation, and Job Design Requirements

In the relationship of technology to people systems, we are primarily concerned with three issues:

- the skills needed to apply the technology,
- the motivating nature of the technology including the potential for deskilling of work,
- how job designs affect and are affected by the technology.

The need for skills to operate the technology is obvious, so we need not say much about it. The primary motivational issue is that people may fear that technology will "deskill" their jobs, i.e., take a relatively skilled, craft-like job and put all or much of the decision-making and craft into automation, leaving the job holder with little to do but load and watch a machine. For people who hold such jobs, deskilling is extremely demoralizing, even if there is a significant skilled element left to the job. One common fear has been that the traditional draftsman would be replaced by engineers who will create the design on a CAD system. This fear was particularly prevalent in the 1980's when many organizations were first introducing CAD on a widespread basis. Our research then (Liker, Fleischer, and Arnsdorf, 1992; Fleischer and Liker, 1992) suggested this fear was largely overstated. In fact, in most cases it was the draftsmen who were taught how to use the CAD systems. In the process, many companies soon discovered that CAD could not automate many key aspects of drafting work. There are many skills, conventions, and general know-how associated with drafting that were not known by the engineers. In addition, engineers in the U.S. were generally not interested in learning how to use CAD and getting involved in the detail necessary to create the product model. Higher level designers did not always want to develop the level of detail in the product model needed to generate fully documented and dimensioned views of the product. Over time, most of the routine work of the draftsmen of old has been incorporated into CAD, but this was usually gradual enough to transition draftsmen to work as designers or to reduce the size of the drafting workforce through attrition.

A second motivational issue is the extent to which the technology helps to motivate the individual using it to maintain interest in his/her work. Technology that is difficult to use (e.g., a computer application that is difficult to navigate, or which often freezes) will tend to be very frustrating and hence be demotivating. In a survey of CAD users in the U.S. and Japan (Liker and Fleischer, 1992a) we found that, in both countries, one of the most important factors in predicting both the use of high level CAD features and the perception that CAD was superior to the old manual methods was "system quality and support." This referred to whether the hardware and software were well supported by technical personnel, whether the system was fast, and whether the hardware and software were reliable. If users had to wrestle with the system or the system often went down, this greatly reduced both their motivation to experiment with higher level features and their satisfaction with the technology.

The people systems do not tell you what technologies are needed, but they do suggest who should have access to different technologies and what skills and training are required so people can use them. The job designs are a way of allocating tasks to people. Based on this, people must be supported with the appropriate technologies to carry out those tasks. For example, if your company separates the role of design engineer and designer, and only the designer uses CAD, it may not be necessary to buy expensive CAD workstations for your design engineers. But if you want your design engineers to use CAD regularly, they will need a workstation. Similarly, if the designers are going to be heavily involved in direct communication with other functional groups, they may need appropriate communication media such as local area networks and electronic mail to communicate with those groups. This can only be determined after you have defined the job designs of each person. The skills and training required for each job type should reflect the technology assigned to that job. Thus, any decisions about skill and training needs should follow decisions about who will use which technologies. Equally, technical advances may make changes in some jobs possible. For example, in the past it may have been infeasible to ask designers to communicate regularly with customers. With the World Wide Web and new conferencing technologies, the cost of doing so is greatly reduced so that it can become a regular part of the designer's job.

Of course, anyone who has attempted to implement new technology knows that technology does not only affect people, but people affect how the technology will be used. If the people are not properly trained and prepared for the technology, if it does not fit into the way jobs are designed, and if it is not introduced sensitively, then the technology may not get used at all or it will be used at a level far below its potential. In the design of the people systems, the parts most relevant to technology are the job designs and the skills and training needs. Later, in planning implementation, you will be forced to confront all of the social, political, and emotional issues associated with changes in people's jobs and lives.

THE LIMITS OF TECHNOLOGY-PUSH

We have visited many companies that seem to think that they can automate their way out of their product development failures. These companies hope that smart technology and the "paperless office" will reduce design costs, improve design quality, and get all those "incompetent people" who make foolish errors out of the loop. We have our doubts.

A few years ago, we (Liker, Fleischer, and Arnsdorf, 1992) examined the state of use of various CAD tools (defined to include CAD/CAM and CAE). In the 1980s and early 1990s, companies were scrambling to modernize their design technology by replacing manual drawing boards with CAD tools. They succeeded in doing this but also found the results disappointing. The expected productivity benefits simply did not materialize.

In 1992 we concluded that the main problem was *how* CAD was being used. CAD was being used as a direct replacement for the drawing board — as a stand alone tool. We argued that the true power of CAD was in its integrative ability, to integrate across versions of the design, across designers, across product and process engineers, across designers and analysts, etc. We visited automotive companies and their suppliers, major appliance manufacturers, and aerospace defense contractors who did largely electronics design — all were large companies considered leading edge users of CAD. While this integration was generally the intent of the companies we visited, they had generally failed to achieve their goals for two reasons. First, the companies often did not think through whether the CAD applications they had in mind were appropriate for their design tasks. Second, the companies were installing CAD systems without changing the organization to encourage the coordination and communication that is at the heart of integration.

In our 1992 article, we identified five promises typically made by CAD vendors that were going unfulfilled in the companies we visited. Even several years later we are approached by companies who say that the central message of what we said then is still true, and that they are still having difficulties with integration. Here are the five promises and some reasons why they were not being fulfilled.

1. ***Automation of routine design tasks will increase productivity of designers and design engineers.*** The real power of CAD for design comes when the design is modified or when it is built from pieces of old designs already in the system or when using parametric design models. We found that designers were typically functioning as islands, developing their designs from scratch and borrowing little from others or from their own previous designs. They reported that they did not have the time to store and retrieve primitives for future designs. This had a lot to do with the way they were organized to work on individual pieces of the design from their functional perspectives. Also, there were no disciplined processes developed to capture the history of designs. Each design project essentially started from scratch.
2. ***The ability to design in three dimensions will increase the designer's conceptual capacity and creativity, improve design quality, and create a geometric database useful for downstream analysis and production operations.*** We found that the predominant use of CAD was to produce 2-D drawings that were essentially electronic renderings of blueprints. Even when a CAD database was developed, it was usually used to then create 2-D printouts. Seldom were different parties all using the 3-D database on a screen to study the

design and make decisions based on the 3-D image. Nor were 3-D models being used to integrate across designers who would fit them together for interference checks. Engineers rarely even looked at the CAD tube and felt that use of CAD was below their status as engineers — they were too busy going to design meetings and meeting with suppliers to sit down and learn how to use CAD. They wanted 2-D printouts to look at and carry with them to meetings. Despite the relative lack of use of the 3-D capabilities, in some companies management had set as policy that all design would be done in 3-D even though building these models was an onerous task and took a lot more time.

3. ***Design databases will be electronically transferred to manufacturing or CAD/CAM, improving quality and productivity and reducing lead time.*** There was limited compatibility between the CAD and CAM programs in mechanical design. The 3-D model had to be "flattened" with appropriate cutter paths generated. In some cases, the manufacturing people responsible for generating the CAM programs found little value in starting with the CAD database and started from scratch from the drawings. And it was rare that suppliers were using CAD data from their customers to generate programs for numerically controlled equipment. Often they did not have compatible systems, and many complained that their customer's designers did not prepare the data with sufficient accuracy to enable them to generate NC tapes directly.
4. ***Design databases will be electronically transferred to engineering analysis and simulation programs, which will alleviate costly testing of prototypes.*** This promise assumes that the CAD database can be directly uploaded to an analysis module, and that CAD and CAE are iterative processes in which analysis results form design decisions and design changes are fed back to CAE for further analysis. It also assumes that computer models reduce or even eliminate the need for physical prototypes. Both assumptions were suspect. What we found in the case of structural design was that engineering analysis was done on a relatively small number of designs and that it was done in parallel to CAD, by different people who were experts in CAE (generally who held graduate-level degrees), on different computer systems, and they generally worked off of blueprints rather than databases. The CAE analysts said that the CAD data were not in a form useful to them — they needed certain simplified cross-sections for structural analysis. Some of the CAD systems had automatic mesh generation capabilities for Finite Element Analysis, but the analysts said these were based on

crude algorithms which yielded inaccurate results. The CAE analysts were often backlogged. Design engineers generally found they could not wait for the analysis results and had to go ahead with decisions, looking to the analysis results only for later confirmation. Few of the design engineers or their managers were willing to believe the engineering analysis results unless they were consistent with tests of physical prototypes, so prototypes were not being eliminated. As we argue in the Toyota case in the Appendix, physical prototypes can be an excellent point of integration as different parties (engineers of different subsystems, suppliers, manufacturing) come together to see how their pieces fit together. Often the physical prototypes reveal many unanticipated interactions which it seems unlikely engineering analysis could adequately capture.

5. ***The design process will become paperless as databases, rather than drawings, are sent to customers and suppliers.*** As we have already made clear, the CAD database was hardly replacing paper. If anything, the ability to make rapid (detailed) changes to the design led to more frequently reissued drawings and the use of more paper than ever.

In several instances above, we qualified our statements by giving mechanical design examples. We observed a much more natural use of CAD databases to link design, analysis, testing, and manufacturing in electronics design than in mechanical design. Thus, when the circuit was designed it could be almost instantly tested and analyzed. Also, the design database could be used to drive the creation of prototype circuit boards naturally and seamlessly. This may be partly the result of cultural differences between electrical and mechanical engineers. Many of the mechanical engineers we know still don't quite trust computers, while the electrical engineers have been brought up to believe in electronic devices as part of their fundamental belief systems. It may also be that CAE for electronic applications is more advanced, but we suspect this itself is the result of this cultural difference.

On the other hand, even in electronics firms we saw serious problems of integration. One coordination problem was between circuit designers and packaging engineers. The circuit designers just laid out the functional requirements but did not have to worry about how it fit together on a set of physical circuit boards. Often they did not have the data in a form that was useful to the packaging engineers. Circuit designers also had to coordinate with each other (if several circuit designers were developing subsystems that needed to work together) and packaging engineers had to work with mechanical engineers designing the physical housing for the circuits. There were frequent complaints about problems of coordination at these interfaces.

Obviously some things have changed since 1992 in the capabilities of design technologies and the sophistication of use. Some of what we described above might be dismissed as learning pains associated with introducing a new technology and learning how to use it. Regardless, we feel there is another more important message. The message lies in the fit between three areas — design tasks, technology integration, and organizational integration. Technology integration and organizational integration go hand in hand, and the appropriate levels of technology integration and organizational integration depend on the design task — what we refer to as the "work process" in this book. The methodology developed in this book is designed to get you to determine your task needs first, analyze your organizational requirements, and then consider how technology will support the design tasks and fit with your new organization. Starting in the reverse order by selecting technology for its own sake may be a recipe for failure.

Conclusions

In this chapter we have emphasized the need to find an appropriate fit between work processes, organization, people, and technology. Technology should be one of the last considerations — selected and implemented to support the work processes and organization. It is important to resist the seductiveness of buying technology for its own sake. We provided a framework that shows the relationship of technology to each CE element and specific subelements. You will find in our methodology in Part II that decisions are made about technology concurrently as you make decisions about work processes, organization, and people. In fact, there is no separate section in the design chapter dealing with technology generically. Instead, there is a number of small subsections integrated with the other CE elements.

CHAPTER 8

•

WHERE AM I? — THE CE PROFILE™

Going to work for a large company is like getting on a train. Are you going 60 miles an hour or is the train going 60 miles an hour and you're just sitting still? • J. Paul Getty

So far we've covered a lot of ground, but it's all been on the conceptual level — lots of ideas and examples, but no specific way for you to figure out where you stand. This chapter will introduce a short questionnaire (the CE Profile™) that you can use to determine how far along your company is toward having the structures and processes needed for doing good CE. At a very rough level, it will assess strengths and weaknesses and, in general, give you some ideas about where you should start working toward providing the context for doing CE better.

The CE Profile in this chapter serves two purposes.

1. It provides a first opportunity to apply the concepts of Part I of this book to your organization. The quality of this profile will depend on how well you have understood the concepts presented up to this point.
2. It serves as a screen for the much more detailed assessment in Part II of this book. The CE Profile will give you a rough idea of your strengths and weaknesses, and you will want to pay particular attention to assessing more specifically your areas of weakness in Part II.

THE CE PROFILE™

Once you've read a book like this, your natural reaction is probably to want to know how your company stacks up against the ideal. The CE

Profile is just what it sounds like — a checklist that can be completed in about 20 minutes that will give you a quick answer to the question, "How are we doing?"

We created the CE Profile from the structure of the previous chapters, starting with the Work Process. We took each of the main topics in each chapter and created a set of questions about them. With only a few exceptions, each question is scored Yes/No. There are only four requirements for you to do a good CE Profile.

1. ***Be informed.*** If you don't have a clue about how to answer a question, take a little time to get some information. Better yet, work on the Profile with a cross-functional team, in order to maximize the chances that someone sitting around the table will know the answer.
2. ***Live with uncertainty.*** For a Profile, you can't take the time and effort to really know *all* the answers. Do this fairly quickly with whatever information you can gather at a relatively low level of effort. Although this may appear to conflict with #1 above, it really doesn't - it's just a matter of balance.
3. ***Be conservative.*** If you're unsure about how to answer because your reality is not quite so black and white as the Yes/No answers, make your answers conservative. In most cases, the reality of what's happening in your company is *worse* than you think it is, because few organizational cultures permit the honest admission of weaknesses.
4. ***Be honest.*** Don't kid yourself — it won't help you to stay in business if you cheat to get a high score.

Any Yes/No type of assessment can never be extremely accurate. In Chapter 11 we provide you with a process for conducting a very detailed assessment that can take you weeks to complete. An in-depth assessment inevitably takes lots of time and effort, and you have to find a balance between accuracy and the effort involved. Here we take the side of low effort. But that's OK; the purpose is to get you started, to help you decide if you need to do anything at all, and, if so, where to start.

We've tried to make the CE Profile as easy to use as possible. This means that we've avoided "customizing" it for specific situations. So, if your situation differs from what the "norm" appears to be, you'll need to interpret the scores as best you can. The "normal" situation we've defined is a company that manufactures some product, has one or more suppliers of something other than raw materials, and has customers who use what they make either as end users or in manufacturing as part of some larger product. We expect that whatever you make is complex enough to have been "designed," whether that design was done by you or some other company, such as your customer. Also, if a concept is confusing to you, then you should be willing to go back and review the chapter in which it is discussed.

You can find the actual CE Profile questionnaire at the end of this chapter. Feel free to make copies of the CE Profile to use within your company or to get input from multiple people.

Scoring the CE Profile

Write the "raw" scores in the appropriate column of Table 8.1 for each CE Element. Your raw score is the number of times you checked "yes" under each element. Multiply your raw score for each element by the weight in its row. This gives you a weighted score for each CE Element. The reason we weight the scores is that we believe each of the CE Elements is about equal in importance. Some of the elements have more or fewer questions, depending on how easy it is to break the element down into small sections. There's no logical reason to rate something as more important just because we could think of more questions to ask about it. The "weight" corrects for this imbalance.

CE Elements	# items	Raw Score x	Weight =	Weighted Score
Work Process	9		2.22	
Internal Organization	46		0.43	
Supply Chain Involvement	15		1.33	
People Systems	18		1.11	
Technology	8		2.50	
Total	96			Σ = ______

Table 8.1 • CE Profile Scoring Table

Sample Ratings

After you've calculated your weighted scores for each CE Element, you can either use the scores in numerical form or you can plot them on a bar chart or spider diagram. In Table 8.2 we provide an example set of scores and charts that have been plotted for them.

The maximum possible total score is 100. We suggest you use a very familiar method for interpretation of this type of scoring:

90-100 = A — outstanding structure and processes for CE
80-90 = B — good structure and processes for CE

CE Elements	# items	Raw Score x	Weight =	Weighted Score
Work Process	9	*4*	2.22	*8.88*
Internal Organization	46	*38*	0.43	*16.34*
Supply Chain Involvement	15	*8*	1.33	*10.64*
People Systems	18	*11*	1.11	*12.21*
Technology	8	*7*	2.50	*17.5*
Total	101			$\Sigma = 65.57$

Table 8.2 • Example Scoring Table

70-80 = C — marginal structure and processes for CE

60-70 = D — could do a whole lot better at structure and processes for CE

below 60 = well, what can we say . . .

You can also look for strengths and weaknesses with this rating system using either the bar chart (Figure 8.1) or the spider diagram (Figure 8.2). The weights in the scoring table were selected so that in each of the five areas there is a maximum possible weighted score of 20 points. Very low scores may suggest the first or second thing you should work on to improve your CE Profile. Note that the greatest weakness

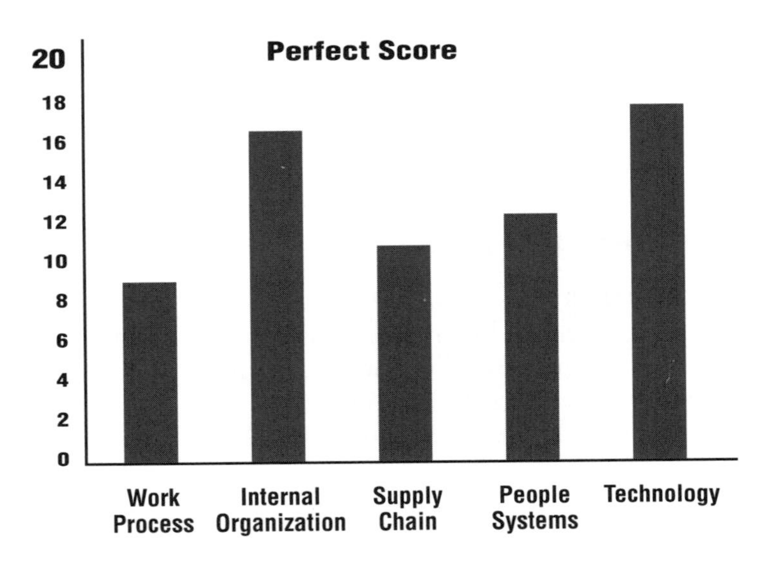

Figure 8.1 • Bar Chart for CE Profile

for our sample firm is in supply chain involvement and work process. They could probably just focus on those areas first and harvest a big improvement.

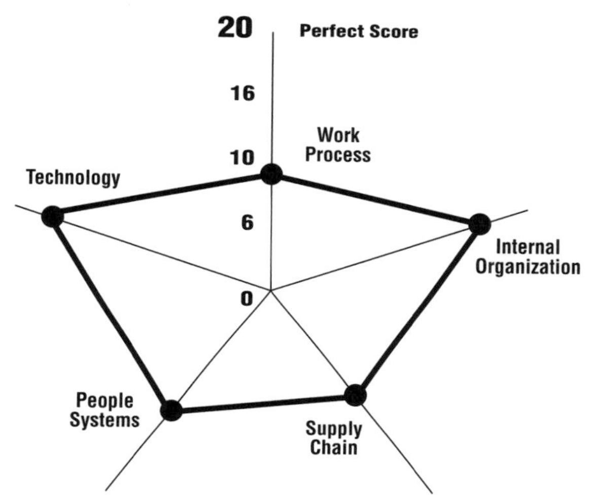

Figure 8.2 • Spider Diagram for CE Profile

How to Use the Results of the CE Profile

There are two types of scores you will get out of the CE Profile. The first is the total CE score. This does not tell you how well you're doing at CE or product development. What it tells you is how effective your *context and support systems* are for CE. If you have a low total score, your systems to support CE are weak. If you improve them, you'll likely improve the effectiveness with which you can do CE and sustain that success in the long run. This type of score tells you *if* you need to change at all, but it doesn't tell you anything about *what* should change.

The second type of score is the CE Element score, those for Work Process, Internal Organization, etc. Low element scores indicate weaknesses in the five specific areas. For example, in the sample scores (Table 8.2), the score for Supply Chain Involvement was quite low, as was the score for Work Process. Working on these two issues would probably result in the biggest improvement in support for CE. You can look at the individual items in the category for specifics on which you should work. But a much more specific look at areas for improvement is described in the in-depth assessment in Chapter 11.

Be careful before you act, however. The CE Profile is highly subjective and not very accurate. We strongly recommend you do a more in-depth assessment targeted to the specific areas you choose to work on. In Chapter 11 we provide you with methods and tools for conducting such an assessment.

CE Profile™

INSTRUCTIONS

Check Yes or No for each of the questions below. After you complete each section, total the scores as indicated in the scoring box for that section. Then use the Scoring Tables to calculate your total weighted CE Profile score.

WORK PROCESS (CHAPTER 3)

1. Do you have a formal product development process? Yes - ❑ No - ❑

2. For whatever process you follow (formal or informal), does it include:
   a. activity precedence? Yes - ❑ No - ❑
   b. schedules? Yes - ❑ No - ❑
   c. exit and entry rules? Yes - ❑ No - ❑
   d. controls? Yes - ❑ No - ❑
   e. tools and resources? Yes - ❑ No - ❑

3. Is your process easy to use? Yes - ❑ No - ❑

4. Is your process customized as appropriate for each division or product line? Yes - ❑ No - ❑

5. Is your process general enough to allow flexibility, but detailed enough to be useful? Yes - ❑ No - ❑

> ***WORK PROCESS SCORING***
> *How many times did you check "Yes" under Work Process?* _____

INTERNAL ORGANIZATION (CHAPTER 4)

1. Does your *organization architecture* promote:
   a. sufficient coordination within functions? Yes - ❑ No - ❑
   b. sufficient coordination across functions? Yes - ❑ No - ❑
   c. economical use of talent? Yes - ❑ No - ❑

d. ease of performance evaluation and control? Yes - ❑ No - ❑
e. focus on goals that are customer-driven? Yes - ❑ No - ❑
f. coordination of effort toward your customer? Yes - ❑ No - ❑

2. Does your *project management* structure promote:
   a. sufficient coordination within functions? Yes - ❑ No - ❑
   b. sufficient coordination across functions? Yes - ❑ No - ❑
   c. focus on goals that are customer-driven? Yes - ❑ No - ❑
   d. focus on goals that are company-driven? Yes - ❑ No - ❑
   e. responsiveness & speed of development? Yes - ❑ No - ❑

3. Do your *Standard Operating Procedures* promote:
   a. coordination within functions? Yes - ❑ No - ❑
   b. coordination across functions? Yes - ❑ No - ❑
   c. focus on goals that are customer-driven? Yes - ❑ No - ❑
   d. focus on goals that are company-driven? Yes - ❑ No - ❑
   e. responsiveness and speed of development? Yes - ❑ No - ❑

4. Do your *planning, scheduling, and monitoring systems* promote:
   a. coordination within functions? Yes - ❑ No - ❑
   b. coordination across functions? Yes - ❑ No - ❑
   c. focus on goals that are customer-driven? Yes - ❑ No - ❑
   d. focus on goals that are company-driven? Yes - ❑ No - ❑
   e. responsiveness and speed of development? Yes - ❑ No - ❑

5. Do your *tools, techniques and standards* for design promote:
   a. coordination within functions? Yes - ❑ No - ❑
   b. coordination across functions? Yes - ❑ No - ❑
   c. focus on goals that are customer-driven? Yes - ❑ No - ❑
   d. focus on goals that are company-driven? Yes - ❑ No - ❑
   e. responsiveness and speed of development? Yes - ❑ No - ❑

6. Do your *performance metrics* promote:
   a. coordination within functions? Yes - ❑ No - ❑

b. coordination across functions? Yes - ❑ No - ❑
c. focus on goals that are customer-driven? Yes - ❑ No - ❑
d. focus on goals that are company-driven? Yes - ❑ No - ❑
e. responsiveness and speed of development? Yes - ❑ No - ❑

7. Does your use of *credentials and training* promote:
a. coordination within functions? Yes - ❑ No - ❑
b. coordination across functions? Yes - ❑ No - ❑
c. focus on goals that are customer-driven? Yes - ❑ No - ❑
d. focus on goals that are company-driven? Yes - ❑ No - ❑
e. responsiveness and speed of development? Yes - ❑ No - ❑

8. Are you using *teams* to promote:
a. coordination within functions? Yes - ❑ No - ❑
b. coordination across functions? Yes - ❑ No - ❑
c. focus on goals that are customer-driven? Yes - ❑ No - ❑
d. focus on goals that are company-driven? Yes - ❑ No - ❑
e. responsiveness and speed of development? Yes - ❑ No - ❑

9. Does your *culture* promote:
a. coordination within functions? Yes - ❑ No - ❑
b. coordination across functions? Yes - ❑ No - ❑
c. focus on goals that are customer-driven? Yes - ❑ No - ❑
d. focus on goals that are company-driven? Yes - ❑ No - ❑
e. responsiveness and speed of development? Yes - ❑ No - ❑

INTERNAL ORGANIZATION SCORING
How many times did you check "Yes" under Internal Organization? ______

SUPPLY CHAIN INVOLVEMENT (CHAPTER 5)

Before you attempt to answer the following questions, read these directions. If your suppliers only provide raw materials or commodities (items you could order out of a catalog), skip to the next section (People Systems). If none of your suppliers provides "custom" parts or components (customized to some degree for your product), skip to the next section (People Systems). But, if that is the case, ask yourself whether this is because they should not be doing custom work or because you are not using their capabilities.

If you're still with us, think about the two or three non-commodity suppliers in your supply chain who provide the highest value-added custom components for your product. Answer the following questions about them.

1. Do you have staff located at any of those suppliers' sites?
   Yes - ❑ No - ❑

2. Do any of your suppliers have staff collocated at your site?
   Yes - ❑ No - ❑

3. Do these suppliers participate in your product development process? Yes - ❑ No - ❑

4. Do you give these suppliers minimum critical specifications for their components as opposed to detailed specifications?
   Yes - ❑ No - ❑

5. Do these suppliers have influence over the setting of specifications for their components? Yes - ❑ No - ❑

6. Do these suppliers get involved in the concept stages of the design of your product? Yes - ❑ No - ❑

7. Do these suppliers have major testing responsibilities for prototypes of their components? Yes - ❑ No - ❑

8. Do these suppliers have a strong enough technical capability to take responsibility for product development? Yes - ❑ No - ❑

9. Do your purchasing policies support buying complete subsystems from suppliers when appropriate? Yes - ❑ No - ❑

10. Do your purchasing policies support involving suppliers in the early stages of product development when appropriate?
    Yes - ❑ No - ❑

11. Do you have long-term partnership relationships with suppliers who should be involved early in product development?
    Yes - ❑ No - ❑

12. Do these suppliers provide innovative product technologies customized for your needs? Yes - ❑ No - ❑

13. Do you have project management systems in place to ensure that suppliers involved in product development get their contributions in on time? Yes - ❑ No - ❑

14. Do these suppliers actively work on ideas for modifying the product to reduce costs and pass part of those savings on to you as a customer? Yes - ❑ No - ❑

15. Do these suppliers have appropriate technologies to exchange product data with your company? Yes - ❑ No - ❑

SUPPLY CHAIN INVOLVEMENT SCORING
How many times did you check "Yes" under Supply Chain Involvement? _____

PEOPLE SYSTEMS (CHAPTER 6)

1. Do your job designs support:
   a. working with people from other functions? Yes - ❑ No - ❑
   b. working closely with suppliers and customers? Yes - ❑ No - ❑
   c. a focus on goals that are customer-driven? Yes - ❑ No - ❑
   d. a focus on goals that are company-driven? Yes - ❑ No - ❑
   e. responsiveness and speed of development? Yes - ❑ No - ❑

2. Do most people have the technical skills needed to do their jobs and contribute to cross-functional work? Yes - ❑ No - ❑

3. Do most people have the social skills needed to work in a cross-functional team environment? Yes - ❑ No - ❑

4. Do you have a skills profile (or the equivalent) for most positions? Yes - ❑ No - ❑

5. Do you have a record of the skills profile for most individual job holders? Yes - ❑ No - ❑

6. Do you provide sufficient training for all employees? Yes - ❑ No - ❑

7. Do all departments and teams understand their goals? Yes - ❑ No - ❑

8. Do all departments and teams understand the goals they have in common? Yes - ❑ No - ❑

9. Are goals and metrics consistent? Yes - ❑ No - ❑

10. Are rewards consistent with goals? Yes - ❑ No - ❑

11. Are individual and group goals consistent? Yes - ❑ No - ❑

12. Is the goal setting process viewed as legitimate? Yes - ❑ No - ❑

13. Is the appraisal system viewed as fair? Yes - ❑ No - ❑

14. Is the appraisal system viewed as accurate? Yes - ❑ No - ❑

PEOPLE SYSTEMS SCORING

How many times did you check "Yes" under People Systems? _____

TECHNOLOGY (CHAPTER 7)

1. Does your current technology adequately support:
   a. work process modeling? Yes - ❑ No - ❑
   b. core design work? Yes - ❑ No - ❑
   c. project management? Yes - ❑ No - ❑
   d. communication across functions? Yes - ❑ No - ❑
   e. communication with suppliers and customers? Yes - ❑ No - ❑
   f. data exchange with suppliers and customers? Yes - ❑ No - ❑

2. Does everyone have the technology they need? Yes - ❑ No - ❑

3. Are you looking ahead to future technologies?

Yes - ❑ No - ❑

TECHNOLOGY SCORING

How many times did you check "Yes" under Technology? _____

SCORING

In the Scoring Summary below, list your total number of "Yes" checks for each CE Element. Place those scores in the CE Profile Scoring Table and multiply by the appropriate weights. Sum the weighted scores to find your total CE Profile Score.

SCORING SUMMARY

Work Process ______

Internal Organization ______

Supply Chain Involvement ______

People Systems ______

Technology ______

CE Elements	# items	Raw Score x	Weight =	Weighted Score
Work Process	9		2.22	
Internal Organization	46		0.43	
Supply Chain Involvement	15		1.33	
People Systems	18		1.11	
Technology	8		2.50	
Total	96			Σ = _______

CE Profile Scoring Table

PART II

•

Changing to Support Concurrent Engineering

Chapter 9

•

A Structured Methodology for Change

You never had time to learn. They threw you in and told you the rules and the first time they caught you off base they killed you. • Ernest Hemingway[1]

In the quotation at the top of the chapter, Hemingway is talking about war. While business is a lot less deadly, it can still be, shall we say, *competitive*? In fact, it's so competitive these days that many people feel much the way Hemingway did in World War I — things are so hectic that you never really have time to learn what to do, but if you make a mistake, you're dead. The purpose of Part II is to provide some guidance and, more importantly, to provide a structure in which you take the time to learn.

In this chapter we provide an introduction to the Concurrent Engineering Effectiveness (CEE) Change Methodology. The CEE Change Methodology is a process and a set of tools for systematically changing the context in which your company does product development to make it easier to do Concurrent Engineering (CE). By *process* we mean a structured sequence of steps to follow with detailed directions for what to do in each step. The details about what to do in each step of the process are given in Chapters 10–13 of this book. Here we will provide a brief overview so you can get a sense of what the process is about.

One of the problems with many structured processes is that they force you into a pattern that all but determines the outcome. We have tried

[1] From *A Farewell to Arms*

not to do that in the CEE Change Methodology. Instead, we have provided a *structure for discovery and learning*. By this we mean that the activities and tools do not lead to answers which we have determined in advance. Rather, each activity will lead you to learn or discover something valuable. The model we presented in Part I will guide the topics of discovery, but it will not determine what you learn. If you use that model with the process and tools we provide, when you put the discoveries together you will have a coherent picture of how to change the structure and process of your company toward being able to do CE.

What is a Change Methodology?

There are any number of definitions of a change methodology, but for our purposes it includes activities (and directions for how to perform them), an order in which those activities should take place, and a set of tools for helping you perform the activities. We have provided all of that in the CEE Change Methodology.

Activities

These are the things you need to do in order to accomplish your objectives. The methodology should tell you what you need to know in order to carry out each activity. Ideally, it would tell you *everything* you need to do. Realistically, this is impossible. All we can hope to do is to provide a set of activities that most organizations will need to carry out, most of the time. Our experience suggests that these activities are usually good ones, that most people learn a great deal from doing them, and that they usually lead to useful outcomes. You may find you can get away without doing some of these things, although most of the time that will be because you have a substitute activity which provides a similar result. You will also find that there are other things you need to do depending on your specific situation.

Order

Sometimes the order in which you do things matters. It may matter because an activity formally requires the output of a prior activity as input. Or, it may matter if the output of an activity will get stale if it waits too long (e.g., if you plan training as a result of a requirements analysis today, but don't deliver it for another year). At other times order doesn't formally matter, but since you can't do everything all at once you do them in an order that seems to make some kind of sense. In describing our process we will provide you with a suggested order in which activities should be done. We will try to indicate where order matters because of input-output considerations (e.g., the results of the as-is assessment are input to the design phase), and where we're providing an order simply for convenience.

Tools

Tools make it easier to perform the activities in the process. Without a tool, an activity may seem abstract and hard to accomplish, despite terrific directions. A tool often makes the activity seem easier to accomplish, resulting in a greater likelihood of it being done. Other times, the tool is essential to the activity. While we usually think of a tool as being something hard (like a hammer) or as being highly structured (like computer software), a tool can also be relatively soft and unstructured. For example, a simple communication matrix (see Exhibit 11.12 for an example) has little structure, but it can be an invaluable aide to analyzing communication patterns. In Chapters 10–13 we provide a variety of tools that make it easier to carry out the activities we have described. There will be at least one tool for each activity.

The Change Process

The process we recommend for moving to CE has four main phases (Figure 9.1):

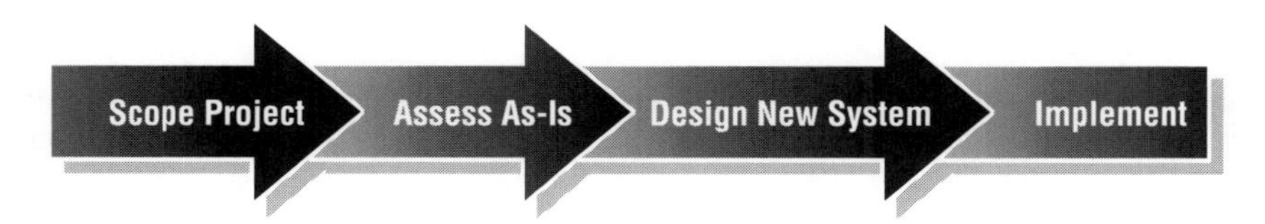

Figure 9.1 • CEE Change Process

Scope the Project

Here you define what you are trying to do, clarify your objectives, and make sure you have enough support inside your company to make it worth proceeding. This phase includes developing a vision and action plan, gathering support and commitments within your company for moving to CE, and benchmarking. You do a lot of selling here which you will continue to do throughout the total effort. You can think of the activities in this phase as figuring out just what you want to do and getting support for doing it.

Assess As-Is

In this phase you figure out just where you are now, so you know what your problems are and so you can understand your company well enough to design a new system. You learn in some detail about your current processes and contexts for product development. This is very different from the CE Profile we introduced in Chapter 8. In the CE Profile you used your global judgments about each of the CE Elements to rate your company. In the detailed As-Is Assessment (see Chapter 11), you will explore each CE Element in depth. Typically this phase takes several weeks to several months, depending on how much detail you want to get into and how many of the CE Elements you decide to assess.

Design New CE System

Based on your As-Is Assessment, as well as principles you develop and your benchmarks, you design the revised CE Elements for your company so you can do concurrent engineering and take advantage of new technologies for product development. You need not revise all of the CE Elements, just the ones that appear to have the biggest payoff in terms of the objectives you set for yourself in the Scoping phase. You may notice that we have not called this "planning." We distinguish "design" from "planning" because planning is often the result of looking backward and thinking about a short-term future. Design is a creative process in which you create new systems for the long-term future.

Implement

In this phase you decide where, when, and how to implement your design, how to sell it, evaluate it, and continuously improve it. We particularly emphasize the need to make decisions about how widespread and how rapid your rollout should be. We talk about the need to do one or more pilots, and how you can use those to help you continue to sell CE to management and to those who have to get the work done.

While in Figure 9.1 the four phases appear to be completely sequential, in fact, just as with any CE project, the phases overlap considerably. For example, making sure you have support for CE, which begins in the first phase, actually continues throughout the process. Design work may begin before the assessment is complete and, indeed, implementation may begin at any point after a particular new element has been designed.

Needed: An Analytic Problem-Solving Mindset

For the CEE Change Methodology to be useful, you need to have an analytic mindset. Some people approach problems from an "intuitive" point of view, in which they look for a "whole" solution to pop out of their minds. Other people are happy to simply "copy" the way others have solved a problem. Still others, and these are our target audience, take an engineering or analytic perspective. They prefer to gather information, systematically analyze it, and derive solutions from that analysis. Any of these approaches will work some of the time; none of them will work all of the time. In this book we have emphasized the analytic/engineering approach.

We should note that there is some overlap between the approaches, and indeed it is probably undesirable to rely on only one of them. For example, although we emphasize the analytic approach and provide tools to help with the analysis, we also suggest that you benchmark other companies so you can make use of solutions which have already been proven, rather than trying to reinvent the wheel. However, we caution against slavishly trying to copy what someone else has done in another time and in other circumstances. Equally, any creative problem-solving activity will rely on human beings "intuitively" coming up with innovative new ideas about how to solve a problem. However, we believe that this intuition is best served by information that is presented in a complete and understandable form, rather than attempting to invent things from thin air. Again, the benefits of information collection and analysis are essential.

Is This Book Too Analytic?

The very best problem-solving takes place in a balanced way, with a balance struck between analysis, intuition, and copying others. Some readers may look ahead in this volume and jump to the conclusion that our approach is unbalanced, that we're too heavily weighted to the analytic. And, to some extent, we admit this is true. We did it this way for two reasons. First, we felt there was a strong need to provide analytic tools concerned with the organization, people, and process issues related to concurrent engineering. Second, in Part I of this book we address the topics needed to build the holistic concepts for thinking intuitively about CE. In several places in our process we provide hooks for the reader to use these concepts. Finally, there is a huge literature about benchmarking that will tell you everything you ever wanted to know about how to copy from others. We urge you to benchmark as much as necessary, and use that information within the analysis.

Some readers also may look at the number of analytic steps in the following chapters and say that there is just too much to do — you could spend your life doing all that analysis and never get any real work done. We have two answers to that criticism. First, we know from experience that, *with discipline*, you can march through each of these steps in a reasonable period of time, with a reasonable level of effort. We also know from experience that, without discipline, you can take any analytic method and spend your life on it. The second answer is that you don't have to do *everything* that we describe. We provide a complete *menu* from which you can choose. Just like you don't have to eat everything on the menu when you go into a restaurant, you don't have to do everything we suggest in this book. After your initial analysis, you're going to pick those things which are going to provide you with the most benefit and work on those in detail. So, there is a lot of material in this volume, but you probably aren't going to use it all; or at least you aren't going to use it all at once.

What Will You Have When You're Done?

If you go through each step in the methodology, you will have implemented the support structures for concurrent engineering and people will be doing their work differently. You will also have engaged in a lot of assessment and design, and you will probably have a number of reports and presentations to show for that work. But none of that activity is really to the point. None of it is of any value if you haven't also implemented your designs.

A well-known commercial for athletic shoes says, "Just do it!" When your goal is just to have fun, then it's probably OK to just lace up your shoes and go out to "do it." When your goal is the survival of your company and hundreds or thousands of jobs, then more careful planning is called for. If you have done your job well, the use of CE will mean that your company will achieve many of the same kinds of results we described in Part I.

When Can You Use This Methodology?

We have tried to make as few assumptions as possible in developing this methodology. After all, we don't know you or your company or what you're trying to do. But we had to make some assumptions; so here they are.

- Your organization is a manufacturing enterprise that has some involvement in developing new products.
- You (or your company) are interested in implementing concurrent engineering (or some variation). This assumption is fundamental. The CEE Change Methodology is not designed to *convince* you to adopt CE, although you might look at the Methodology to see what's required to fully do so.
- You don't mind hard work and analytic thinking. The CEE Change Methodology is a highly analytic process. If you operate on a more

intuitive basis and don't find much benefit from analytic approaches, this methodology probably isn't for you.

How to Use the CEE Change Methodology

Form a Team

The process we describe in this volume is best performed by a cross-functional team rather than by an individual. In theory, an individual could do every activity in the book, and use every tool. However, most of the activities require significant information collection and analysis. Both of these are much better performed by a group. A cross-functional group has access to much more information than any individual. A well-functioning group also has far greater analytic capability than an individual. Finally, a group typically has much greater creative potential than does an individual. This last point is very important since most of the activities involved in the Methodology require creative problem solving.

Treat It as a Project

When you have to deal with the details of implementing CE (once you're past the strategic level), we find it most useful to think about moving to CE as a *project*. A project should have an *objective* and a *plan* for achieving that objective. Your objective for your CE project might be to reduce time-to-market and cost for new products by instituting CE. Your plan needs to define the activities and the resources needed to accomplish this objective.

Most people who develop plans do it for projects with which they are pretty familiar. For example, an experienced design engineer would likely be the one to develop a plan for developing a new widget; an experienced plant manager would develop the plan for putting together a new manufacturing system. Not many people have experience moving a company to CE. Even those who have done this successfully have done it in a different company than yours, with different assumptions, different contexts, and different objectives. While someone with such experience may be able to help (and there are many consultants available with this type of experience), they're really not in the same category as an experienced design engineer planning a new product development project. What's needed is a *generic method* which you can use as the basis for building your plan — this is the CEE Change Methodology.

Use the Process to Sell CE

We indicated above that the CEE Change Methodology shouldn't be used until a decision has been made to go ahead with the move to CE. However, the fact that there is a structured approach to implementing CE may make it easier to sell both management and staff that the risk of change is low. After all, if the change process is so well understood that there is a structured process for doing it, then the risks involved with implementation must be relatively small.

Enter the Process in the Middle

If you feel you've already done parts of this, you should enter the process at the appropriate point. For example, if you've already developed a vision and objectives, and have a champion, then you can probably move right along to the As-Is Assessment phase. If you do this, you should make sure to at least read the sections you think you've done so you can tell if you missed anything important; and do a jargon check to make sure you understand the technical terms used in later chapters.

Cherry-Pick Activities and Tools

You may have your own methodology, but there are parts of your process that are incomplete. Go ahead and pick out those activities and tools you think will be useful and use them. While we think that using the whole methodology provides the greatest benefit, many of the activities and tools provide significant benefit on a stand-alone basis.

Use the Tools and Methods for Problem-Solving

After you become familiar with the methodology, you may find that you lack the interest or the sanction to go ahead with a full-blown project to plan and implement CE. Our experience shows that many of the tools we describe in this book are excellent problem-solving tools. This is a form of "cherry-picking," but it's targeted on the problems to be solved, rather than the process. If you do this, let us know what you used and how it worked.

The CEE Change Tools

Throughout this volume you'll see examples of various analysis tools we use in the Coseat Inc. examples. Each of these will be identified as an Exhibit, rather than as a Figure or Table. These Exhibits take the form of matrices, tables, diagrams, forms, and the like. We have not provided *blank* copies of these tools in order to keep this book to a more manageable size.[2] None of the forms is really very complicated. We recommend that you develop computer forms of those tools you choose to use. For example, any of the matrices can be built as a computer spreadsheet. This is our preferred way of working with the matrices.

We advise using the matrix forms as the focus of discussion for your work group. You can do this by using computer projection equipment to project the matrix (or other tool) on a screen and making changes in the matrix dynamically while you work. This not only provides an excellent focus for the group, but it means that the group leader (or facilitator) has much less work than usual to do after the meeting is over!

The alternative to working with the computer is to work on a whiteboard or to use large sheets of paper (newsprint). This is the tried and true method that has worked for facilitators for many years. And it will work using the matrices and other forms. But, if you do it that way, we still suggest that you later translate these into computer form and use that as the *archive* for the group. Storage on the computer will be easier, and presentation back to the group at the next meeting will be much clearer.

Do You Need a Consultant?

Lots of companies use a consultant, or indeed many consultants, to help them work through a process such as this. From what we've seen, companies that do this kind of thing *without* an external consultant seem to have a higher probability of success. In fact, one of our motivations in writing this book was to provide access to the rest of the world to some of the tools typically used by consultants.

Why would companies do better without consultants? There are several reasons.

1. ***Diffusion of responsibility.*** Too often the consultant seems to get the responsibility for making change happen, when that responsibility needs to be with everyone in the company, from top management on down. We see this in companies where there really isn't an internal drive to change, but someone in charge feels like they ought to change, perhaps to imitate their neighbors or competitors. In cases like this, no one in the company wants to do the really hard work of change, so they hire an outsider to do it for them. Unfortunately, unless the people inside get motivated to change, it won't happen.

[2] We are planning to provide free copies of these forms on the Industrial Technology Institute's World Wide Web site. Use your Web browser to go to the following address to see what's available at the time you read this: http://www.iti.org/cec/index.htm.

2. ***Learning is lost.*** If the consultant does all the planning and changing of systems, then no one inside understands what was done. When the consultant leaves (as they all do), then deep understanding of the new system is lost — no one knows why they're doing things the way they are, so they tend to slip back into the bad old ways of doing business.
3. ***Hammers in search of nails.*** Many consultants have their "bag of tricks" which they use in their practice. Sometimes the tools in their bag are a few things they've invented, sometimes the tools are those they used when they last held a job in industry. Regardless, there is a strong tendency to look for tasks to which they can apply their tools. Sometimes they find a nail for their hammer, but sometimes they find a flower. Regardless, they have to use the hammer. Of course, this is not true of all consultants, but it is true of enough that it causes plenty of problems.

We urge you, if possible, to try and run the change process using internal people, starting with top management. If you must use a consultant, try using one primarily for the following purposes.

- ***Information and training.*** A good consultant has had experience with a large number of firms, many of which have some similarities to you. In contrast, the people inside your company have probably had experience with at most a few others. The consultant's greatest asset may be bringing the benefit of this experience to you.
- ***Assessment and analysis.*** A good consultant has a variety of methods for making judgments about your company, and has a degree of objectivity that someone inside will probably be lacking. Both assessment and analysis should not be done independently by the consultant, but should be done jointly with internal staff to make sure the learning stays inside. In this volume, we have tried to provide you with the tools such a consultant might use.
- ***Facilitator.*** Change processes can be difficult to manage amidst the chaos of everyday management. A consultant can assist by helping to run meetings or even to help run the whole change process. In general, we think it far better to use internal people to do this, but it can be done by a consultant as long as he/she doesn't actually take responsibility for making change happen.

The Case Studies

In the Appendix you will find nine detailed case studies of real companies which we believe provide examples of really good practice in CE. These companies are *Chrysler*, *Eaton*, *Ford*, *Motorola*, *Newport News*

Shipyard, *Sun Microsystems*, *Texas Instruments*, *Toyota*, and *Whirlpool*. We will refer to those cases from time to time in the text, and indeed we referred to some of them several times in Part I. In a sense, these are the results of some of *our* benchmarking. We will also use these cases as the benchmarks for Coseat Inc. We have substantial experience with many other companies, but these companies either provided a particularly good story or they demonstrated something that we needed to describe to you. While we urge you to read the cases, we also urge you to do your own benchmarking and write your own case studies. At least for us, the act of writing up the case is a great learning experience — you're forced to think about what you've observed. Try it.

A Running Example — Coseat Inc.

As we describe the CEE Change Methodology in the next four chapters, we will provide many examples, some of which are from real companies. However, we felt a strong need to provide an example that runs throughout the description of the process, so you can get a sense of its continuity. In order to make it a "good" example, i.e., one that demonstrates the principles involved really well, it had to be fictional. We have labeled this running fictional example *the Coseat Inc. example*. We developed the example firm based on several companies we know quite well. None of the real companies which served as the basis for the example is quite so "bad" as Coseat is in the early examples, nor is any quite so "good" as Coseat becomes by the end of the process. We provide the following description of Coseat Inc. as the scenario on which subsequent exhibits are based.

Coseat Inc. is a manufacturer of seating systems and seating components that supplies the automotive, aerospace, and heavy equipment industries. A seating system is a complete seat with cover, frame, control system, and rails that is ready to be placed into the vehicle. Coseat makes many of the components of a seat, such as the automatic control system, the frame, and the trim. They also buy many of the same components from external suppliers, depending on special requirements from their customer.

Coseat has about 4000 employees located in 5 plants in the U.S. and an R&D Center, located in a suburb of Detroit. About 70 percent of Coseat's $400 million in sales is to the automotive industry, 15 percent is to aerospace (mostly makers of business aircraft), and 15 percent is to heavy equipment makers, such as those that make tractors and earth movers. Most of the automotive seating business (about 60 percent) involves complete seat systems, while the rest involves component sales to *automotive OEMs for programs where the OEM assembles its own seats.*

Coseat pioneered the concept of building complete seating systems for the automotive industry, and for several years this has been its competitive advantage. Now several other companies which formerly produced only components are getting into the business. In addition to this renewed competition, Coseat's customers in all industries are looking for new relationships with suppliers, particularly ones in which the supplier acts in a full-service supplier role. This means not only building a complete seating system, but also determining customer needs and designing the seat system, as well as supporting it after the sale. Coseat has a history of excellent manufacturing and part design, but has little experience working directly with consumers or designing (as opposed to building) whole seating systems.

Along with all of these changes, Coseat's customers are under very heavy pressures to reduce costs and to speed their responsiveness to market changes by reducing the time they take to develop and bring new products to market. These pressures, of course, come to bear on Coseat as well.

The CE Profile for Coseat Inc.

In order to give a better sense of Coseat Inc., we have completed a CE Profile for the company. Here is the completed questionnaire for the company, as well as the summary analysis (Exhibit 9.1).

Exhibit 9.1 • CE Profile for Coseat Inc.

CE Profile™

Instructions

Check Yes or No for each of the questions below. After you complete each section, total the scores as indicated in the scoring box for that section. Then use the Scoring Tables to calculate your total weighted CE Profile score.

Work Process (Chapter 3)

1. Do you have a formal product development process?

   Yes - ✔ No - ❑

2. For whatever process you follow (formal or informal), does it include:

   a. activity precedence? Yes - ✔ No - ❑

b. schedules? Yes - ❑ No - ✔
c. exit and entry rules? Yes - ❑ No - ✔
d. controls? Yes - ✔ No - ❑
e. tools and resources? Yes - ❑ No - ✔

3. Is your process easy to use? Yes - ❑ No - ✔

4. Is your process customized as appropriate for each division or product line? Yes - ❑ No - ✔

5. Is your process general enough to allow flexibility, but detailed enough to be useful? Yes - ❑ No - ✔

WORK PROCESS SCORING
How many times did you check "Yes" under Work Process? 3

INTERNAL ORGANIZATION (CHAPTER 4)

1. Does your *organization architecture* promote:
   a. sufficient coordination within functions? Yes - ✔ No - ❑
   b. sufficient coordination across functions? Yes - ❑ No - ✔
   c. economical use of talent? Yes - ✔ No - ❑
   d. ease of performance evaluation and control? Yes - ✔ No - ❑
   e. focus on goals that are customer-driven? Yes - ❑ No - ✔
   f. coordination of effort toward your customer? Yes - ❑ No - ✔

2. Does your *project management* structure promote:
   a. sufficient coordination within functions? Yes - ❑ No - ✔
   b. sufficient coordination across functions? Yes - ❑ No - ✔
   c. focus on goals that are customer-driven? Yes - ❑ No - ✔
   d. focus on goals that are company-driven? Yes - ❑ No - ✔
   e. responsiveness & speed of development? Yes - ❑ No - ✔

3. Do your *Standard Operating Procedures* promote:
   a. coordination within functions? Yes - ✔ No - ❑

b. coordination across functions? Yes - ❑ No - ✔
c. focus on goals that are customer-driven? Yes - ❑ No - ✔
d. focus on goals that are company-driven? Yes - ❑ No - ✔
e. responsiveness and speed of development? Yes - ❑ No - ✔

4. Do your *planning, scheduling, and monitoring systems* promote:
a. coordination within functions? Yes - ❑ No - ✔
b. coordination across functions? Yes - ✔ No - ❑
c. focus on goals that are customer-driven? Yes - ❑ No - ✔
d. focus on goals that are company-driven? Yes - ❑ No - ✔
e. responsiveness and speed of development? Yes - ❑ No - ✔

5. Do your *tools, techniques and standards* for design promote:
a. coordination within functions? Yes - ❑ No - ✔
b. coordination across functions? Yes - ✔ No - ❑
c. focus on goals that are customer-driven? Yes - ❑ No - ✔
d. focus on goals that are company-driven? Yes - ❑ No - ✔
e. responsiveness and speed of development? Yes - ❑ No - ✔

6. Do your *performance metrics* promote:
a. coordination within functions? Yes - ❑ No - ✔
b. coordination across functions? Yes - ✔ No - ❑
c. focus on goals that are customer-driven? Yes - ❑ No - ✔
d. focus on goals that are company-driven? Yes - ✔ No - ❑
e. responsiveness and speed of development? Yes - ✔ No - ❑

7. Does your use of *credentials and training* promote:
a. coordination within functions? Yes - ❑ No - ✔
b. coordination across functions? Yes - ❑ No - ✔
c. focus on goals that are customer-driven? Yes - ❑ No - ✔
d. focus on goals that are company-driven? Yes - ❑ No - ✔
e. responsiveness and speed of development? Yes - ❑ No - ✔

8. Are you using *teams* to promote:
a. coordination within functions? Yes - ❑ No - ✔
b. coordination across functions? Yes - ❑ No - ✔

c. focus on goals that are customer-driven? Yes - ❑ No - ✔
d. focus on goals that are company-driven? Yes - ❑ No - ✔
e. responsiveness and speed of development? Yes - ❑ No - ✔

9. Does your *culture* promote:
   a. coordination within functions? Yes - ✔ No - ❑
   b. coordination across functions? Yes - ❑ No - ✔
   c. focus on goals that are customer-driven? Yes - ❑ No - ✔
   d. focus on goals that are company-driven? Yes - ✔ No - ❑
   e. responsiveness and speed of development? Yes - ❑ No - ✔

INTERNAL ORGANIZATION SCORING
How many times did you check "Yes" under Internal Organization? _11_

SUPPLY CHAIN INVOLVEMENT (CHAPTER 5)

Before you attempt to answer the following questions, read these directions. If your suppliers only provide raw materials or commodities (items you could order out of a catalog), skip to the next section (People Systems). If none of your suppliers provides "custom" parts or components (customized to some degree for your product), skip to the next section (People Systems). But, if that is the case, ask yourself whether this is because they should not be doing custom work or because you are not using their capabilities.

If you're still with us, think about the two or three non-commodity suppliers in your supply chain who provide the highest value-added custom components for your product. Answer the following questions about them.

1. Do you have staff located at any of those suppliers' sites?
   Yes - ❑ No - ✔

2. Do any of your suppliers have staff collocated at your site?
   Yes - ❑ No - ✔

3. Do these suppliers participate in your product development process? Yes - ❑ No - ✔

4. Do you give these suppliers minimum critical specifications for their components as opposed to detailed specifications?
Yes - ❑ No - ✔

5. Do these suppliers have influence over the setting of specifications for their components? Yes - ✔ No - ❑

6. Do these suppliers get involved in the concept stages of the design of your product? Yes - ❑ No - ✔

7. Do these suppliers have major testing responsibilities for prototypes of their components? Yes - ✔ No - ❑

8. Do these suppliers have a strong enough technical capability to take responsibility for product development? Yes - ✔ No - ❑

9. Do your purchasing policies support buying complete subsystems from suppliers when appropriate? Yes - ❑ No - ✔

10. Do your purchasing policies support involving suppliers in the early stages of product development when appropriate?
Yes - ❑ No - ✔

11. Do you have long-term partnership relationships with suppliers who should be involved early in product development?
Yes - ❑ No - ✔

12. Do these suppliers provide innovative product technologies customized for your needs? Yes - ❑ No - ✔

13. Do you have project management systems in place to ensure that suppliers involved in product development get their contributions in on time? Yes - ❑ No - ✔

14. Do these suppliers actively work on ideas for modifying the product to reduce costs and pass part of those savings on to you as a customer? Yes - ❑ No - ✔

15. Do these suppliers have appropriate technologies to exchange product data with your company? Yes - ✔ No - ❑

SUPPLY CHAIN INVOLVEMENT SCORING
How many times did you check "Yes" under Supply Chain Involvement? 4

PEOPLE SYSTEMS (CHAPTER 6)

1. Do your job designs support:
   a. working with people from other functions? Yes - ❑ No - ✔
   b. working closely with suppliers and customers? Yes - ❑ No - ✔
   c. a focus on goals that are customer-driven? Yes - ✔ No - ❑
   d. a focus on goals that are company-driven? Yes - ✔ No - ❑
   e. responsiveness and speed of development? Yes - ❑ No - ✔

2. Do most people have the technical skills needed to do their jobs and contribute to cross-functional work? Yes - ✔ No - ❑

3. Do most people have the social skills needed to work in a cross-functional team environment? Yes - ✔ No - ❑

4. Do you have a skills profile (or the equivalent) for most positions? Yes - ❑ No - ✔

5. Do you have a record of the skills profile for most individual job holders? Yes - ❑ No - ✔

6. Do you provide sufficient training for all employees? Yes - ❑ No - ✔

7. Do all departments and teams understand their goals? Yes - ✔ No - ❑

8. Do all departments and teams understand the goals they have in common? Yes - ❑ No - ✔

9. Are goals and metrics consistent? Yes - ❑ No - ✔

10. Are rewards consistent with goals? Yes - ❑ No - ✔

11. Are individual and group goals consistent? Yes - ❑ No - ✔
12. Is the goal setting process viewed as legitimate? Yes - ✔ No - ❑
13. Is the appraisal system viewed as fair? Yes - ✔ No - ❑
14. Is the appraisal system viewed as accurate? Yes - ✔ No - ❑

PEOPLE SYSTEMS SCORING

How many times did you check "Yes" under People Systems? 8

TECHNOLOGY (CHAPTER 7)

1. Does your current technology adequately support:
   - a. work process modeling? Yes - ❑ No - ✔
   - b. core design work? Yes - ✔ No - ❑
   - c. project management? Yes - ✔ No - ❑
   - d. communication across functions? Yes - ✔ No - ❑
   - e. communication with suppliers and customers? Yes - ❑ No - ✔
   - f. data exchange with suppliers and customers? Yes - ❑ No - ✔
2. Does everyone have the technology they need? Yes - ❑ No - ✔
3. Are you looking ahead to future technologies? Yes - ❑ No - ✔

TECHNOLOGY SCORING

How many times did you check "Yes" under Technology? 3

Scoring

In the Scoring Summary below, list your total number of "Yes" checks for each CE Element. Place those scores in the CE Profile Scoring Table and multiply by the appropriate weights. Sum the weighted scores to find your total CE Profile Score.

Scoring Summary	
Work Process	*3*
Internal Organization	*11*
Supply Chain Involvement	*4*
People Systems	*8*
Technology	*3*

CE Elements	# items	Raw Score x	Weight =	Weighted Score
Work Process	9	3	2.22	6.66
Internal Organization	46	11	0.43	4.73
Supply Chain Involvement	15	4	1.33	5.32
People Systems	18	8	1.11	8.88
Technology	8	3	2.50	7.50
Total	96			$\Sigma = 33.09$

CE Profile Scoring Table

[end of Exhibit 9.1]

Exhibits 9.2 and 9.3 show the results for Coseat Inc. graphically. Before we try to make sense out of the scores, it is important to realize that these are really low scores for all of the CE elements, and the total score of 33 is terribly low. However, we said we wanted to create a "bad" example, and so we did. Regardless, the scores do suggest that internal organization and supply chain involvement would be the most likely places to start for Coseat Inc., and that people systems are a relative strength (compared to other elements). What this means is that in the subsequent in-depth as-is assessment, Coseat might initially focus primarily on internal organization and supply chain issues, with only cursory attention paid to the other elements until a later time.

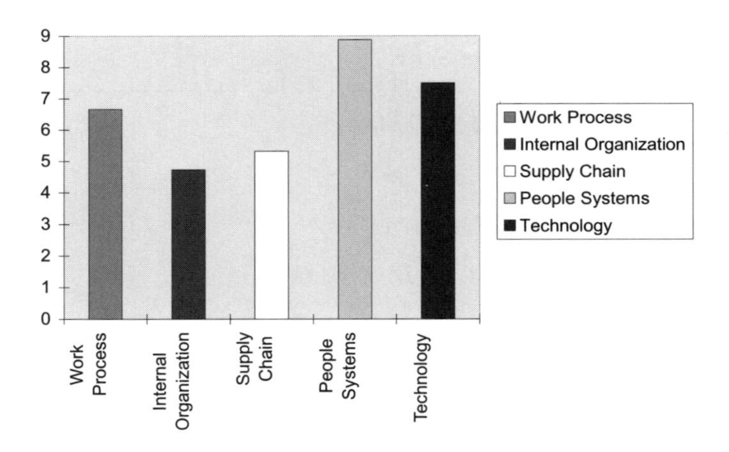

Exhibit 9.2 • Bar Graph for Coseat Inc. CE Profile

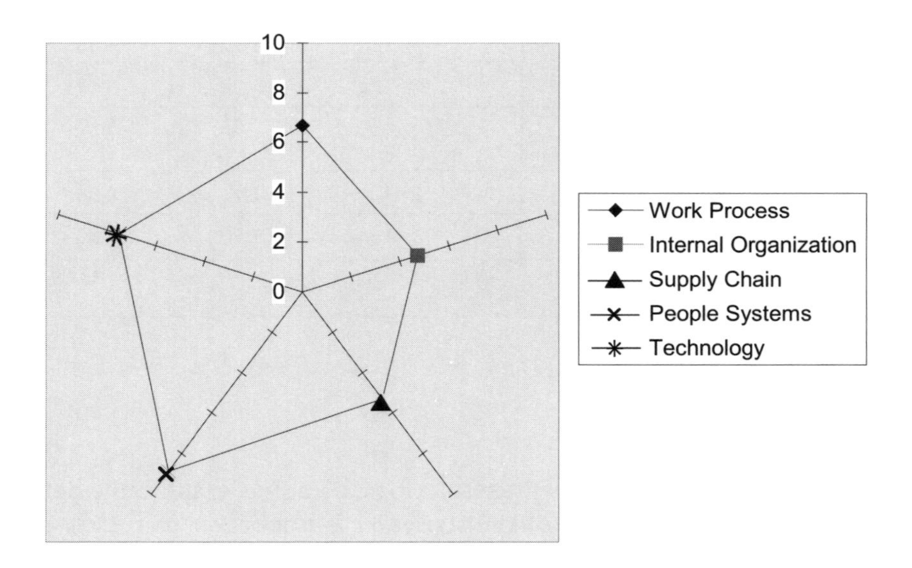

Exhibit 9.3 • Spider Graph for Coseat Inc. CE Profile

CONCLUSION

In Part II of the book, we've provided you with the means to *do something* about the organization, people, and process issues around concurrent engineering. We've provided a process, methods and tools for analysis, and an approach to designing and implementing a new CE system. We urge you to look through the material here and try some things. To a great extent, change is about learning — learning new ideas, learning new ways of doing things, learning about people and the way they react to those new ideas. Each of the steps in the CEE Change Methodology is designed to facilitate learning as well as doing. Go for it!

CHAPTER 10

•

SCOPING THE CHANGE EFFORT

...in war, before the battle is joined, plans are everything, but once the shooting begins, plans are worthless. • Dwight Eisenhower

This first phase of the change effort (Figure 10.1) is dedicated to selling a broad vision of Concurrent Engineering (CE) to top management, defining the boundaries of the change effort, and laying the groundwork for all other activities. It includes making sure you know what you are going to do in the next phases, and that you know there is a high level of commitment to the effort. It includes making sure that all necessary people are involved, and that they know what they will have to do. These kinds of steps are absolutely necessary if implementation is to succeed. In effect, this is the early plan for the "project" to design and implement CE.

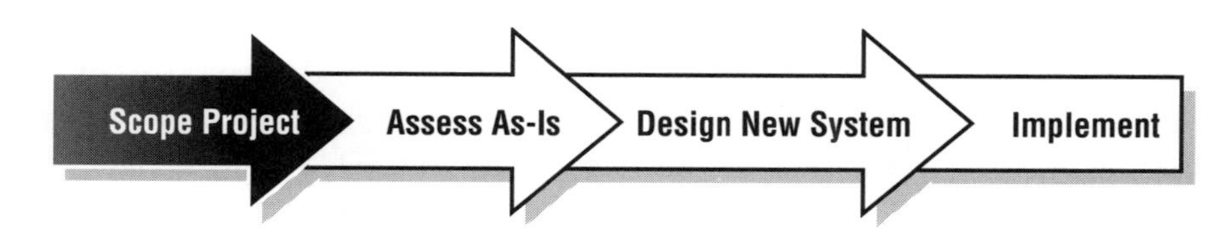

Figure 10.1 • Four Phases of the CEE Change Process

Consider the situation at Coseat Inc., the fictional company we introduced in Chapter 9.

Ed Cazotsky, Director of Design Engineering at Coseat Inc.'s Detroit Technical Center, has been concerned about Coseat's product development process. He considers it to be too slow and expensive. While Coseat's niche

of developing complete seating systems has been a competitive advantage for several years, Coseat has been getting a reputation in the automotive industry of sometimes being the component supplier who makes the whole car program late.

Furthermore, Coseat's profits have been lagging on recent seating programs, and Cazotsky attributes this to a combination of high development costs and high supplier costs.

Coseat's customers have been implementing concurrent engineering over the past decade with considerable success. Coseat has taken advantage of this by participating in its customer's product development processes. However, the lesson has not migrated to Coseat's internal development practices nor to its relationships with its suppliers. In other words, Coseat's account managers and product engineers participate on their customer's product development teams, but practices at home are unchanged over the past decade.

Cazotsky hasn't been blind to this. He recognizes that Coseat has to change. Having seen how long it took his automotive customers to change, he also knows that this won't be easy. He knows enough to realize that he can't just announce that everyone will do CE tomorrow. He needs high-level support and he needs a plan. But, of course, planning can just be an excuse to put off doing anything, and he doesn't want to do that. He needs to plan and act concurrently, just as he wants his engineers to do.

Ed Cazotsky needs a vision of where he's headed; he needs to collect some high-level supporters; he needs to pull the affected parties into the process; and he needs to figure out what he's going to do next. The activities in this first phase of the CEE Change Process are designed to do all of that. Note that this phase is not designed to create a corporate strategy. In Coseat Inc.'s case, the corporate strategy involves building on a niche it has already created (manufacturing seating systems) by increasing Coseat's value added to its customers in design — in effect becoming a "full service" supplier. Concurrent engineering supports this strategy and is therefore not likely to be in conflict with any major efforts the company is attempting to make. In the absence of strategy, or in the presence of a strategy which cannot easily accommodate CE, then the whole process may not be appropriate.

There are four major steps in the Scoping phase (Figure 10.2):

1. clarify project concept
2. secure commitment and sponsorship from top management
3. develop project vision and boundaries
4. develop project action plan.

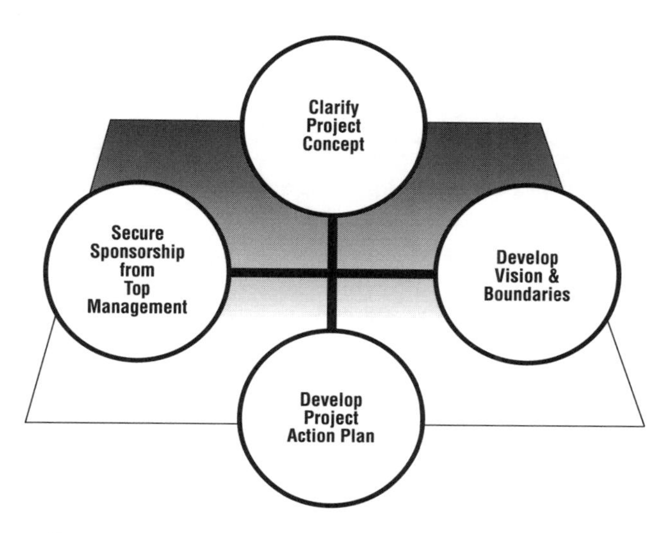

Figure 10. 2 • The Scoping Phase

These four steps are done more or less concurrently. You will almost certainly need a concept and vision before you can begin to secure sponsorship or develop an action plan. But, more than likely, the process of obtaining sponsorship and action planning will result in significant changes in your concept and vision. Furthermore, the process of gathering support and refining the vision will continue all the way through implementation.

CLARIFY PROJECT CONCEPT

The early advocates are those who are already convinced that CE is an important innovation to adopt within the company. However, they may not necessarily have the decision-making authority to follow through. By developing a clear project concept that can be shared with others, and by identifying a high-level "champion," the project increases its chances of obtaining the full organizational support that will be required. This is true regardless of *your* level in the company. If you are a high-level executive, you may choose to be the champion, but the role of champion must be filled. If you are lower in the organization, then you will need to find someone of a higher level than yourself to serve in that role.

We recommend three activities in this step (Figure 10.3):

1. form a group of early advocates
2. identify a CE champion
3. clarify the project concept.

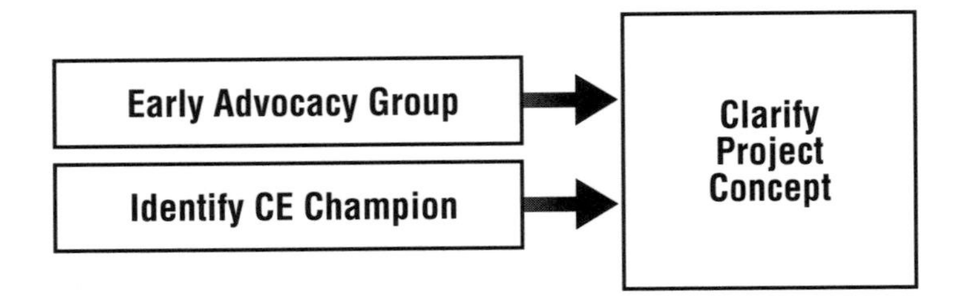

Figure 10.3 • Activities to Clarify Concept

You will almost certainly have to do the first two activities before you can get very serious about the third, since you need to gather the people together in order to develop the project concept.

Form Early Advocacy Group

The early advocacy group may be only a few individuals most committed to CE, or at least concerned about a problem that might be solved with CE. Although it is not necessarily a committee per se, these people may have discovered through mutual networking that they have a shared interest and vision in introducing CE to the organization. The early advocacy group will ideally be able to do some background work in order to gather broader commitment and support from others in the organization.

The role of the early advocacy group will be to clarify the project concept sufficiently to take it to a broader decision-making group (laterally as well as vertically) in the organization. Its purpose is also to review, at an early stage, the potential linkages between implementation of CE and overall company strategy.

Identify a CE Champion

The champion's key role is to make an initial case to top management that CE should be pursued. Ideally, the champion should have the following characteristics:

- sufficient familiarity with the strategic direction of the firm to determine whether a fit might exist between CE and the firm's overall direction;
- capability of obtaining resources required in the initial phases of the project to plan and implement CE, until a project sponsor is found;
- high enough position in the organization that he/she can make a strong case to the manager with authority to allocate resources to the project throughout its project life (the project sponsor).

Clarify CE Project Concept

The project concept puts all your ideas about the project together in one place. It includes your vision for CE for the company, as well as information about the environment, both internal and external. Some of this comes from other steps in the Scoping phase (e.g., the project vision has its own step). This is also a good time to do a quick, low-cost self-evaluation, such as the CE Profile (Chapter 9).

To muster broader support in the company, a clear, concise CE project concept document, often called a white paper, may be useful. It also helps the early advocacy group to clarify its goals early on in the process. Remember, the "project" in this case is the activities you will undertake to plan and implement CE. The CE project concept white paper might be based on the outline in Exhibit 10.1.

1. A project vision (later to be refined by a larger, more representative group).
   a. Brief description of project vision.
   b. Relationship between project vision and company vision, mission and strategy.
2. Why change is needed: triggers for change.
   a. Internal dynamics triggering need for change (e.g., drop in sales, supplier or client demands).
      1) Review company strategy for fit.
      2) Review company situation (could include CE Profile).
   b. External dynamics triggering need for change (e.g., competitors are moving to this approach, keep up with the times).
      1) Immediate competition.
      2) Industry trends.
      3) General environments.
3. The expected outcomes for the project if it is carried out (e.g., more business, lower costs, faster delivery times).
4. Resource needs: initial design, long-term projections.

Exhibit 10.1 • Outline for CE Project Concept White Paper

Secure Commitment and Sponsorship from Top Management

Once a core group of advocates has developed a CE project white paper and a CE champion has been identified, it is necessary to develop broader commitment to CE. This needs to happen both laterally within the organization and among the functional groups affected by the change, as well as vertically, in top management. This can be done by identifying a sponsor and setting up a guidance team.

Identify a Project Sponsor

The sponsor should have general management oversight as a line executive at or near the top of the organization (e.g., general manager or vice-president of division, CEO, COO, or CFO). This sponsor will be the key to assuring continued support throughout the life of the project, not only for the resources required, but for the changes necessary across the organization and/or division. A sponsor is someone who can obtain:

- initial start-up support,
- necessary approvals to start work,
- "blessings" to establish a Guidance Team.

In some companies, the champion and sponsor may be the same individual, especially in fairly small firms. In the very small firm, the champion may even be a member of the early advocacy group. The larger the firm, the more likely it is that effective change will require individuals who bridge the distance between the early advocacy group and the management who has the authority to release the resources needed to carry out the CE project.

Set Up a Guidance Team

The purpose of the Guidance Team is to oversee the project from this point forward, to provide planning direction, and to assure support from appropriate individuals and groups within the organization.

Membership. Criteria for membership on the Guidance Team might include the following.

- Representatives from internal groups likely to be affected by the plan. These might include functional managers and/or representatives from marketing, manufacturing, and engineering. These should be decision-makers and/or those directly linked to decision-makers.
- At least some members of the early advocacy group, since they have the interest and expertise needed for the project.
- Optionally, the champion and even the sponsor may serve on the team. The larger the company, the less likely the champion and/or sponsor are to be involved in the day-to-day planning, although both should be kept informed of what is going on and should be contacted for necessary approvals.
- Someone with expertise about CE and about organizational change. This could be people from the early advocacy group or just someone who either knows about CE or is committed to learning about it during the course of the project.

External Input. It may be important to get input from suppliers or others from outside the organization. Depending upon the number of suppliers and their relative importance to the firm, they might be included on one or more subcommittees that could have overlapping membership of Guidance Team members, and/or others in the organization who report back to the team.

Roles. Roles of the Guidance Team include:

- clarify project concept, including its vision, objectives, and project boundaries;
- oversee completion of the plan;
- continue to provide guidance to the project over the course of the project's life to assure cross-functional/cross-divisional input and commitment to implementation.

Training. Unlike the members of the Early Advocacy Group, the Guidance Team members will not necessarily know much about CE. They should be trained in some depth so they can act as a resource for others in the company. The training should take two forms. The first is some form of traditional training and reading. Training is available from a variety of sources — you probably get brochures all the time. Reading resources can be found obviously in this book, as well as in the References section at the back. The second form of training is to visit other companies that are doing CE well. Some suggestions are in this book; others can be found in the trade press for your industry.

Team Operating Procedures. For effective team functioning, the following issues need to be resolved.

- ***What decision-making authority does the team have?*** In general, the more empowered a team is to actually make decisions and get things done, the better. This is especially true if the team includes wide representation from the relevant functions. If the team is truly representing the managers of those functions, then it is possible to get a great deal done.
- ***Within the team, who has what authority?*** Is there a team leader who has the power to act on the team's behalf when the whole team is not present? Depending on the company's culture, this may or may not be feasible. Indeed, it may not be desirable if the leader tends to act in ways contrary to the wishes of the members of the team.
- ***Who will keep a record of team activity?*** Someone needs to record what the team decides and does, both during the team's meetings and to make the rest of the company aware of what the team is doing. This can be a position that rotates among the members.

- ***Who will schedule and coordinate activities and how will this take place?*** Someone usually needs to act as the central contact to facilitate the team's activities — to set up meeting space, order lunch, etc. Again, this can be rotated if necessary.

Develop CE Project Vision, Objectives, and Boundaries

Once a Guidance Team and sponsor have been identified, it is helpful to develop a more detailed project vision and objectives by this more broadly defined group. Although the Project Concept White Paper should be used as input, the more broadly defined group may modify or substantially alter the original version at this point.

Develop CE Vision. The CE Visioning Exercise in Exhibit 10.2 provides an example of an exercise that could be used to rapidly develop a vision. Note that this exercise won't work with a group that needs to learn what CE is all about. A vision can only be created by an informed group that knows what the building blocks of their preferred future can be. As a minimum, all members of the group should have read Part I of this book. Preferably, they would know even more, based on the additional reading, training, benchmarking visits, and experience.

Develop Project Objectives. Once the vision for CE is agreed upon, specific broad objectives for the CE project should be identified. In particular, what are the key project objectives and how do they fit in with the overall company vision/mission/strategy? It is usually helpful to think of these as general outcomes to be achieved by the project, such as decreasing time-to-market by 50 percent.

Develop Project Boundaries. The guidance team should also identify the boundaries of the project. Which parts of the organization will be involved? If the company is a large one, perhaps only one division might be involved at first. Are there particular areas that are out of bounds, such as the union contract or the reward system? These can also be identified later; but the more explicit that assumptions are stated at this point, the more smoothly implementation is likely to follow.

Project Action Plan

Once the project vision, objectives, and boundaries are identified, a specific action plan can be laid out to plan and implement CE. The project action plan should reiterate the project vision, objectives, and boundaries as well as spell out action steps to carry out. The Project Action Plan can be built on the Project Concept White Paper, except

Purpose of the exercise: At the end of the exercise, participants will have outlined general goal statements for the CE Project and a vision of CE that is shared. This is an open forum encouraging creativity and team participation. This exercise should only be carried out after group members have undertaken a period of education about CE and related topics.

Time for exercise: 75-85 min.

Materials: Note pads, pens and markers, flip chart, handout example of future scenario, masking tape.

Features and instructions

5 min – Explain exercise objectives and instructions.

15-20 min – Develop common information base among participants. Information such as business requirements, understanding technology (including CE), the system model, and reasons why the company needs to change should be discussed among participants.

5-10 min – Distribute scenario and allow participants to make a list of characteristics of the organization they would expect to see after it has adopted CE successfully.

15 min – Go around the group allowing each person to call out one idea at a time and record each idea on a flip chart. Continue going in a round-robin fashion until a group has exhausted all the ideas. Post the flip chart pages on the wall.

10 min – Each individual gets three stars to vote for characteristics that must be present, and five check marks for ones that should be present. Each person can place their votes on a single characteristic or distribute them among several. Consolidate characteristics and choose the 5-7 biggest vote-getters. Rewrite the largest vote-getters on a new sheet of newsprint.

15 min – Share results with the group and ask to convert the results into goal statements. These should be broad goals for the CE Project. Seek consensus on the critical elements of how they wish the project to proceed.

10 min – Finalize vision statement.

SCENARIO EXAMPLE

It is _____ (fill in date — e.g., 5 years from now). You are in a time balloon hovering over your organization. You can see, think, and feel all that is going on. You are pleased and proud of how effectively your company has implemented Concurrent Engineering. What do you see happening? What are people doing that lets you know you have reached your goals?

Exhibit 10.2 • CE Visioning Exercise

that the Action Plan provides much more detail about specific activities. The outline in Exhibit 10.3 provides guidance for developing such a document.

Outline for Project Action Plan

1. **Project vision and objectives.**
2. **Project boundaries.**
3. **Milestones:** Specific activities required to complete the project, evidence that such activities are complete, and a timetable to complete them.
4. **Assumptions:** Things we believe to be true, including general trends as well as those specific to your markets, competitors, customers and your own company that support the value of this project.
5. **Allocation of responsibility for various activities.**
   a. Project management.
   b. Data collection and analysis for As-Is condition.
   c. Design activity (might be decided later).
   d. Implementation (might be decided later).
   e. Project sponsor, link to top management.
6. **Key participants:** Decide which division(s) will participate, and/or be affected.
7. **Budget:**
   1. Design phase.
   2. Estimation, for implementation phase.

Exhibit 10.3 • Outline for Project Action Plan

Concurrent Planning

The "steps" in this phase are not the kinds of things that can be done once and then forgotten. All will be modified and updated as you go along. While it would be nice to have a sponsor stick with the project over its lifetime, we all know that people come and go. Hence, you may well have to find a new sponsor. When that happens, it's likely that the vision and ultimately Project Action Plan will need to change. Equally, the action plan will probably need to change over time as you learn new things during the course of the As-Is Assessment, or when things change around you. Expect it to happen, and when it does, be adaptable.

Just a Word of Warning: We've described the CEE Process as nice and orderly and rational. For better or worse, the organization you want to use it in is probably none of these things, and the same can be said for the people in it. While you're going to have to live with that, we hope that the tools and methods provided here will furnish some structure to help you work through the problems you are likely to face.

Potential Problems and Solutions

Can't Find a Champion or Sponsor

There are at least a couple of ways this could happen. One is that you might be able to find someone at a high level who gives the go-ahead, but is not really committed and won't provide any resources. The other is that you can't find anyone at a high level with any interest at all. Let's look at each case separately.

- ***Weak Support without Resources.*** The strategy in this case is to do things that have little or no cost — things that can be supported from whatever budgets you can access or which are effectively free. If you can make a success at the little, no-cost activities, you can begin to build an aura of overall success, which then might allow you to do something a little larger, and which might begin to garner support from your sponsor. You would, of course, also do the usual selling that you would do normally.
- ***No Support at All.*** In some organizations, it is possible to proceed without high-level support at all, but it's usually risky. Your best bet — if you have no support at all — is to continue to try and get some. Frankly, if your ability to argue your case is so weak with your management that you can't even get weak support, the chances that you will succeed in later more difficult steps is very low. Assuming you have tried and failed to win support with higher levels of management, you might consider trying to sell some of your peers on this, in an effort to set them up as the sales agent to higher management. There can be any number of reasons why you haven't succeeded, and there's certainly some chance that letting someone else try might do the trick.

Some Parts of the Organization Won't Participate

Ideally, all of the relevant parts of the organization would participate in helping to plan and implement CE. But, unless you have strong support from a very high level, there's a reasonable chance that someone isn't going to participate. A pretty common example of this is when engineering initiates the CE effort, but can't bring manufacturing or other functions directly on board. There are at least two ways this non-participation can take shape: passive or active resistance.

- ***Passive Resistance.*** In passive resistance, the non-participant refuses to be part of any planning efforts, but doesn't actively oppose those efforts that do take place. More important, members of the non-participating organization are probably willing to be part of the actual work of doing CE as individuals. For example, if manufacturing was a passive non-participant in an effort led by a product development engineering department, then an individual manufacturing engineer, if asked to participate on a cross-functional team, might well agree. The change strategy for passive resistance is to bring individuals from the resisting organization into active CE work as much as possible. Then document the benefits gained from this participation. If you're reasonably successful, you'll build enough momentum so that the other functions will find themselves drawn into the effort.

- ***Active Resistance.*** In active resistance, the non-participant actively works to keep its personnel out of the CE activity. This may not take the form of an actual refusal to participate, but personnel from the non-participating unit are never available when needed. After a while you'll start to get the picture. In a case like this, you will probably have to proceed without the resistant organization. For some non-participating functions, you may be able to do reasonably well without them, but for others (e.g., manufacturing if CE is initiated by engineering), you may have more of a problem. A short-term solution (if you have the resources) is to create temporary resources within your own organization. For example, it's quite common for an engineering department to bring manufacturing engineering resources or even accounting and estimating resources inside the department when those resources can't be obtained from other places in the company. Of course, this is creating redundant resources for the company as a whole — and that's why we referred to this as a short-term solution. Sooner or later the active resistance of the relevant functions will have to be broken down, but that may only come after you have been able to demonstrate the benefits of CE (albeit partial) through the use of internal resources.

The Coseat Inc. Example

Now let's revisit the seating company, Coseat Inc. (You will remember Ed Cazotsky, Director of Design Engineering at Coseat's Technical Center.)

Ed has been talking to some friends of his at the Technical Center. These include Esther Maxim, a senior project engineer; Harold Gates, Director of Manufacturing Engineering; and Leon Chan, Manager of Testing. Mostly they talk over lunch and before or after another Meeting they've been attending. Over the past few weeks each member of this group has been getting increasingly more concerned about the position Coseat Inc. seems to be in. Finally, over lunch, they decide to have a formal meeting to discuss their concerns. They invite representatives of each of the major functions of the Technical Center, as well as their boss, Robert Ferry, General Manager of Product Design.

The meeting commences with an introduction by Ed Cazotsky about what he sees as the source of Coseat Inc.'s problems. He talks about what he's read about CE and what some of their automotive customers are doing in the area. After an hour of discussion, there's general agreement that CE is worth investigating further. Ferry appoints a committee consisting of Cazotsky, Gates, and Chan to develop a white paper that can be used to stimulate discussion both within the Technical Center and, later, to the company at large. They do this, and two weeks later at another meeting of the group, Ferry agrees to act as champion to take the idea to other parts of Coseat Inc.

Ed and his small group of friends are an Early Advocacy Group. Chances are they all did some reading on their own, and talked about their ideas with their co-workers. Robert Ferry is going to serve as the CE Champion. In Exhibit 10.4 you can see excerpts from the white paper developed by the group.[1]

Vision
Coseat will use the concurrent engineering (CE) approach to involve engineering, manufacturing, marketing and other functions in all product development activities. Coseat will work more closely with both suppliers and customers. Coseat's employees will be both involved and empowered to make these things happen.

Why Change is Needed
Internal Drivers. Coseat wishes to maintain its niche as the premier developer of automotive seating systems. However, profits and sales have been declining as a result of frequent late delivery of new systems.

External Drivers. Both customers and competitors are increasingly using CE to reduce time-to-market with new seating systems. Customers are increasingly looking for best practice in product development from their suppliers.

Expected Outcomes
Significant reductions in product development lead time; improved responsiveness to customer requirements, reduced product cost.

Resources Needed
Cross-functional planning team from the Technical Center, product engineering, manufacturing, marketing, procurement, and account management. This relatively high level team would meet for a total of about 15 hours over a three month period, and do about 30 hours of work off-line to plan the initial effort into CE.

Exhibit 10.4 • Excerpts from CE Concept White Paper

[1]We have not included a whole white paper because it would be too lengthy for the space we have available. We've tried to show you the argument being made, rather than the language that would be used, since that will vary tremendously from company to company.

After several meetings at which the project concept white paper is refined and clarified, Ferry agrees to act as a champion for the concept with his boss, the VP of Operations, Ken Leopold. As it happens, Leopold has been giving this very same idea some thought lately and was trying to figure out how to get started. Ferry's white paper provides some organizing concepts, as well as the impetus to get something started.

Leopold now does very much what Cazotsky did at his level — he starts talking to his peers, in this case, other vice-presidents. Each expresses some degree of uncertainty about the idea, but is willing to go along with a limited planning effort. A Guidance Team is established with representatives of each of the major functions involved in product development: engineering, manufacturing, marketing, finance, human resources, and supplier management. The team develops a refined vision, a set of objectives for CE, and boundaries for the planning effort (Exhibit 10.5).

Vision

Coseat will use the concurrent engineering (CE) approach to involve engineering, manufacturing, marketing and other functions in all product development activities. Coseat will work more closely with both suppliers and customers. Coseat's employees will be both involved and empowered to make these things happen.

Project Objectives

Background. Project objectives were derived from a benchmarking process and from listening to the voice of our customers. For example, we learned from Motorola that it is possible to seek and achieve a ten times quality improvement. On the other hand, our customers told us about their needs in time-to-market and late deliveries. Not all of these goals can be achieved by the project alone; for example, reducing late deliveries might not seem like a product development process goal. However, the manufacturing process at Coseat is influenced by product design and is the cause of many of the late deliveries and we need to keep that goal in mind as we develop manufacturing systems. We derived the following objectives:

1. Reduce time-to-market by 50%
2. Improve quality by 10×
3. Reduce engineering changes 25%
4. Reduce warranty costs by 30%
5. Reduce scrap rate by 25%
6. Reduce late deliveries by 90%

Project Boundaries

The project will focus on the customer interface, suppliers, product design, and manufacturing engineering. Plant organization will not be directly affected or involved in the process. Plant operations should not be affected except when new products are introduced and when new manufacturing systems are installed. Major reorganization of the company structure is in scope for consideration, but of course will require involvement and review at the highest levels.

Exhibit 10.5 • Example Vision, Objectives, and Boundaries

With the vision, objectives, and boundaries defined, the team then developed an action plan, parts of which can be found in Exhibit 10.6.

Milestones
Assessment Complete (Report to top management) - start+8 weeks
Go/no go decision about design - start+10 weeks
Initial Design Complete - start +16 weeks
Go/no go decision about continuation - start+18 weeks
Revisions to Design Complete - start+22 weeks
Go/no go decision about Phase II - start+24 weeks
Implementation begins - start+25 weeks

Responsibilities
1. Project management – Ed Cazotsky from Design Engineering will manage the CE Planning Project.
2. Data Collection for As-Is Assessment – representatives from all participating departments will be responsible for data collection in their departments. Analysis will be a joint responsibility of the whole team.
3. Design – TBD
4. Implementation – TBD
5. Project Sponsor – Ken Leopold, VP Operations

Participants
Engineering, R&D, manufacturing, marketing, finance, human resources, and supplier management.

Budget

Exhibit 10.6 • Excerpts from Project Action Plan

SUMMARY

We've discussed a process for getting everything in place to get ready to do an As-Is Assessment, design a new CE system, and implement it. This includes an approach for assembling an early advocacy group, and selecting a project champion, sponsor, and Guidance Team. We've also presented the broad outlines for a CE White Paper, a visioning activity, and a Project Action Plan. We haven't gone into these in great detail because what you actually do will be very idiosyncratic — in each company there will be only specific individuals who are able to play the roles we've described. How you find these people, approach them, and convince them to help you really depends on you and them, and on the specific situation you're in.

You need to cover these bases before you launch into any kind of assessment or design activity. Without the necessary vision and support, you could easily be wasting your time, as well as the time of those you work with.

CHAPTER 11

•

ASSESSMENT OF THE AS-IS SITUATION

To order is to recognize. To know that, in an endless, unknown sea, there is an island upon which you have set before. • Peter Hoeg

The second phase of the change effort (Figure 11.1) is designed to assess the current situation in your company and its supply chains so you know where you are starting from.

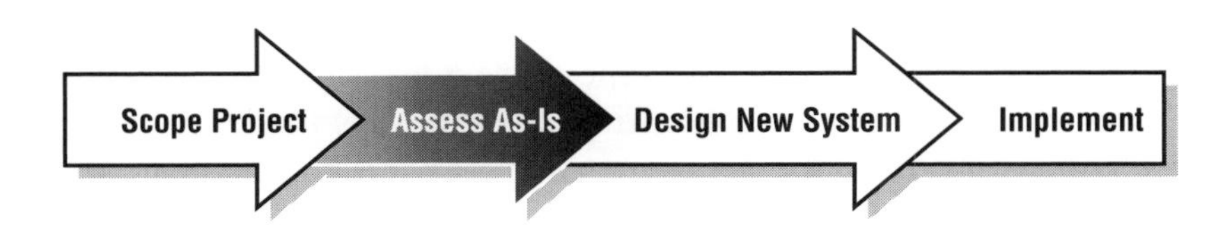

Figure 11.1 • Four Phases of the DAO Change Process

Ed Cazotzky, Director of Design Engineering at Coseat Inc.'s Technical Center, has just been appointed to lead Coseat Inc.'s efforts to plan and implement CE (see Chapter 10). He has a vision; he has a team; he has the general outline of a plan, but what should he do next? Ed has done a lot of reading and he's visited a couple of his customers' sites where they're doing CE. He and his team members think they have a pretty good idea about what CE is and what Coseat would look like if they did CE. The problem is getting from here to there. Even if they know generally where "there" is, do they know where they're starting from? Can they develop a detailed plan without understanding their current system? While the members of the guidance team are smart, knowledgeable folks, they quickly realize there's an awful lot about the company and the way it does busi-

ness that they don't really understand. Ed and his team decide they need to investigate Coseat's current state of affairs before they move too quickly into detailed planning for CE.

PURPOSES OF AN AS-IS ASSESSMENT

An As-Is Assessment provides information about your current state of affairs. It describes how things are done now and how well they are being done. The As-Is Assessment serves two very important purposes.

1. ***Baseline.*** By telling you how things are going now, the As-Is Assessment provides a baseline against which you can compare future performance after you have introduced CE. One obvious reason for wanting to do this is so you can demonstrate what a wonderful thing CE is for your company. If you can demonstrate that CE resulted in performance improvements, and if you can quantify those improvements, you can make yourself look pretty good (and you can feel pretty good about yourself as well!). Another, less obvious reason for doing this is so you can find out if anything has indeed changed after you introduce CE. After all, you can "roll out" a program like CE, but it doesn't necessarily mean that people's behavior will change. Without a baseline you would have no formal way of knowing if practices were indeed different.
2. ***Identify Opportunities.*** By looking in detail at existing practices and structures, the As-Is Assessment provides you with the detailed information you need to identify opportunities for improvement which will provide the greatest benefit to your company. Frankly, it might be wonderful to have some grand CE plan and to just impose it. But the world rarely works that way — you have to work with existing people and skills, and you have to change existing structures and practices. If you are going to change something that currently exists, then you have to pick your spots. Where are the greatest opportunities for improvement? Where can change be introduced with the least disruption to existing work?

We can't emphasize enough how vital the assessment will be for the success of CE. For many "action-oriented" managers and engineers, it can initially seem like a waste of time; after all, they say, "We know how this company operates — we work here." We can respond to that with three comments. First, how well do you really know how things work? Even in very small companies (as small as 75–100 people) we find that there is often no one who really understands how things work in all areas of the company. In a large company it is all but impossible to get a small team together that fully understands how things work. Second, even when you get a team of functional experts together, they usually find it very difficult to communicate to people in other functions about what is happening in their area. Thus, all the needed information may be in the heads of the people sitting around a table, but getting it out for use by the whole team can be very difficult. Third, a model is needed to help classify and analyze the data in a useful way. The As-Is Assessment gets all of the needed information out around the table, in a single place, and in a form understandable to the whole team.

An Assessment Model

Just as with the model of your work process (see Chapter 3), your As-Is Assessment can be as simple or as complicated as you are willing to make it. Make it too simple and you don't learn anything; make it too complicated and you never get to do anything else. We will present a relatively complicated model — we want you to push the bounds of complexity so you learn a lot. However, we will also provide advice to help you keep it from getting out of hand. If the assessment takes much more than 4–6 weeks, you're either looking at too much detail or you're not really focusing on the task.[1]

There are four primary components to our As-Is Assessment Model (Figure 11.2).

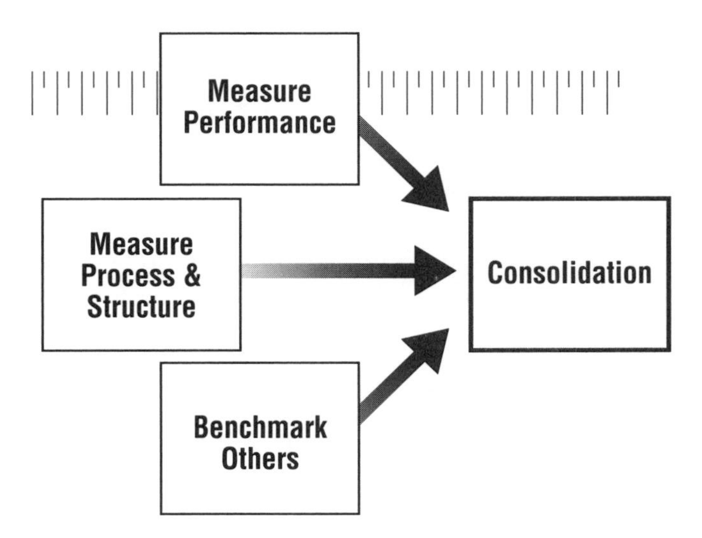

Figure 11.2 • As-Is Assessment Model

[1] This time frame works if your team is devoted to this on a mostly full-time basis. Obviously the time required will depend on what proportion of your effort is devoted to the task. Our recommendation is to focus on it and get it over with as quickly as possible. Your company isn't getting any better as long as you're just doing the assessment.

1. ***Measure Performance***. Measure your product development performance so you can assess how well you are doing as compared to others and as compared to yourself when you have implemented CE.
2. ***Measure Process and Structure***. Measure processes in detail so that you understand how each process is performed. Assess structures so that you understand how they affect and are affected by processes.
3. ***Benchmark Others***. Find out how others do CE (in terms of performance, process, and structure) so that you can learn from their example. In a few cases, learning may mean copying, but in most cases it will mean *adapting* a lesson from a very different setting.
4. ***Consolidation***. None of these components can stand alone. At the end of the chapter, in the consolidation component, we will tie the many pieces together so you can make sense of them and make decisions about how to design your CE system.

For each of the components described below we will provide you with a set of instructions, one or more tools to help you actually carry out the instructions, and a discussion of what to do with the information you've gathered.

THE ASSESSMENT TEAM

We recommend the As-Is Assessment be carried out by a cross-functional team of individuals who are knowledgeable in their functional area, and have strong social connections to others in their area. The reason for the social connections is that no one knows everything, and therefore every member of the team will need to feel comfortable going around and asking for information. The cross-functional team will probably be the Guidance Team (Chapter 10), but it could be a subcommittee of people delegated by the Guidance Team to do the assessment.

"AW GEE, MAW"

Again, do you really have to do everything we say? Do you really have to collect and analyze exactly the way we say to? On the one hand, the answer is of course not; you don't really have to do everything according to some formula. On the other hand, we urge you to accept the *spirit* of our method. By this we mean that you need to seriously consider each of the issues we ask you to assess. If you already have deep expertise in an area and don't believe you need to use the specific method we suggest, that will probably be all right. If you have a method for assessing something that you prefer to ours, by all means use the method you prefer! But you should be sure to address, in some way, all of the main issues we present in this chapter.

Benchmarking

Benchmarking is a complex activity and many excellent books have been written about it.[2] We will not attempt to duplicate their efforts in a few pages. Rather, we will provide a few suggestions on how to take maximum advantage of benchmarking in the context of the Concurrent Engineering Effectiveness Methodology.

Match Benchmarking to Your Assessment

When you look at how other companies do CE, focus on the issues we suggest in this As-Is Assessment. In fact, you can think of a benchmarking visit as being similar to an assessment, except you won't have the opportunity to evaluate it in quite as much detail. As you will see later in this chapter, we use benchmarking data in the Consolidation activity. The closer in form your benchmarking data are to your performance and process measures, the more useful they will be. However, remember that other companies may measure things differently than you do. For example, their approach to measuring costs may be entirely different from yours, if, for example, they use activity-based costing and your company does not.

Look at Other Industries

Don't restrict yourself to your own industry. While you obviously want to know about best practices in your own industry, chances are you won't find real breakthroughs that way.[3] Look for excellent practices in other industries by focusing on specific processes, rather than the details of what's being made. For example, information of use to plastic moldmakers can be found and studied in industries as diverse as automotive, toys, furniture, electrical equipment, and computers.

Visit Other Companies

Don't depend on written case studies (such as the ones in this book) for all of your benchmarking. While we would be the last to say that case studies are unimportant, they are not a replacement for actually talking with people who are doing CE. We developed our case studies by spending a lot of time with such people, and we learned what seemed

[2] See, for example, Camp (1989), Pryor (1989), or Besterfield, et al. (1995).
[3] Although it's always possible — just look at the lessons Toyota has presented for the rest of the automotive industry.

important to us. You will certainly learn other things that we may have never thought about on such visits.

Assessing Performance

The essence of assessing performance is measurement — what is success and how do you measure it? For the most part, these measures should relate back to your objectives. For example, if one of your objectives is to reduce time to market with new products, then your performance metric will be something like months from some start point you have selected (e.g., beginning of concept design) to delivery of the first product to a customer. These measures of performance will usually be fairly obvious — they're mostly related to time, cost, and quality. There is also a good chance that you are already collecting at least some data about them. Even if you aren't currently collecting such data, the need to do so is quite clear.

The importance of collecting and using performance measures is usually very obvious to most managers, so we won't spend much time on it here. We want to stress that it is important to get and use the best data available. This may mean creating your own sources of data and databases if the existing information systems don't provide what you need. As an example, existing information systems may only track a project when it has reached a certain level of development (e.g., when you have a formal contract from a customer). You may need to track the total time of development from some earlier point in time (e.g., when unpaid, but

Objective	Measure	Data Source
Reduce time to market by 50%	Months from assignment of task to project manager to delivery of first seat to customer	Project tracking system
Improve quality 10x	cpk	Plant quality tracking system
Reduce ECNs by 25%	# changes submitted after release	Tracked by project manager
Reduce warranty costs by 30%	Total warranty cost for first year	Accounting system
Reduce scrap rate by 25%	Scrap rate at factory	Factory information system
Reduce late deliveries by 90%	Seat deliveries to customer as reported by customer, compared to weekly schedule	Factory information system

Exhibit 11.1 • As-Is Performance Measures for Coseat Inc.

very crucial, concept work is being done). You would need to develop some way of tracking both calendar time and person-hours during these early stages of development. While this may be time-consuming and costly for you, unless you do it you run the very severe risk of being unable to demonstrate the benefits of CE to your management.

In Exhibit 11.1 we show the objectives Coseat Inc. set for themselves, and the measures identified to track progress toward the objectives prior to the start of their effort to introduce CE.

Assessing Processes and Structures (General Approach)

The bulk of this chapter describes many detailed procedures for assessing processes and structures. The reason for all of this detail is that there is no one assessment approach for all elements of your work process, organization, and technologies for concurrent engineering. Each element needs a tailored approach. However, there are some general guidelines on how to approach the task of collecting data and using the assessment tools. These suggestions provide a start at planning the assessment.

- ***Cross-Functional Team***. The assessment should generally be carried out by a cross-functional team. This is to ensure multiple perspectives on both the data collection and the analysis.
- ***Multiple Data Sources***. When we suggest collecting information, collect it from more than a single source. This will help to ensure more accurate and unbiased information. When selecting people to talk with about any given topic, try to make your selection as similar as possible to the larger group of people they are representing.
- ***Objectivity***. When you collect information try to be as neutral as possible with the people with whom you are speaking. In other words, try to avoid influencing them with your own biases.
- ***Speed***. Try to move through any method as quickly as possible. You may want to use a neutral facilitator to help you move along quickly in any given exercise.

We will divide this part of the assessment into the same structure as we provided in Part I: work processes, internal organization, supplier relations, people systems, and technology. We have not attempted to assess strategy, although consideration of strategy and its fit for CE were discussed in Chapter 9.

The elements of the As-Is Assessment should be done in approximately the order we provide below (Figure 11.3). This is because infor-

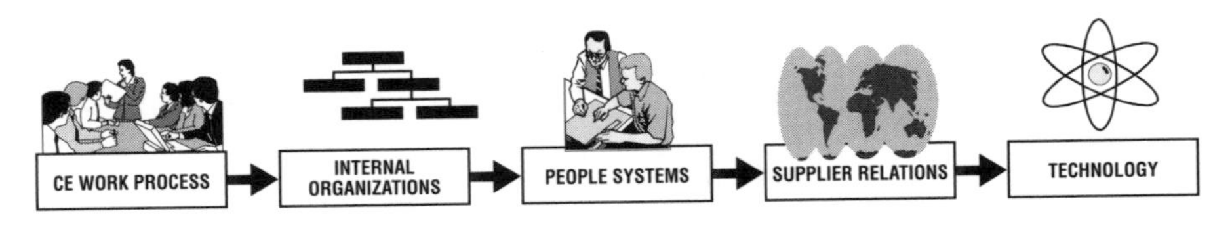

Figure 11.3 • Elements of the As-Is Process and Structure Assessment

mation about work process is needed for the internal organization assessment, and information from the internal organization assessment is needed for the people system assessment. For the other elements, order is not so important.

Assessing Work Process

There are four steps involved in assessing your work process. Figure 11.4 shows how they are connected to the overall structure of the work process assessment.

1. ***Map As-Is Process.*** Using some form of work process modeling like a flow chart, PERT, or IDEF0, map the process of product development as it is actually practiced. Review Chapter 3 of Part I to help you determine the appropriate modeling approach for the level of detail you want to get into.
2. ***Document Formal Processes.*** Document any formal product development processes which are supposed to be used according to company policy.

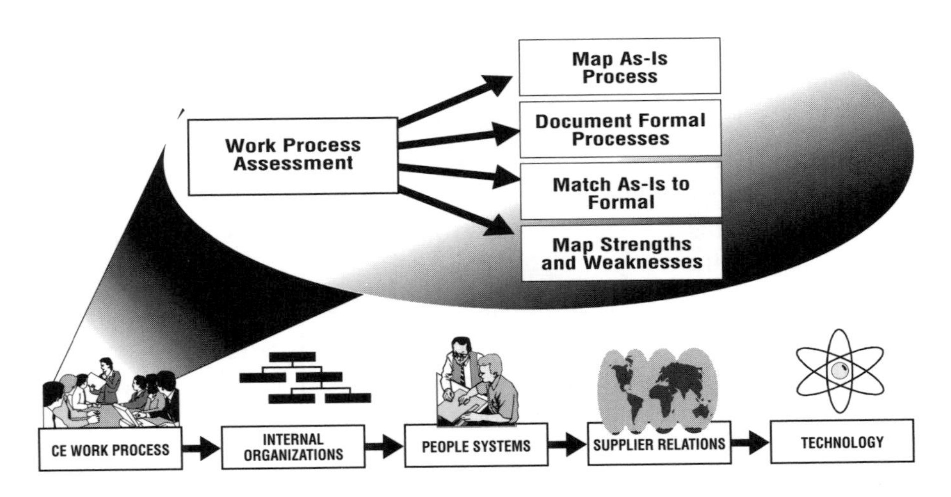

Figure 11.4 • Work Process Assessment

3. *Match As-Is to Formal.* Document any differences between the actual as-is processes and the formal processes.
4. *Map Strengths and Weaknesses.* Assess strengths and weaknesses of the as-is system, including documentation of non-value-added steps in the process.

MAP AS-IS PROCESS

Use one of the modeling methods described in Chapter 3 of Part I or a similar method to map out your existing processes in product development at a relatively high level. It is important to remember that the purpose of this exercise is not to have a finely developed map of the process; rather it is to help you understand how you currently do product development. Not only would the development of a finely detailed map be too expensive and time consuming, but we doubt whether it would really help you all that much. Most companies find the high-level mapping exercise really helpful; most companies find a very detailed mapping a big waste of time.[4]

You can develop this type of mapping in either of two ways. One approach is to get people from different functions associated with product development in a room for a few hours and ask them to map out their process using the guidelines we have given you. If you give them a time limit and an example (see the Exhibits below) they should be able to do this without too much trouble. The alternative is for you (or your team) to interview these same people individually (or in groups) in order to collect the information about processes. You can then develop the process maps, have them review your work and then revise. We have used both approaches and both work reasonably well. In general, it's probably much more efficient in terms of everyone's time to get them all together in a room. You may run the risk that some individuals in the group will dominate the discussion, leaving you with a somewhat distorted view of the process; but, in general, that risk is small.

One thing to watch out for is the existence of multiple processes. You may indeed have several processes for developing new products. It may be that every design team has a different process (and creates a new one each time a project begins); or it may be that different kinds of products have different processes. The presence of multiple processes is not a bad thing, in fact, different types of products generally need different development processes. Try to document the few most dominant pro-

[4] This isn't to suggest that detailed mapping is always without value. If you have a highly structured process that can be finely mapped, then it may be worthwhile to do so in order to remove all the little process inefficiencies. Our experience with product development processes is that they are so fluid as to make detailed mapping overkill.

cesses within the limits of your scope. If you keep things at a moderately high level, it is usually the case that minor differences between processes (e.g., small differences between project teams developing similar products) will disappear. On the other hand, there may be major differences in design philosophy which you need to document, since these may have serious implications later on.

Ed Cazotzky and the Guidance Team from Coseat Inc. have decided to figure out how the company does product development. They'll develop a process map using IDEF0 with a cross-functional team. Ed asks his peers in other parts of the company to assign representatives to attend a half-day meeting to plot it out. Rather than having members of management do this work, he asks for senior people who perform actual project work. He figures these people will understand how things really get done much better than their managers do. As representatives, he's got the following:

Esther Maxim — Senior Project Design Engineer (reports to Ed)
Ted Franshaw — Senior Manufacturing Engineer (reports to the Director of Manufacturing Engineering)
Lester Smith — Purchasing Agent (responsible for electrical components)
Lee Wistan — Account Manager for Coseat's General Motors Sales
Miles Kendall — Manager for Market Research (Miles is the only market research professional in a very small department)
Mary Totin — Senior Tooling Engineer
Harry Andrews — Test Engineer

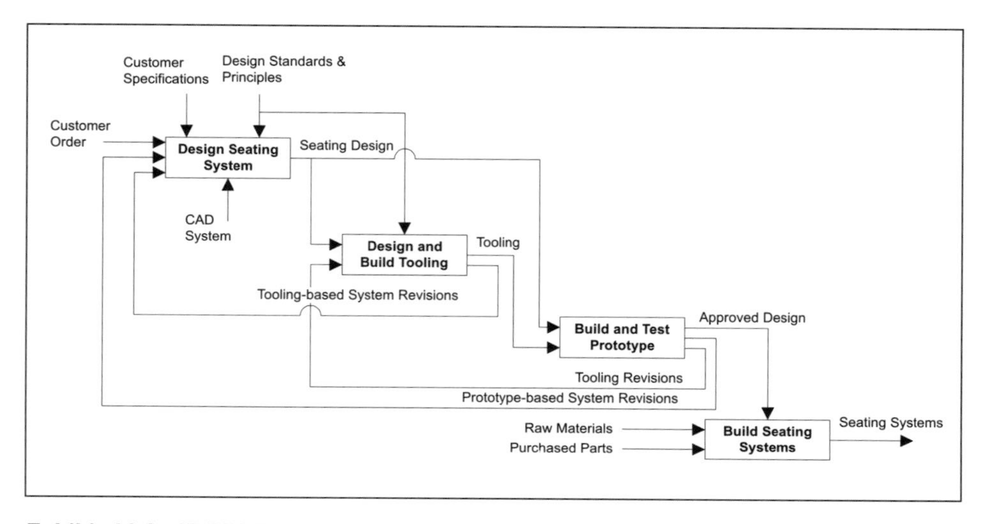

Exhibit 11.2 • IDEF0 Flow Diagram for Coseat Inc. Product Development Process

The Guidance Team suggests that Ed act as the facilitator for the group. Prior to the meeting, Ed meets individually with each member of the team to explain the purpose of the group and the larger objective management has in moving toward CE. Ed starts the meeting itself with a round of introductions, since several of the members have never met. He explains the purpose of the meeting again to give everyone a chance to ask questions. Ed briefly explains a simplified approach to IDEF0 process modeling, including activities, inputs, outputs, resources, and mechanisms. He asks that the members of the group not get caught up in the details of the modeling approach — he'll worry about that later when he develops the formal model.

He starts things off with a question — "What are the high level steps we follow when we develop a seating system?" The group initially develops something that looks a bit like the model shown in Exhibit 11.2 (the exhibit itself is a more formal model that Ed developed later, based on what the group said it does). Ed then takes each box in Exhibit 11.2 and asks the group, "What are the steps we follow to design a seating system (or design and build tooling, etc.)?" The final model for Design Seating System can be found in Exhibit 11.3. Ed makes careful notes about the meaning of terms and what is meant by each activity.

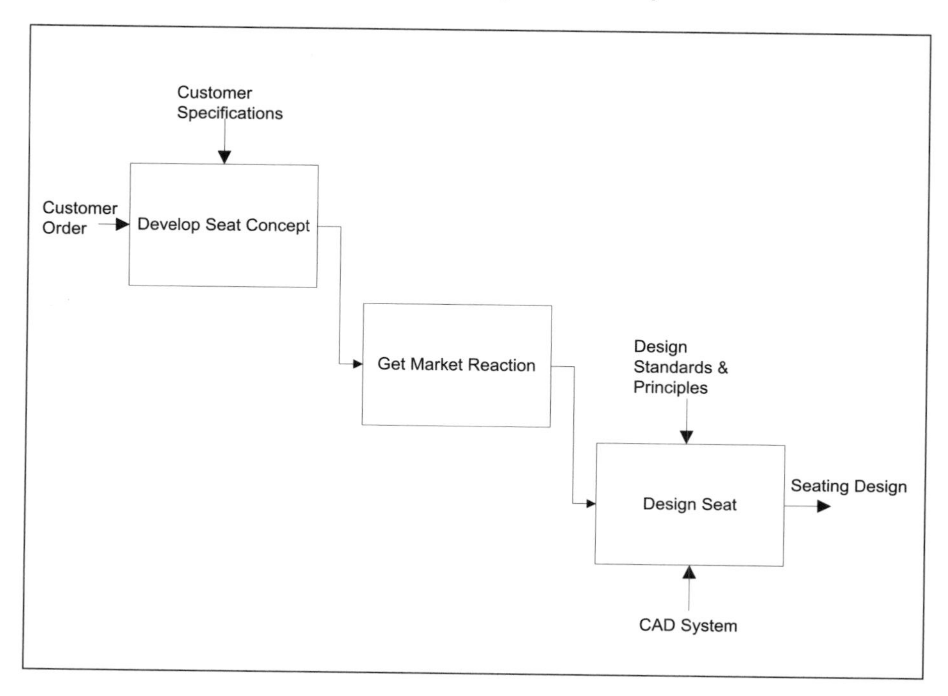

Exhibit 11.3 • IDEF0 Diagram for Coseat Inc. Design Seating System

One problem with the flow diagram is that, while it shows task precedence, it does not show timing. Since timing is very important in CE, the group also develops a Gantt chart which shows the approximate timing of each major activity (Exhibit 11.4). They note that the timing is very sequential.

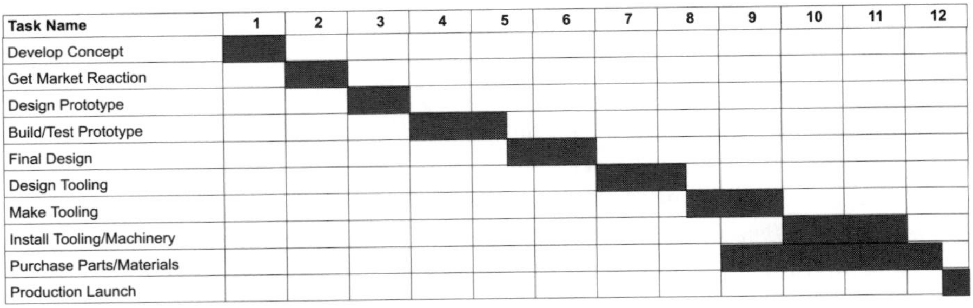

Task Name	1	2	3	4	5	6	7	8	9	10	11	12
Develop Concept												
Get Market Reaction												
Design Prototype												
Build/Test Prototype												
Final Design												
Design Tooling												
Make Tooling												
Install Tooling/Machinery												
Purchase Parts/Materials												
Production Launch												

Exhibit 11.4 • Gantt Chart for Coseat Inc. Work Process

Document Formal Processes

Many companies will have a formal product development process which specifies how development projects should be run. Here we're asking you to find out if there is such a process. If there isn't, or if it is very old (more than a few years) and obviously ignored, then you probably shouldn't bother documenting it. On the other hand, if it is still considered to be current (regardless of whether everyone uses it) you should go ahead and document it.

For our purposes, documenting means to summarize the process and assess its use. Even if the process is embodied in a 1000-page document, find the relevant parts (such as the high-level process we just discussed) and put it in a short summary, just so you have some idea about what the process is. The second thing you should do is to ask around and find out if the process is used. You should probably do this while you are collecting information that will enable you to map the process.

In case you're looking for a Coseat Inc. example here, there isn't one. Coseat was one of the many companies that lacked any formal process.

Match As-Is to Formal

Here we want you to compare the actual as-is work process with the formal process. Are they a lot alike? If they are alike, it tells you that people in the company are used to following a formal process. This will be good to know later on if you propose that they create a new process. If they are not alike, it might suggest people are generally unwilling to cooperate in such an endeavor; or, more likely, it suggests the formal process didn't meet their needs — so they do what they have to do instead. It is rarely the case that a formal process is literally applied in each product development process. In fact, engineers are often not even aware of the formal product development processes in their companies. It

Work Process Observations Worksheet

1. **Identify and describe the observed weaknesses and strengths of the identified activities (right ones, wrong ones, too many, too few, wrong order, etc.).**
   - Getting Market information too late in the design process. Our products need to be more customer driven from the start. Need some sort of customer needs identification process.
   - Need a clearer identification of the tooling design and build processes. The existing processes are too poorly defined.
2. **Identify and describe the observed weaknesses and strengths of the interfaces between activities, the items that move from one activity to the other (including adequacy of content, quality of the match in expectation between two communicating activities, missing elements, etc.).**
   - Do not have common understanding of the information flowing from one activity to another. The senders and receivers disagree on what the content of the various information packages really should be.
   - Basic product design process is well understood and consistently followed. However, there is too much "throw it over the wall" built into the process.
   - One exception: the information flowing from tooling design to toolmaking is well defined and complete.
3. **Identify and describe the observed weaknesses and strengths in the sequencing and coordination of the various activities (such as timing issues, unrecognized dependencies, well-handled dependencies, good and bad assumptions, etc.).**
   - Prototype design actually begins before marketing has completed its work. Bad decisions result from this lack of customer voice.
   - Tooling design should start earlier. Preliminary tooling design work could begin before the final design work is complete.
4. **Identify all terms that need to be carefully defined. They should include any that have been shown to have significantly different interpretations by different people or groups who need to have a common understanding.**
   - Release - used four different ways in the company, generally without clarification.
   - All terms in process model need clearer definition (e.g., concept, design standards, prototype concept revisions, prototype, prototype design).
5. **Identify any other strengths and weaknesses you may have spotted.**
   - Serious lack of understanding of what the company actually does on a day-to-day basis by the average engineer. There needs to be a lot more top-down information dissemination.

Exhibit 11.5 • Coseat Inc. Work Process Observations Worksheet

is not worth spending a lot of time documenting in detail all the gaps between the formal system and actual practice. This should only be done at a high level—a superficial analysis in this case is usually good enough.

Map Strengths and Weaknesses

Here we want you to think about what's good and bad with the actual current process (not the formal process). Are there non-value-added steps? Are steps in the right logical order? Is there concurrency built into timing where you think it might be useful? Don't draw any conclusions at this point (save that for later). Ed Cazotzky at Coseat Inc. completed the Work Process Observations Worksheet (Exhibit 11.5) to summarize this information. Ed's comments are in italics.

Assessing Internal Organization

Next we will look at the internal organization of the people involved in product development. There are six activities in the internal organization assessment as shown in Figure 11.5.

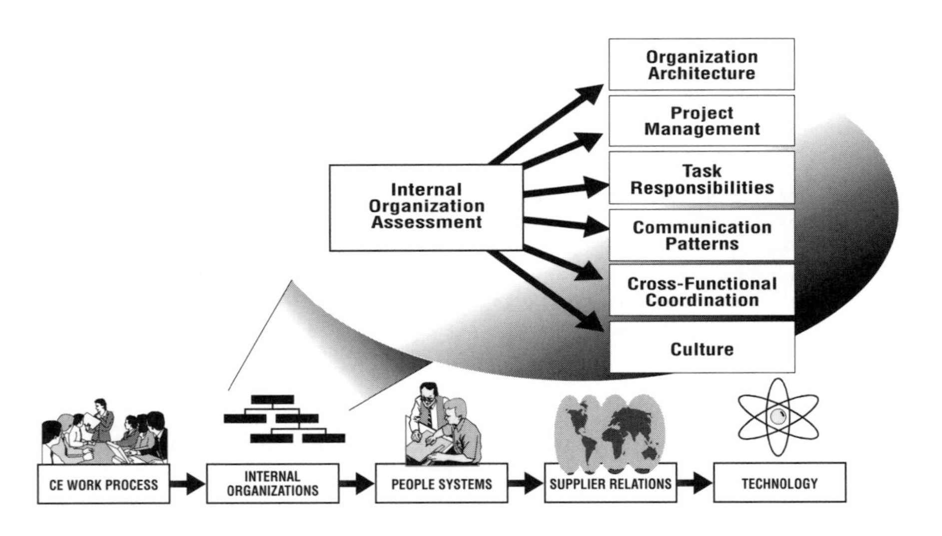

Figure 11.5 • Internal Organization Assessment

1. ***Organization Architecture.*** Document the current formal organization structure.
2. ***Project Management Approach.*** Document the current approach to project management.

3. ***Task Responsibilities.*** Document which functions in the organization are responsible for which parts of the work process.
4. ***Communication Patterns.*** Understand how different functions communicate with each other both formally and informally.
5. ***Cross-Functional Coordination.*** Understand how different mechanisms are used to facilitate coordination for different parts of the work process.
6. ***Culture.*** Understand how your company's culture affects your ability to coordinate across different functions and with suppliers; also understand how it affects your ability to introduce change.

Organization Architecture

Based on the discussion of organization architecture in Chapter 4 of Part I, describe the architecture of your organization as either functional, product line, matrix, or hybrid. If it is a hybrid, describe which parts fit into the different categories. Describe strengths and weaknesses of the current structure as you see them. As you can see in Exhibit 11.6, Coseat Inc.'s structure is very much functional.

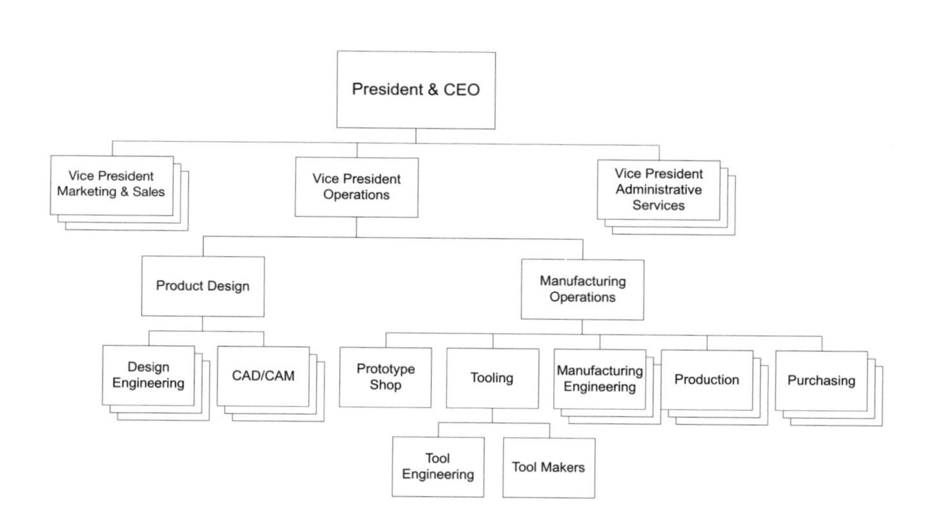

Exhibit 11.6 • Coseat Inc. Organization Architecture

Ed Cazotzky and his team used the Organization Architecture worksheet (Exhibit 11.7) to document their observations.

Organization Architecture Worksheet

1. **What type of formal organization characterizes your company?**
   Functional form
2. **What type of formal organization characterizes your own division? (skip for smaller companies)**
   Functional
3. **What benefits do you perceive from your company's (or division's) formal organization?**
   Synergy within functions
   Economical use of talent
   Ease of performance evaluation/control within departments
4. **What drawbacks do you see created by your company's formal organization?**
   Goals are not product or customer driven
   Difficult coordination across functions for product development
   Design and Tooling are in separate departments, reporting in turn to different mid-level managers. Poor coordination leads to inefficiencies and tooling that needs to be reworked.

Exhibit 11.7 • Organization Architecture Worksheet for Coseat Inc.

PROJECT MANAGEMENT APPROACH

Based on the discussion in Chapter 4, determine the form of project management most common in your organization. In Exhibit 11.8, Ed Cazotzky has used the Project Management Worksheet (see Table 4.3 for additional explanation) to show which project management approach is used at Coseat Inc.

Next, Ed filled out the "your priorities" column in the "Engineering Priorities Worksheet."[5] *Ed is using it to rate his organization's priorities. He rates on a scale from low to high how important each will be in his new concurrent engineering approach. This requires jumping ahead from the as-is and thinking about the to-be goals. Ed looks at each engineering priority, and in each case believes it rates very high for the priority of his organization. Looking across the columns he sees that the best project management structure to address each of these priorities lies between "heavyweight" and "autonomous." Each has advantages and disadvantages which Ed will have to consider in the design phase. Neither*

[5] You may recognize this form from Chapter 4 where it was called the "Strengths and Weaknesses of Project Management Structures" and used to describe how each project management structure satisfied a set of engineering priorities.

Project Management Responsibilities	**Absent**	**Liaison**	**Light-weight**	**Heavy-weight**	**Autonomous**	**Your Company**
Distribute and share technical information among project members and facilitate problem solving		X	X	X	X	*yes*
Distribute reports, minutes of meetings		X	X	X	X	*yes*
Set project goals			X	X	X	*yes*
Schedule and coordinate project activity			X	X	X	*yes*
Allocate funds and equipment for project			X	X	X	*yes*
Select staff for project (or significant influence)				X	X	*no*
Evaluate performance of project members (or significant influence)				X	X	*no*
Evaluate overall performance of project members					X	*no*
Long term professional development of project members					X	*no*

Exhibit 11.8 • As-Is Project Management Worksheet for Coseat Inc.

is strong enough on "synergy within functions" to satisfy Ed, so it is clear the project management structure will need to be supplemented with other coordination mechanisms to get people within functions to coordinate.

Engineering Priorities	Absent	Liaison	Light-weight	Heavy-weight	Autonomous	Your Priorities
Synergy within functions	High	High	High	Moderate	Low	*High*
Synergy across functions	Low	Low	Moderate	High	High	*High*
Achieve customer goals	Low	Low	Moderate	High	High	*High*
Achieve company goals	Low	Low	High	High	Moderate	*High*
Responsiveness and speed of development	Low	Low	Low	Moderate	High	*High*

Exhibit 11.9 • Engineering Priorities Worksheet for Coseat Inc.

Finally, Ed summarizes his observations on the Project Management Comments Form (Exhibit 11.10).

Project Management Comments Form

1. **What Project Management approach does your company mostly use (Exhibit 11.8)?**
   Lightweight
2. **How well does that approach fit with the items in the Engineering Priorities Worksheet (Exhibit 11.9)?**
   The lightweight project manager is clearly inadequate to address our engineering priorities for synergy across functions, getting people aligned to meet company goals, and the responsiveness and speed of product development. We need a heavyweight or autonomous project management structure. To maintain the synergy within functions that we need the new project management structure will have to be supplemented with other coordination mechanisms.
3. **What problems do you currently have meeting customer needs?**
   We are often late and fail to meet exact specs.
4. **What problems do you currently have regarding cross-functional conflict and/or missed opportunities?**
   We still have a "throw it over the wall" mentality in design engineering and in marketing.
5. **What seems to work well in terms of project management?**
   We have very skilled, respected people as project managers.

Exhibit 11.10 • Project Management Comments for Coseat Inc.

Task Responsibilities

The purpose here is to identify how different functional groups in your organization are involved in the main activities you described in the work process model. In other words, we want to attach the "who" to these different activities. Knowing the "who's" will help us in the next step of the process to examine how different functions communicate and coordinate to accomplish an activity.

Functional groups are generally groups of people with similar skill sets such as engineering analysis, electrical engineering, mechanical engineering, and tool making. These functional groups may or may not actually be grouped together organizationally in departments. It is possible that one department will actually serve multiple functions, such as design, engineering analysis, and manufacturing engineering. Nonetheless, design, engineering analysis, and manufacturing engineering are three distinct functions that require different skills and expertise. Therefore, they should be treated as three functions for this analysis. It is also possible that one individual will be the sole engineering analyst and thus be the "functional group" for engineering analysis.

There are many ways to describe how functional groups are involved in activities. We have identified the following four different types of involvement (sometimes identified by the acronym *RASI*).

1. ***Responsibility.*** The group has major responsibility for the conduct of that activity. They may delegate some of the task to others or actually do it themselves — in either case, they are responsible.
2. ***Approval.*** The group has authority to approve or not approve key decisions in this activity. They are decision makers.
3. ***Support.*** The group provides important support for this activity. This support might be in the form of providing information, providing resources, or performing specific delegated tasks that are part of the activity.
4. ***Informed.*** The group is informed about the progress or outcomes of the activity.

The task here is to identify how different functional groups are involved in your key design-related activities. We have developed a tool to help you do this: the Functional Involvement Matrix.

Functional Involvement Matrix. The Functional Involvement Matrix (see Exhibit 11.11a, for an example) classifies the involvement and responsibilities of functional groups for different activities. You should begin with the activities from your high level analysis of your work process. Determine each function's responsibility for each activity using the RASI scheme. After documenting who is responsible, you should make a preliminary assessment of your strengths and weaknesses. For example, are the right people being informed? Are the people with the most knowledge of the product sufficiently involved in decisions about the product? If one function is responsible and the rest are informed, that suggests the one function is making decisions alone and then passing on completed decisions to the others. Is this appropriate, or should multiple functions be more involved in making the decisions up front?

In order to complete the Functional Involvement Matrix, Ed Cazotzky assembles the same team he used to develop the process map. Although all functions in the company are not represented on the team, there are team members who know enough about the unrepresented functions to do the analysis. Ed facilitates the two hour meeting at which the team completes the matrix. He starts things off by asking the question, "What are the primary functions in our company involved in the product development process?" He then shows the team the process map and leads a discussion in which the team selects steps to include in the matrix. Finally, he moves across the rows cell by cell, asking, "Is this function involved, and if so, at what level?"

Note that Ed could have done a fairly elaborate study to collect this kind of information. For example, he could have done a survey of all managers or of everyone involved in the product development process. We believe this would have been unnecessary. Rather, we suggest you take a "quick and dirty" approach. In general, a "rough cut" is all you need for the kind of planning we suggest in this book. At times you may find you have a need for much more detailed analysis in a narrowly focused area — by all means do so if you have the time and resources.

The interpretation of the matrix in Exhibit 11.11a is quite straightforward. We can see that Marketing and Sales, for example, has authority over the concept and is responsible for getting the market reaction jointly with program management. After that, they completely drop out of the process. In contrast, design engineering is responsible for developing the concept jointly with the program manager. But they are only informed about the market reaction, after which they design the prototype, have authority over the build and test of the prototype, and have responsibility for the final design. They drop out after that except for being informed about launch. We can see from this a series of real "over the wall" hand-offs. Just what should be done about this is not at all clear at this point, however, nor should we expect it to be without the full set of evidence. Ed Cazotzky has made his comments in the Functional Involvement Comments Form (Exhibit 11.11b).

		Project Management	Marketing and Sales	Design Engineering	Prototyping	Test Engineering	Tool Engineering	Tool Production	Manuf. Engineering	Purchasing	Suppliers
DESIGN SEATING SYSTEM	Develop Concept	A	A	R							
	Get Market Reaction	R	R	I						I	
	Design Prototype	A		R	I						
	Build/Test Prototype	A		A	R						
	Final Design	A		R			I	I	I	I	I
DESIGN & BUILD TOOLING	Design Tooling						R	I	I	I	I
	Make Tooling							R	I		
INITIATE PRODUCTION	Install Tooling/Eqiupment	I							R	S	R
	Purchase Parts & Tooling	I								R	S
	Production Launch	I		I			I		R	S	S

KEY: R Responsible; A Approval; S Support; I Informed

Exhibit 11.11a • As-Is Functional Involvement Matrix for Coseat Inc.

When using a tool like the Functional Involvement Matrix, it is good practice to keep notes about the entries in each cell. Without these notes, an entry such as "Support" in a cell is much less meaningful several weeks later when the group is trying to interpret the markings on the matrix. There are several good ways to record such notes. Probably the best is to record minutes of meetings with notes included in the minutes as an appendix. The notes might also be recorded on a flip chart during the meeting and filed for later use. A more "high-tech" approach would be to record the notes as part of a computer spreadsheet version of the matrix. We have found it very useful to use a computer projection version of the matrix as a tool during group discussions. You could record notes off to the side of the matrix within the spreadsheet. Alternatively,

Functional Involvement Comments Form

1. **List any cases in which people who should be involved are currently not in the loop** (consider blank cells in Functional Involvement Matrix).
   Project management should be involved in the tooling phase and are currently out of the loop.

   Marketing & Sales drop out after market reaction and should be supporting prototype development and approving the final design. They should then be informed of any changes to the final design in the tooling and production stages.

   Design engineering does not have sufficient involvement after final design and should be involved in some way consistently through launch.

   Test engineering has no direct involvement in the process and should be supporting prototype design, build, and test.

   Tool engineering and manufacturing engineering should have input earlier in the process in the prototype stage to give their input on manufacturability. Tool engineering should also support installation of tooling and equipment and support production launch.

   Purchasing and suppliers are involved too late in the process. Suppliers involved in product development should be brought in earlier depending on their roles in product development and purchasing should be involved in presourcing agreements.

2. **Consider obvious cases where a function does not have the level of involvement warranted by their expertise** (consider the cells in Functional Involvement Matrix that have some letter in them).
   Project managers do not have sufficient responsibility (given that we are considering moving to a heavyweight or autonomous project manager role).

   Design engineers should not simply be informed about production launch but should have some responsibility.

   Manufacturing engineers should have a stronger role in final design and tooling.
   Some suppliers should have a stronger role in final design, tooling design and production launch.

Exhibit 11.11b • Functional Involvement Comments for Coseat Inc.

some programs (such as MS Excel®) provide a "notes" function associated with each cell. We have found Excel's version of this to be awkward to use in practice, although later versions of the software might make this feature more practical.

Communication Patterns

The purpose of this activity is to document your current patterns of communication across functional groups. These are the same groups used in the Functional Involvement analysis. The focus here is on communication related to your product development across these functions.

The primary questions that are traditionally defined about communication are who, what, how, how often, and to what purpose? The *who* in this case is defined as the functions. The *what* are the activities (e.g., design, analysis, etc.) that the functions perform. The *how* are the mechanisms used for communication. In this section we will focus on the *patterns* of communication, which includes issues related to who, how, and how often. We will address the other communication questions in the next section under "coordination."[6]

Communication patterns have three critical dimensions: direction, synchronicity, and frequency.

- ***Direction***. Direction can be communication flowing in one direction only; from party A to party B, or B to A; or flowing in both directions.
- ***Synchronicity***. Synchronous communication is almost immediate, two-way "give and take" across parties. Asynchronous is when a significant lag exists from the time a communication is sent until a response is received (e.g., design engineering sends purchasing a request for quotation and several days later purchasing sends back the quotation).
- ***Frequency***. There may be frequent communication between functions, or there may be very little.

The amount and type of communication that is necessary depends on the activities that need to be coordinated across groups. If several groups are all involved in an activity, they will need to communicate about it. If it is a complex activity that requires a lot of intense joint problem solving, then two-way synchronous communication is needed. If one group can do the job by themselves and they simply need to keep other groups informed, then one-way communication of low frequency may be sufficient. After documenting patterns of communication, you should make a preliminary assessment of your strengths and weaknesses in communications. For example, is information flowing directly to the point of action? Do groups get the information resources they need to perform their core tasks? Are the people with useful knowledge for the task providing that knowledge at early stages of decision making?

We have developed a Communication Matrix for analyzing communication patterns across functions within your organization. Exhibit 11.12 shows the As-Is Communication Matrix filled out for Coseat Inc.[7]

In Exhibit 11.12, for each pair of organizational units, each cell indicates whether *product-related communication* is primarily one-way or two-

[6] The difference between communication and coordination is too subtle to worry about here. In our coordination section, we will focus more on the questions of how and to what purpose than we do in this section.

[7] The processes for completing the Communication Matrix and the Coordination Matrix in the next section are so similar to that for completing the Functional Involvement Matrix that we haven't included scenarios to tell the story about how Ed Cazotzky and his team completed them. You can see the results of their efforts in the completed exhibits.

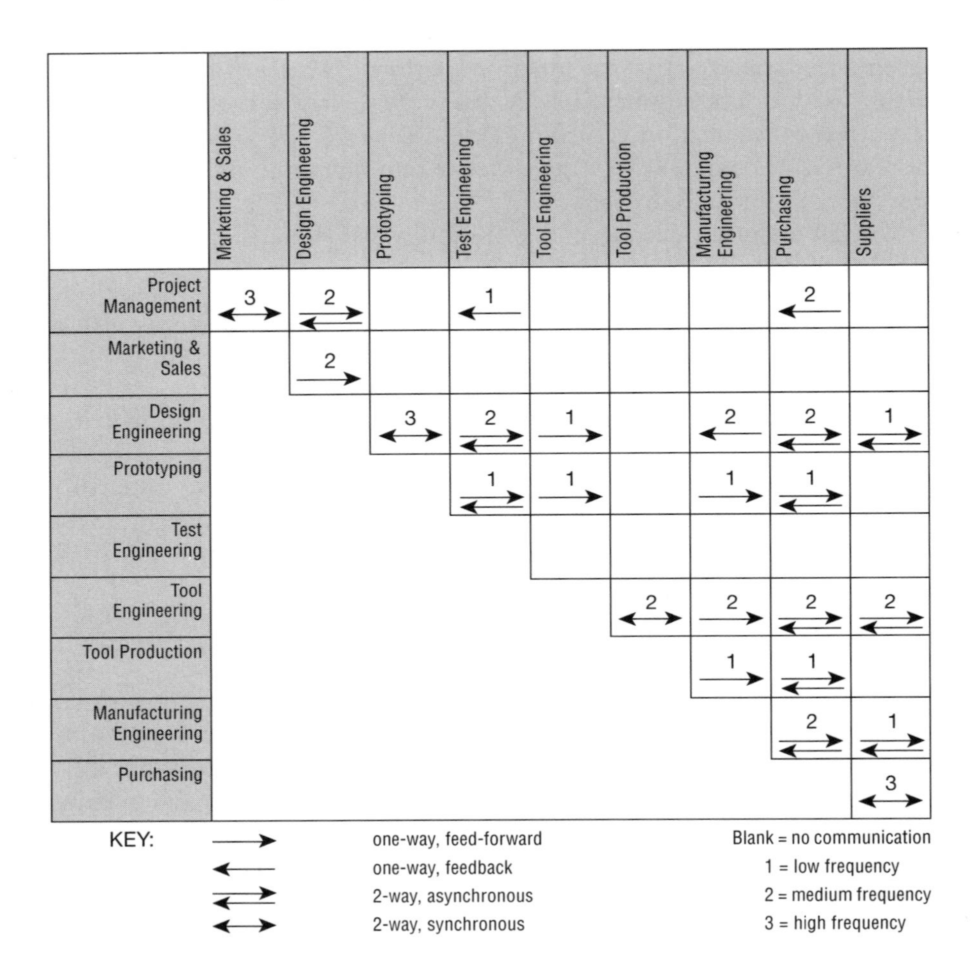

Exhibit 11.12 • As-Is Communication Matrix for Coseat Inc.

way. Note that some information about the direction of communication can be obtained by inference from the As-Is Process models and the As-Is Functional Involvement Matrix.

We can see that design engineering communicates frequently and reciprocally with the prototype shop. Tool engineering has a moderate level of reciprocal communication with tool making. Purchasing communicates very frequently and reciprocally with suppliers. However, most other links are either one-way/low frequency, completely blank, or asynchronous. For example, marketing and sales communicates one-way with design engineering; purchasing communicates asynchronously with most functions other than suppliers. Information is received and sent in a bureaucratic way with little joint decision making and problem solving. Note also the absence of communication from design engineering to manufacturing engineering. Manufacturing engineering sends communications to design engineering, but only when they have problems with the design — in a reactive mode. The result of this kind of throw-it-over-the-wall communication pattern is likely to be stilted communication, little joint problem solving, and problems which will not be detected until production has already started. Coseat mostly has reactive fire-fighting rather than proactive problem solving.[8]

Cross-Functional Coordination

In this step you document the *methods* used to coordinate between different functions, as well as the *quality* of coordination they provide. In Chapter 4 we described five Cross-Functional Coordination mechanisms (mutual adjustment; direct supervision; and standardization of work process, outputs, and worker skills). The Cross-Functional Coordination Matrix is used to document both the types of mechanisms used and the quality of coordination they provide.

Exhibit 11.13 shows an example of the As-Is Functional Coordination Matrix for Coseat Inc. Each cell shows the mechanisms used for coordination between two functions, as well as a number (from 1 to 3) indicating the quality of coordination. Since the matrix is symmetrical, only the non-redundant cells are filled in. Exhibit 11.13 shows that most coordination is done with or by program managers and design engineers. Much of the coordination that takes place with the label "Mutual Adjustment" is in fact only ad hoc meetings which take place as needed. If we look at the cross between design engineering and marketing and sales (which we examined earlier under communication and functional involvement), we can see that only "Standardization of Output" is used for coordination — which effectively means the market researchers send a standard report to the design engineers. The number given for quality of coordination should be based on this scale:

- ***1*** = inadequate coordination - *frequent and significant problems arise due to failure to coordinate*
- ***2*** = barely adequate coordination - *occasional problems arise due to failure to coordinate; these are usually not significant*
- ***3*** = very active and useful coordination - *few problems arise from failure to coordinate and these are almost never significant*

[8] We have not reproduced these comments in an exhibit with an "As-is Communication Comments Form." It should be essentially the same format as the "Functional Involvement Comments Form" (Exhibit 11.11b).

Functional Coordination Matrix	Project Management	Marketing & Sales	Design Engineering	Prototyping	Test Engineering	Tool Engineering	Tool Production	Manufacturing Engineering	Purchasing	Suppliers
Project Management	2 DS, Plans	3 MA, O	3 MA, DS, S	3 DS, S	2 DS, S	3 DS, S	3 DS	3 DS	3 WP	2 O, MA
Marketing & Sales			1 O							
Design Engineering			2 O, S, MA	3 MA, WP	3 MA, WP, O	2 MA, WP, O	1 MA, WP, O	2 MA	2 MA, WP	2 MA, O
Prototyping					3 MA, WP			2 MA		
Test Engineering						2 O		1 O		
Tool Engineering							3 MA, WP	2 MA		
Tool Production								2 O		
Manufacturing Engineering										2 MA, O
Purchasing										2 MA, O, WP
Suppliers										

KEY:
MA = Mutual Adjustment
DS = Direct Supervision
WP = Standardization of Work Process
O = Standardization of Outputs
S = Standardization of Worker Skills

1 = Inadequate Coordination
2 = Barely Adequate Coordination
3 = Very Active And Useful Coordination

Exhibit 11.13 • As-Is Cross-Functional Coordination Matrix for Coseat Inc.

CULTURE

In some ways, the Cultural Assessment cuts across all five of the high level CE Elements. That might suggest that it be done last, after you've assessed all of those elements. On the other hand, we also described it as part of the Internal Organization in Chapter 4. So, structurally we've placed the Cultural Assessment here with the other parts of the Internal Organization assessment. Nonetheless, you can do this assessment at any point in the process.

In the Cultural Assessment, we ask that you (and your team) learn about and describe artifacts, values, and assumptions in each of the five CE Elements. This is not an easy thing to do, especially when you are embedded in the culture you're trying to observe. The problem of course is that the "assumptions" part of the assessment is particularly difficult for an internal observer to perceive objectively. Some amount of "psychological distance" is probably needed to do a really good Cultural Assessment. If you use only internal staff, at best you will only be able to perform an incomplete and perhaps superficial assessment. Indeed, if you can afford to do so, you might well hire an external organizational consultant (e.g., an anthropologist or organizational psychologist) to conduct the Cultural Assessment for you.

Having said that, you can still learn a great deal about your culture by doing the assessment yourself. Even if it is incomplete from a scientific point of view, if you learn a number of useful things about your organization, then it is worth doing. We will provide you with a simple method for conducting a Cultural Assessment that can be done quickly and relatively painlessly. As long as you recognize up front that it is less complete and less objective than such an assessment done by a qualified outsider, you need not feel too defensive about the results.

The basic approach we recommend is to form a cross-functional team to observe and discuss the five CE Elements in terms of artifacts, values, and assumptions. You will then draw conclusions about your culture across the three different levels of culture. For this purpose, a cross-functional team is essential. This is because it will allow people from the different functions to have some objectivity in observing each others' functional area. You can, if you wish, just take the team into a room and fill out the Cultural Assessment Matrix (as in Exhibit 11.14a). However, you will do a much better job at this if you try the following suggestions.

- ***Collect Artifacts.*** As part of your assessment in all of the other areas, you will be collecting a lot of information, such as process maps, organization charts, human resource policies, etc. These are all artifacts. You already have them, so use them as part of your cultural assessment. One of the nice things about this is that there's only a little extra work involved.
- ***Observe Artifacts.*** As you go around collecting the artifacts we referred to above, you will see things, such as how offices are arranged, what kinds of things are displayed on the walls, and who comes to meetings. Write these observations down.

- ***Collect Evidence of Value Statements.*** Most companies have plenty of statements of what they value and what the people in them value. Some companies have these things framed and put up on the wall, most have them embedded in formal policy statements (e.g., a "quality" policy). There are also value statements embedded in other kinds of documents and in things people say. Record some of these statements — whether you see them in a document or hear them in a meeting.
- ***Talk With People About Their Assumptions.*** Sitting around over coffee or a beer, ask people what they believe about how the world works. You don't have to do something like this as a formal interview, nor do you have to ask some ridiculous question (e.g., "Hey Fred, what assumptions do you make about what motivates people?"). We find the most useful question to ask to get at assumptions is, "Why do you think that?" Another similar question is, "What makes you think _____ works that way?" Asked in a friendly, inquisitive manner, these kinds of questions aren't threatening and are often perceived as complimentary to the person you're asking. Do this if you're comfortable with it; otherwise you may get some strange stares. These kinds of discussions can be held with individuals or with groups, as formal interviews or not.
- ***Use Artifacts and Values to Infer Assumptions.*** Look at both the artifacts you collect (and observe) and the values you learn about, and try to make some guesses about what assumptions underlie these things. For example, if you see design engineers with window offices and reserved parking spaces, and manufacturing engineers with cubicles and a mile walk to their cars, then you can make some pretty good guesses about the importance of status in the company as well as the difference in status between the two functions.
- ***Schedule Informal Time for Discussion.*** Allocate a certain amount of time in your team to discuss cultural issues. In particular, try to explore the differences in assumptions between individuals on your team in terms of the assumptions they make and the values they hold.

Ed Cazotzky and his team at Coseat Inc. get together to discuss how to do their Cultural Assessment. They divide their task into pieces. Their first task is to fill in the first three columns of the Cultural Assessment Matrix — artifacts, values, and assumptions. First they look at what they've already got in each cell of the matrix (Exhibit 11.14a). In terms of artifacts and formal values, they've got quite a bit in most of the

CE Elements	Artifacts	Values	Assumptions	Cultural Conclusions
Work Process	*no evidence of a formal process; complaints about how information never seems to get where it's supposed to*	*many statements about the importance of the customer and meeting their needs; "Quality is the most important thing we do" prominently displayed*	*people should work independently and not have to ask others for help; smart people are the main reason for high quality*	*People may not believe that teams are a good idea or that team work is better than individual efforts.* *May resist efforts to introduce formal processes.*
Internal Organization	*strong sense of obeying the chain of command; physical layout is very orderly; each function has its own suite of offices separated from others by locked doors; meetings are very formal, quiet*	*many statements from top management about how this is a team-based company.*	*everything must be controlled from the top, providing individuals with the least amount of autonomy possible*	*Very authoritarian, may resist participative, team based activity, despite statement from management.*
Supplier Relations	*suppliers located quite a distance away; most meetings take place here; suppliers are very deferential to buyers*	*formal policy statement about how suppliers should be working with us as "partners"*	*suppliers should do what they're told*	*Supplier relations are very top-down; resistance should be expected from buyers to idea of suppliers participating closely with design teams.*
People	*people tend to ignore job descriptions to do whatever is necessary; training certificates, diplomas proudly displayed in offices; individual behavior is rewarded, no group rewards*	*"People are our most important asset" prominently displayed*	*design engineers are higher status than almost anyone; Coseat has better, more qualified people than competitors*	*"Can-do" attitude should help with change in the long run. Emphasis on training will help. Status differences will need to be broken down before team-based culture can be adopted.*
Technology	*workstations on engineers' desks; access to engineering applications either local or over network*	*several engineers stated that if they only had the latest computer software, all their problems would be solved*	*new technology will solve all problems; "we don't need to worry about that 'soft stuff'"*	*Technology per se will not be a problem, but engineer's attitudes to soft issues will be a source of resistance to change.*

Exhibit 11.14a • As-Is Cultural Assessment Matrix for Coseat Inc.

categories. But they also know there are lots of things they should start observing. Ed asks each member of the team to observe behavior at the various meetings they go to during the next two weeks. He also asks them to look around their buildings and make some notes about what seems to characterize the people and the way they interact.

Two weeks later, at their next meeting to discuss the Cultural Assessment, the members of the team write their observations on note cards which they attach to a storyboard with push pins. The team then does an affinity exercise in which the cards get moved around to group them into categories. These categories are then given names which are placed in the Cultural Assessment Matrix where the team agrees they belong. The next day the team meets again to discuss what to place in the cultural conclusions column of the matrix. They look across each row and discuss what seem to be common themes.

The Cultural Assessment yields several issues pertaining to the ability of the company to change. This leads Ed and the team to reconsider which areas they want to emphasize, and whether they have enough top level support for their change effort.

Assessing Supplier Relations

Your company's relations with its suppliers are crucial to its ability to make new products. Most large manufacturing companies have chains of suppliers which provide parts and components that make up the majority of the product. Even smaller manufacturing companies rely on suppliers to provide raw material, parts, and services to help them. Thus, any manufacturing firm's ability to integrate their product development processes requires an understanding of how the suppliers fit into that process, and what the suppliers can do to make it work more efficiently and effectively.

In order to understand your current supplier relations, and to decide how to improve supplier contributions to your overall development process, you need to understand where you are today as a baseline. This assessment will cover five steps: a description of your suppliers, the activities suppliers are responsible for, how you communicate with suppliers, the roles suppliers play in your product development process, and your mechanisms used to manage suppliers. This assessment is not quite as simple as it might seem since you probably don't have the same relationship with every supplier, nor does every supplier matter as much as every other one. We have ordered the five steps to aid the process of data collection and analysis (Figure 11.6).

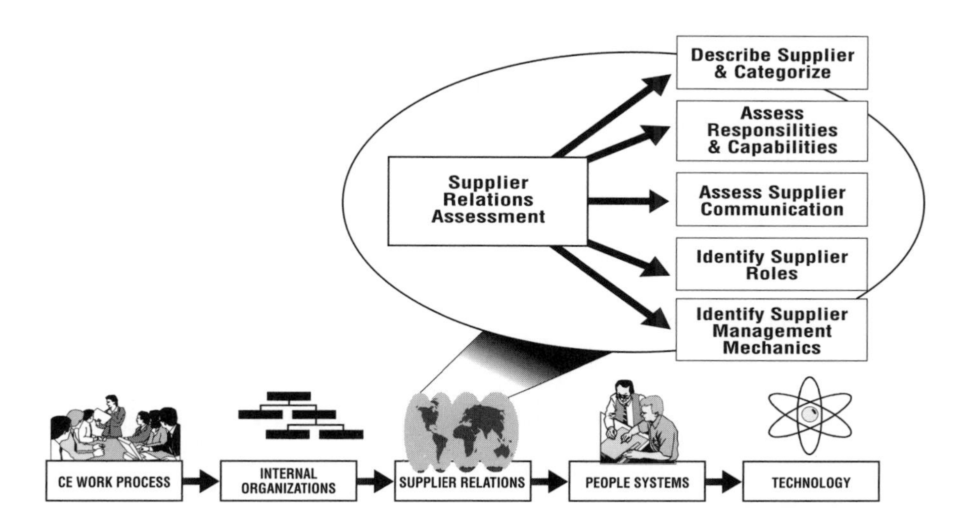

Figure 11.6 • Supplier Relations Assessment

1. ***Describe Suppliers and Categorize***. Assemble information about your suppliers and divide them into categories by their primary function in helping you to make your product.
2. ***Assess Responsibilities and Capabilities***. Determine which activities each supplier is involved in, what they are responsible for in those activities, and how capable they are. Assess if the mix of responsibilities and capabilities is appropriate.
3. ***Assess Supplier Communication***. Determine how suppliers communicate with you for each activity they are involved in. Assess if the communication is appropriate.
4. ***Identify Supplier Roles***. Categorize each supplier according to the role it takes in product development with your company.
5. ***Identify Supplier Management Mechanisms***. Describe the types of mechanisms used to manage suppliers to ensure they are able to perform according to their role.

Describe Suppliers and Categorize

Make an organized list of suppliers you want to assess. This first step in the process makes it easier to do the rest. The first decision you need to make is how many supplier relationships to assess. If you have relatively few suppliers (e.g., 10–15), then it probably makes sense to simply assess all of the relationships. On the other hand, if you have a large number of suppliers, it will be necessary to restrict your analysis in some way. We can recommend three ways to do this, in order of preference.

1. Restrict your analysis to those suppliers who have the greatest impact on your product development process — most likely those suppliers who have the greatest unique value added to your product. This suggests you might exclude commodity suppliers. The analysis process we suggest here works best if done with not more than about 20 suppliers. It can be done with more, but it rapidly gets large and time consuming.
2. Restrict your analysis to a particular part or component that you produce. This cuts the complexity of the task tremendously. The downside is that it may restrict the scope of impact you can have with this activity. It can be combined if necessary with the first recommendation.
3. Restrict your analysis to categories of suppliers. As you will see shortly, we recommend you categorize your suppliers by the type of work they do for you. It is possible to conduct the analysis we describe only within categories. This method of restriction will be less useful than analyzing all categories of suppliers, but will still be helpful.

Describe the Suppliers. Make a list of your suppliers with their significant characteristics. Examples of significant characteristics include:

- nature of work they do for you
- type of contractual relationship
- length of relationship
- value added by supplier
- quality rating
- supplier's technology level
- product development capability.

You can add to this list as you wish, but make sure your additions will be needed later. We have made the list short in order to keep the task manageable. All of the characteristics we list will be used in some way further on in your analysis. In the Coseat Inc. example (Exhibit 11.14b),[8] we have attempted to keep the descriptions of the characteristics as short as possible, so that as much information as possible about each supplier can be viewed at once. A useful way to record all of this information is in a table such as was done for Coseat Inc.

[8] A typical seating system supplier in the automotive industry will have many more suppliers than we have shown in this example. The example is kept to nine suppliers to save space and to make for easier reading.

Categorize the Suppliers. The analysis process we recommend for companies with a lot of suppliers requires that you categorize your suppliers by the kind of work they do for you. For example, you could "clump" together tooling suppliers, small part suppliers, stamping suppliers, molding suppliers, etc. You may prefer to categorize by part or component type rather than process, or you may prefer to do both. The categorization used by Coseat Inc. can be seen in Exhibit 11.14b.

Supplier	Category	Kind of Work	Contract Type	Longevity	Value Added	Quality Rating	Tech Level	Prod Dev Capability
Joe's Tool & Die	tooling	stamping dies	project	10 years	20% of product cost	A+	very modern (CAD/ CAM)	substantial
Fred's Molds	tooling	plastic injection molds	project	6 years	20%	A+	very modern (CAD/ CAM)	very little, has potential
Ann Arbor Spring & Wire	specialty spring	fab seat springs	5 year contract for seat X	15 years	5%	A+	modern (some auto)	very little, has potential
Marshall Wire	specialty spring	fab seat springs	5 year contract for seat Z	6 years	5 %	B	ancient	none, little potential
Burlington	fabric	make fabric	5 year contract for seats X, Z	16 years	8%	A+	very modern	currently involved
Franklin Stamping	make parts	stamps brackets	2 year contract for seat Z	8 years	<1 %	B+	modern	none, has potential
Henry's Molding	make parts	injection molds panels	3 year contract for seats X, Z	4 years	1%	B	modern	none, has potential
Mitch's Switches	make components	mfg seat adjust switches	1 year contract for seats X, Z	2 years	4%	A	varies	does much of current design
Acme Frame	make components	fab seat frames	5 year contract for seat X	6 years	6%	B	ancient	does own tooling, has potential

Exhibit 11.14b • Supplier Description for Coseat Inc.

ASSESS RESPONSIBILITIES AND CAPABILITIES

In the previous section (Describe Suppliers and Categorize), you briefly summarized the kind of work done by each supplier (e.g., make tooling, heat treating). In this next task you will show how each supplier is involved in the activities described in your work process. You do this in much the same way you earlier assessed the responsibilities of different functions, but this time using the Supplier Responsibility Matrix. An alternative to using the matrix might be to simply list the activities each supplier is involved in. We believe the matrix is more useful in the long run since it permits you to make comparisons among suppliers.

The Supplier Responsibility Matrix is similar to the Functional Involvement Matrix in that you use it to document responsibility (using the RASI scheme) for the activities you described in the work process. However, it may be necessary to carry the activities you document down to an additional level of detail. For example, in the Coseat Inc. As-Is Functional Involvement Matrix (Exhibit 11.11a), we showed an activity called "Final Design." In the Coseat Inc. As-Is Supplier Responsibility Matrix (Exhibit 11.15), we found it necessary to take that a level deeper into "Final Seat Design" and "Final Part or Component Design" to show the difference in responsibility between design of the whole product (systems engineering) and design of a part or component.

The Coseat Inc. As-Is Supplier Responsibility Matrix (Exhibit 11.15) shows the patterns of responsibility for our example company. The example shows a traditionally managed relationship which probably does

KEY

R = Takes Responsibility
A = Authority
S = Provides A Support or Resources
I = Is Informed

	Supplier's Primary Product Service or Function								
	Tooling		Specialty Springs		Fabric	Make Parts		Components	
	Joe's Tool and Die	Fred's Molds	Ann Arbor Spring & Wire	Marshall Wire	Burlington	Franklin Stamping	Henry's Molding	Mitch's Switches	Acme Frame
Develop Seat Concept									
Get Market Reaction									
Identify Part/Component Requirements			I		I				
Develop Part/Component Concept			S	I	I			I	
Design Seat Prototype									
Design Part/Component Prototype			R	S	R			S	
Build and Test Part/Component Prototype			R	R	R			R	
Build and Test Seat Prototype			S		S			S	
Final Seat Design									
Final Part/Component Design	I	I	R	S	R	I	I	S	I
Design Tooling/Equipment	R	S							
Make Tooling/Equipment	R	R							
Install Tooling/Equipment									
Purchase Parts and Material (Select Suppliers)									
Production Launch									

Exhibit 11.15 • As-Is Supplier Responsibility Matrix for Coseat Inc.

not involve Concurrent Engineering. Coseat Inc. designs the seat system, as well as many of the parts and components. For the most part, suppliers only get involved rather late in the life cycle, as parts move into prototyping. For example, Ann Arbor Spring & Wire gets involved in part concept development only at the level of providing support, and becomes responsible only when their part is actually being designed. The stamping supplier gets involved (other than receiving information) only when the tooling is delivered to their door. The switch manufacturer (Mitch's Switches), in contrast, has a much earlier involvement, providing support to component prototyping as well as to testing the seat prototype. After you complete this matrix, you should record your observations about suppliers who are under- or over-involved.

Assess Supplier Communication

Communication is the major means by which you coordinate your activities with those of your suppliers. In a concurrent engineering environment, communication with suppliers is even more critical than usual since the linkage between activities is so much tighter. You can assess communication with your suppliers in much the same way you assessed communication inside your firm. The only difference is that all the communication of concern is between you and the suppliers — internally we were concerned with communication between the various functions. We recommend that you assess communication according to the dimensions of direction, synchronicity, and intensity as described above.

After documenting patterns of communication, you should make a preliminary assessment of your strengths and weaknesses in communications. For example, is information flowing directly to the point of action? Do suppliers get the information resources they need to perform their core tasks? Are the people with useful knowledge for the task providing that knowledge at early stages of decision making?

The Supplier Communication Matrix is used to assess communication with your suppliers. Complete it much as you did the Communication Matrix in Exhibit 11.12. Exhibit 11.16 shows an example of the As-Is Supplier Communication Matrix as completed by Coseat Inc. Like the As-Is Supplier Responsibility Matrix, the example shows a set of traditionally managed relationships which probably do not involve concurrent engineering. Coseat designs the seat system, as well as many of the parts and components. For the most part, suppliers only get involved rather late in the life cycle, as parts move into prototyping. This is reflected in their communication. For example, there is little communication with either spring maker in part concept development, none with the parts makers, and only one of the component makers. Even in final part design, communication is infrequent and either one way or asynchronous.

KEY: ↔ 2-WAY SYNCHRONOUS; → ONE WAY FROM COSEAT; ⇄ 2-WAY ASYNCHRONOUS; ← ONE WAY TO COSEAT. 1 = Infrequent Communication; 2 = Occasional Communication; 3 = Frequent Communication	Supplier's Primary Product Service or Function								
	Tooling		Specialty Springs		Fabric	Make Parts		Components	
	Joe's Tool and Die	Fred's Molds	Ann Arbor Spring & Wire	Marshall Wire	Burlington	Franklin Stamping	Henry's Molding	Mitch's Switches	Acme Frame
Develop Seat Concept									
Get Market Reaction									
Identify Part/Component Requirements			1 →		1 →				
Develop Part/Component Concept			1 ↔	1 →	1 →			1 →	
Design Seat Prototype									
Design Part/Component Prototype			1 ⇄	1 ⇄	1 ⇄			2 ⇄	
Build and Test Part/Component Prototype			2 ←	2 ←	2 ←			2 ←	
Build and Test Seat Prototype			1 ⇄		1 ⇄			1 ⇄	
Final Seat Design									
Final Part/Component Design	1 →	1 →	1 ←	1 ⇄	1 ←	1 →	1 →	1 ⇄	1 →
Design Tooling/Equipment	2 ↔	2 ⇄							
Make Tooling/Equipment	2 ↔	2 ↔							
Install Tooling/Equipment									
Purchase Parts and Material (Select Suppliers)									
Production Launch			2 ⇄	2 ⇄	1 ⇄	2 ⇄	2 ⇄	2 ⇄	2 ⇄

Exhibit 11.16 • As-Is Supplier Communication Matrix for Coseat Inc.

IDENTIFY SUPPLIER ROLES

Now you assess the overall roles of your suppliers in product development. The responsibility matrix and communication matrix provided information about particular aspects of the supplier relationship. The purpose of this activity is to provide a more global assessment of the roles your suppliers play. You will classify the role of each of your key suppliers and then compare them to a description of an ideal role based on observations of best practice. We recommend that you base this analysis on the categories we developed in Chapter 5: contractual, consultative, mature, and partner.

The Supplier Roles Matrix is a tool for analyzing these roles. Rate each supplier along the dimensions of Supplier Involvement in Product Development which were described in Chapter 5. Based on an average of these scores, you can see which of the four roles most closely fits each supplier as they are today. An example of the matrix can be found in Exhibit 11.17. Rate each dimension on a four-point scale. The anchor points for the four-point scale are in the matrix. For example, "component design responsibility" could be all inside your company if you do the design and then give it to your supplier (score = 1), or it might be all in your supplier if you give them specifications and they go off and do the design on their own and return it for your approval (score = 4). If the design responsibility is mostly yours, score = 2; if it is mostly the supplier's responsibility, score = 3. A few clarifying points follow.

- ***Earliness of supplier involvement.*** In this case, a score of 1 means the supplier is involved in design late in the cycle, e.g., not early. Thus, if they are not brought in until late in the prototyping process, they get a score of 1. If they are brought in even before the concept stage for the total product, to give you ideas for the concept, they get a score of 4.
- ***Supplier's component testing responsibility.*** We are asking the degree to which this is your responsibility as a customer (score 1) or completely your supplier's responsibility (score 4). Of course, you may test the supplier's part as it works in the total product. But if the supplier has complete responsibility for testing the component/subsystem outside of the total product, they get a score of 4.
- ***Supplier development capabilities.*** This refers to the degree to which suppliers have in-house development capabilities. Autonomous means they can do all aspects of development in-house (e.g., concept development, engineering analysis, prototype build, testing, and manufacturing system development).

Compute the average in each column. Then decide which of the roles best fits each supplier. If the supplier is on the boundary between two roles, you should use your judgment as to which role best fits the supplier, or you might note the supplier is on the border, e.g., C/P. You should also note in the matrix any cases of inconsistent patterns for a given supplier by looking down the column. For example, you may find that the supplier is providing a very complex part but has no component design responsibility and high development capabilities.

In Exhibit 11.17 you can see the As-Is Supplier Roles Matrix for Coseat

Inc. The example shows a set of traditionally managed relationships which probably do not involve concurrent engineering. Coseat designs the seat system, as well as many of the parts and components. Only specialty spring suppliers have major design responsibility, and only one of those is handed complete designs. Suppliers ordinarily have no influence on the specifications, again with the exception of the spring suppliers who have a modest amount of influence. Generally suppliers are brought in late in the design process. When the scores are averaged, it is clear that the suppliers are split between contractual and parental relationships. This is at least somewhat justified by the fact that product complexity is relatively low for many of these parts. However, it seems likely that Coseat Inc. is not doing a very good job of exploiting the capabilities of their suppliers. It is also likely that the designers in Coseat Inc., who are removed from the shopfloors of these suppliers, are not doing a good job of designing the products so they can be best manufactured.

	Supplier's Primary Product Service or Function										
	Tooling		Specialty Springs			Fabric	Make Parts			Components	
Key: Assign each characteristic a score on a 1-4 scale. The anchor points are provided in parentheses. For example, assign 1 if the customer is solely responsible and 4 if the supplier is responsible and 2 or 3 if it is somewhere between.	Joe's Tool and Die	Fred's Molds	Ann Arbor Spring & Wire	Springmasters	Marshall Wire	Burlington	Franklin Stamping	Lenawee Stamping	Henry's Molding	Mitch's Switches	Acme Frame
Component Design Characteristic (1 = Customer Only—4 = Supplier Only)	1	1	3	3	1	2	1	1	1	2	1
Product Complexity (1 = Simple Parts—4 = Complete Subsystems)	2	2	2	2	1	2	2	2	2	2	2
Level Of Specifications Provided (1 = Complete Designs—4 = General Concepts)	1	1	2	2	1	2	1	1	1	1	1
Supplier Influence On Specifications (1 = None—4 = Collaborative)	1	1	2	2	1	1	1	1	1	1	1
Earliness Of Supplier Involvement (1 = Late Prototype State—4 = Pre-Concept)	1	1	2	2	1	2	1	1	1	2	1
Supplier's Component Testing Responsibility (1 = None—4 = Complete)	2	2	3	2	2	2	2	2	2	3	2
Supplier Development Capabilities (1 = None—Complete, 4 = Autonomous)	1	1	2	2	1	2	1	1	1	2	1
Average =	1.3	1.3	2.3	2.1	1.1	1.9	1.3	1.3	1.3	1.9	1.3
Overall Relationship (C = Contractual, P = Parental, M = Mature, PP = Partnership)	C	C	P	P	C	P	C	C	C	P	C

Exhibit 11.17 • As-Is Supplier Role Matrix for Coseat Inc.

Identify Supplier Management Mechanisms

Different mechanisms should be used to manage suppliers depending on the role they are playing in your product development process. For example, you should probably only use Early Involvement with suppliers with whom you have a close, trusting relationship, otherwise you wouldn't trust them enough to share proprietary product information. In this activity, you should simply list the supplier management mechanisms you are using with different suppliers. In the Supplier Management Worksheet (Exhibit 11.18), list the mechanisms you use down the side and place a check by the suppliers with whom you're using them. The mechanisms listed in Exhibit 11.18 are defined in Chapter 5.

	Tooling		Specialty Springs			Fabric	Make Parts			Components		
	Joe's Tool and Die	Fred's Molds	Ann Arbor Spring & Wire	Springmasters	Marshall Wire	Burlington	Franklin Stamping	Lenawee Stamping	Henry's Molding	Mitch's Switches	Acme Frame	Comments
Supplier Data Integration			X									
Supplier Partnership												
Preferred Suppliers												
Target Costing												
Supplier Rating												
Early Involvement			X			X						
Supplier Qualification	X	X	X	X	X	X	X	X	X	X	X	All Suppliers Must Be Qualified
Supplier Development												
Long Term Contracting			X			X				X		
Commodity Team												

Exhibit 11.18 • As-Is Supplier Management Mechanisms Matrix for Coseat Inc.

Based on Exhibit 11.18, Coseat Inc. doesn't use many of the established mechanisms with its suppliers. They do qualify all of their suppliers. In addition, they have long-term contracts with a few suppliers, and early involvement with two of those.

People Systems

In Chapter 6 we defined three People Systems elements as having the greatest impact on concurrent engineering. These were job designs, skill acquisition systems, and motivation systems (Figure 11.7).

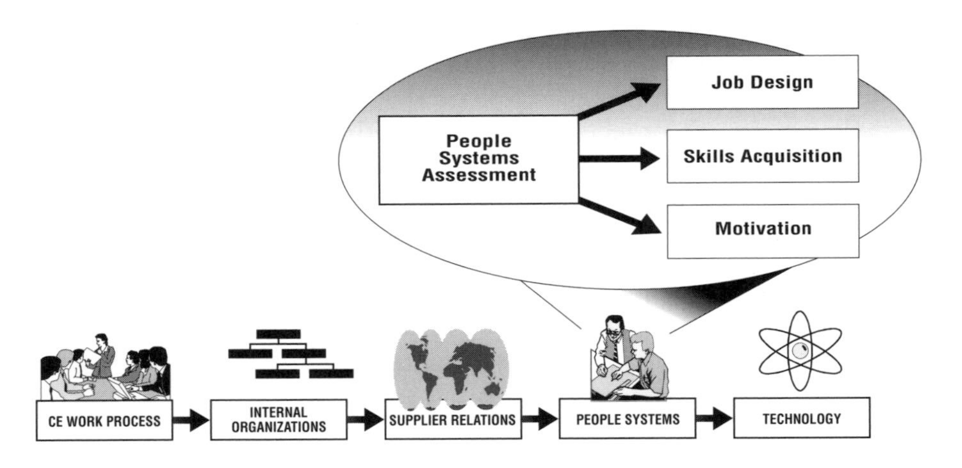

Figure 11.7 • People Systems Assessment

- ***Job Designs.*** Job design is the process of taking a set of work tasks and combining them into a "job" that a person can perform.
- ***Skill Acquisition Systems.*** Both social skills and technical skills are necessary for CE. Do those involved in product development have these skills and is there a process for skill enhancement?
- ***Motivation Systems.*** This includes systems for measuring and rewarding people.

Job Designs

In Chapter 6 we discussed five dimensions of job design (combine tasks, form natural work groups, empowerment, establish environmental relations, and provide feedback on work). As part of the As-Is Assessment, it will be useful to determine if these things have already been "done," so to speak. For example, if some people already perform tasks that are relatively "whole," then there will be little need later (when you design your new system) to try and combine tasks for those jobs any further.

The unit of analysis for this assessment should be the *individual job classification*. By this we mean the distinctly different jobs performed by people in the departments you are assessing. For example, if the scope of this assessment excludes the manufacturing plants, then you need not include those jobs in your analysis. Thus, your analysis might include the following departments: product engineering, manufacturing engineering, marketing, and purchasing. You would assess job designs for all distinct jobs in those departments. Note that this is not necessarily the same as your HR department's classification of jobs. For example, they might have classified four types of designers for pay purposes. Most likely all of those people do basically much the same work; and for your purposes they are all one job.

In Chapter 6 we presented a set of four contingencies that could be used to help choose among the job design options (Table 6.1). These four contingencies included three characteristics of the task(s) for which the job is being designed, and environmental uncertainty (Figure 6.3). We won't address Environmental Uncertainty until the next chapter when we discuss designing the new system. For assessment purposes we will focus on how to measure the task characteristics: task interdependence, task uncertainty, and task skill needs. The "unit of analysis" for all three measures is intended to be an existing job (since this concerns the as-is assessment of job designs).

Task Interdependence. In Chapter 2 we defined three types of Task Interdependence.

1. ***Pooled interdependence.*** This means that units don't actually interact in performance of the task, but rather are dependent on the same "pooled" sources of information or material, to which they all contribute and from which they all draw.
2. ***Sequential interdependence.*** This means one task depends on another to be completed before it can begin. The "downstream" operation requires information or material from the "upstream" operation.
3. ***Reciprocal interdependence.*** This means that two (or more) tasks must be performed together in order for either to be completed.

Task interdependence is only defined based on a relationship between two positions in the context of a specific task. In other words, we can reasonably say that a design engineer has some type of interdependence with a manufacturing engineer. Often that interdependence is sequential (as in "over the wall") and sometimes it is reciprocal (as in CE). However, we cannot say that a design engineer has "x amount of" task interdependence, since it is so context dependent.

What this means is that measuring task interdependence can be a detailed, painstaking task, especially if it is done at the level of individuals and their jobs. While there is some benefit to doing this at the very deep levels of detail necessary for designing individual jobs, at the level we are operating, it is also possible to conduct an analysis of task interdependence at the level of "generic jobs," rather than individuals. This involves a "relationship" matrix, much like we used to assess communication and coordination.

In Exhibit 11.19, the As-Is Task Interdependence Matrix for Coseat

Inc. shows the approximate levels of task interdependence between various jobs in the company. Note that this is filled out for the current process as people within the company see it — not as it "ought to be." We'll get to that later in Chapter 12.

Form Of Task Interdependence P = Pooled S = Sequential R = Reciprocal	Project Manager	Account Manager	Design Engineer	Designer	Draftsman	Prototype Builder	Test Engineer	Tool Engineer	Tool Maker	Manuf. Engineer	Purchasing Agent
Project Manager		S	R			R	S			R	S
Account Manager			S								
Design Engineer				R		R	R	S	S	S	S
Designer					R	S		S	S		
Draftsman											
Prototype Builder							S	S		S	S
Test Engineer								S			
Tool Engineer									S		
Tool Maker										S	
Manuf. Engineer											
Purchasing Agent											

Exhibit 11.19 • As-Is Task Interdependence Matrix for Coseat Inc.

In Exhibit 11.19 you should note that the preponderance of interdependence in Coseat's current system is sequential. For example, we noted earlier that communication between account managers and design engineering is one-way. Our understanding of this relationship is now enhanced by learning that task interdependence is viewed as being sequential. We might argue that it *ought* to be reciprocal, but that would only be the case if the process were different — as it stands, it is sequential.

Task Uncertainty. We've given a set of questions that will provide you with a score for task uncertainty for *each* of the job classes you're rating. Answer the following questions about each job.[9]

1. To what extent is there a clearly known way to perform the major types of work normally encountered?
   1 = Very much so. In almost all cases, it is clear just what will need to be done and how to do it from the start.

[9] These questions are adapted from Hiatt (1992).

3 = ***Somewhat.*** It is usually clear just what will need to be done and how to do it from the start, although we then have to do some improvising as we go.

5 = ***Very little.*** We usually have to figure out a new way to do things on this job as we go—every job is quite unique.

2. To what extent is there a clearly defined body of knowledge or subject matter that can guide you in doing your work?

   1 = ***Very much.*** A clearly defined body of knowledge or standard subject matter covers practically all circumstances.

   3 = ***Somewhat.*** The knowledge that guides my work is only partly defined.

   5 = ***Very little.*** We pretty much have to figure things out for ourselves on a case-by-case basis.

3. To what extent is there an understandable sequence of steps that can be followed in doing your work?

   1 = ***Very much.*** There is always a clear sequence of steps to follow.

   3 = ***Somewhat.*** There is sometimes a clear sequence of steps or there is usually a sequence of steps that I can follow for part of the job.

   5 = ***Very little.*** There is really no clear sequence of steps to follow, we just have to figure out what to do each time.

4. To what extent can you actually rely on established procedures and practices?

   1 = ***Very much.*** Procedures and practices guide me through most of what I have to do.

   3 = ***Somewhat.*** Procedures and practices are somewhat useful in helping me to do my job.

   5 = ***Very little.*** There are few if any procedures and practices to help me do my work.

For *each job*, you can take the average score on the four questions as the Task Uncertainty score for that job.

Task Skill Needs. For each job, make a rough estimate of the level of skill required to perform the tasks needed. For this purpose, a detailed analysis of skills is not required. We suggest you use the following scale for doing this.

- ***Low.*** Skill needs are low for a given job if the tasks can be performed well by a reasonably intelligent person with no specific educational requirements, minimal training (less than a month's worth), and minimal experience (less than two months). Tasks needing low levels of skills include most rudimentary clerical and drafting tasks.

- ***Medium.*** Skill needs are medium if the tasks can only be performed by someone with either high levels of education (e.g., an Associate degree or B.S.) and training (several months might be required) or high levels of experience (perhaps a year or more). Tasks needing medium levels of skills include a designer or a purchasing agent.
- ***High.*** Skill needs are high if the tasks can only be performed by someone with *both* high levels of education (e.g., a B.S. or higher degree in engineering) and training (several months might be required), and high levels of experience (perhaps as much as a year or more). Tasks needing high levels of skills include engineering analysis, design engineering, and project management.

You should now assess each relevant job on each of these characteristics. In Exhibit 11.20 you can see how Coseat Inc. rated their jobs. You don't need to do anything with this information at this time. We're simply collecting data that will be useful later on when we need to think about redesigning specific jobs.

	Task Interdependence	Task Uncertainty	Task Skill Needs
Project Manager	H	H	H
Account Manager	L	H	M
Design Engineer	H	H	H
Designer	H	M	M
Draftsman	H	L	L
Prototype Builder	H	M	H
Test Engineer	H	M	M
Tool Engineer	L	M	H
Tool Maker	L	M	H
Manuf. Engineer	H	H	H
Purchasing Agent	L	M	M

Exhibit 11.20 • Coseat Inc., Job Characteristics Matrix

SKILL ACQUISITION SYSTEMS

The object here is to assess the systems you have in place for acquiring and enhancing skills. It is not primarily to assess the skills of the individuals you have in place right now, although, as you will see, that is one of the systems you ought to have in place. In the skills area there are four basic issues with which you need to be concerned.

1. ***Know what skills you need.*** As we discussed in Chapter 6, you should develop a Skills Profile for each job classification that shows the importance of various skills and where the current job holders are with these skills. Most companies will have job descriptions, and many companies will have a list of technical skills required, but relatively few will in any systematic way consider the social skills needed to really do the job. In Exhibit 11.21 we show a Skills Profile Worksheet for one position in Coseat Inc. — the Design Engineer. The worksheet lists the skills, rates the frequency with which the holder of such a job uses each skill, and rates the current

Job Classification: Design Engineer	Skill Description	Frequency Of Use	Ratings Of Current Holders	Comments
Drafting				
	Ability To Physically Manipulate 3D Structures	1	1	
	Ability To Manipulate Objects And Surfaces	1	1	
	Ability To Draw With A Computerized Tool	1	2	
	Knowledge Of The Elements And Capabilities Of CAD Packages	3	2	DEs Usually Know About 1 Of The 2 Packages We Use
Design				
	Knowledge Of Design Standards	4	3	DEs Often Ignore Existing Standards
	Knowledge Of Product Features And Functionalities	5	5	
	Ability To Interface Designs And Produce Parts Specification	4	5	
	Knowledge Of Product Data Related Electronic Data Interchange	1	1	
	Understanding Of STEP Application Protocols	1	1	
	Ability To Troubleshoot Design Function	5	4	Several DEs Are Weak, Others OK
Manufacturing				
	Knowledge Of Manufacturing Discipline, Including DFA/DFM Capabilities	2	2	
	Knowledge Of Materials Properties	3	3	
	Knowledge Of Production Processes (Preparation, Cutting, Machining, Tooling, Operating Rules And Precedents)	2	2	
Engineering				
	Knowledge Of Procedures For Doing Design Tests (e.g., Instrumentation, Feasibility Testing)	3	4	
	Data Interpretation And Modeling	3	3	
Social Skills				
	Communication	4	3	Several DEs Are Weak, Others OK
	Conflict Resolution	3	4	
	Group Facilitation	1	2	
	Leadership	3	3	

Frequency Of Use Key:
1 = Used Very Little
2 = Used Some
3 = Used A Great Deal

Ratings Of Current Holders Key:
1 = Unable To Perform This Skill
2 = Able To Perform Moderately Well
3 = Able To Perform With High Level Of Performance

Exhibit 11.21 • Skills Profile Worksheet for Coseat Inc., Design Engineer

holders on their skill level for each skill. The point here is to have some kind of comprehensive documentation of skills needed for each job.

There are several noteworthy observations from the worksheet completed by Coseat Inc., in Exhibit 11.21. First, note that the design engineers don't use the CAD system for the most part — they have no skills for use, and they know less than they ideally would even about the system's capabilities. While they know about the company's design standards, they are not fully informed about all. Several engineers are unable to troubleshoot designs that fail to meet the full functional capabilities required. Several engineers seem to have trouble communicating with other people in face-to-face situations, although there are also several who show unusual skill in conflict resolution and group facilitation.

2. ***Know what skills you have.*** The first place you can use the Skills Profile Worksheet is to assess your existing staff's skills against the Skill Profiles for their positions. In most cases a gap means some kind of training is needed. See Exhibit 11.21 for an example at Coseat Inc.
3. ***When you select people to fill jobs, select on the basis of all needed skills.*** Do you use the Skills Profile (or something like it) when creating position announcements? Do you assess the full range of skills of potential new employees? This includes staff who may be transferring to a new position in your company. This doesn't mean that every new hire has to meet every possible skill requirement (that's often impossible), but it does mean you know where the gaps are when you hire them.
4. ***Provide training or educational opportunities to fill skill gaps and to upgrade skills.*** Training needs to be provided in all major areas either to fill skill gaps in individuals or to make sure that existing skills are renewed. The Training Summary Worksheet in Exhibit 11.22 describes current training for Coseat Inc. The picture we get from Coseat Inc.'s Training Summary is that engineers and other professionals get a few external professional development opportunities each year, but no internal training. People who use the CAD system get update training from the vendor as needed. People who work with their hands get no training on a regular basis. Everyone has the opportunity to take advantage of the company's college tuition reimbursement program.

Job Classes	Type Of Training	Training Sources Used	Hours/ Year	Education Opportunities
Design Engineer	1 Workshop/year + 1 Professional Conference	Professional Organizations	30-40	Tuition For Grad Degree
Designer	CAD System Updates	CAD Vendor	15-30	Tuition For BS In Eng.
Draftsman	CAD System Updates	CAD Vendor	15-30	Tuition For AA
Purchasing Agent	1 Workshop/year + 1 Professional Conference	Professional Organizations	30-40	Tuition For Grad Degree
Market Researcher	1 Workshop/year + 1 Professional Conference	Professional Organizations	30-40	Tuition For Grad Degree
Tool Engineer	1 Workshop/year + 1 Professional Conference	Professional Organizations	30-40	Tuition For Grad Degree
Tool Maker	None			Tuition For BS In Eng.
Test Engineer	1 Workshop/year + 1 Professional Conference	Professional Organizations	30-40	Tuition For Grad Degree
Manufacturing Engineer	1 Workshop/year + 1 Professional Conference	Professional Organizations	30-40	Tuition For Grad Degree
Prototype Maker	None			Tuition For BS In Eng.

Exhibit 11.22 • As-Is Coseat Inc. Training Summary Worksheet

In Exhibit 11.23, Ed Cazotzky at Coseat Inc. has summarized the results of his Skills Acquisition Systems Assessment.

Skills Acquisition Assessment Comments Form

1. **Do you have Skills Profiles for all positions?**

   Skills profiles developed for most positions — only exceptions are tooling staff.

2. **Have you assessed current skills against profiles?**

   Existing skills assessed against profiles where we have them.

3. **Is selection done on the basis of all needed skills?**

   Selection is based on technical skill only.

4. **Are training and education opportunities provided?**

   Training is relatively haphazard for senior staff, generally not available for junior staff, except for tuition reimbursement program for college credit.

Exhibit 11.23 As-Is Skills Assessment Comments for Coseat Inc.

Motivation Systems

There are three issues to address under motivation enhancement and maintenance: goals and metrics, appraisal systems, and reward systems.

Goals and Metrics. The basic issue to address here is whether there is consistency[10] among the various goals and metrics operating on the

[10] Note that consistency among goals does not imply that there is no difference between the goals. Goals can be different but still be consistent. For example, one function could have a goal to reduce cost, while another has a goal to increase quality. These are very different, but we also know that they are not only consistent, but in fact mutually supportive.

different people and groups devoted to product development. There are three consistency issues to worry about: horizontal consistency among goals of various functions, vertical consistency between levels of the organization, and consistency between goals and metrics.

Horizontal Goal Consistency: In terms of horizontal consistency, we are interested in two things — common goals and inconsistent goals *between different functions* in the organization. Using the Horizontal Goal Consistency Matrix, Coseat Inc. examined consistency among the goals of several departments (Exhibit 11.24).[11] In this matrix, Coseat Inc. has placed the usual functions in both the rows and columns. In the Goals column of the matrix, the Coseat Inc. team has written a shortened version of the major goals of each functional department. Above the diagonal they have indicated which of these goals each function has in common with others. Below the diagonal they have indicated which goals are in *conflict*. As you can see in Exhibit 11.24, the only goals in common at Coseat Inc. are cost and budget goals, and even then these are not consistent for all departments. In contrast, project management, marketing, and sales are likely to be in conflict when marketing and sales' desire for accuracy (and longer, more detailed marketing studies) conflicts with project managers' need to stay on schedule. Design engineers

	Goals Of Each Function	Project Manager	Marketing & Sales	Design Engineering	Manuf. Engineering	Purchasing
Project Management	Time; Budget; Cost			Budget	Cost	Cost
Marketing & Sales	Accurate Data; Sales	ACCURACY VS. TIME				
Design Engineering	Innovation; Budget	INNOVATION VS. COST				
Manuf. Engineering	Quality; Cost			INNOVATION VS. QUALITY + COST		Cost
Purchasing	Cost					

KEY:
TOP RIGHT = Common Goals (Small Letters)
BOTTOM LEFT = CONFLICTING GOALS (Capital Letters)

Exhibit 11.24 • As-Is Horizontal Goal Consistency Matrix for Coseat Inc.

[11] We haven't included all functions here to save space. You should look at all relevant functions.

and project managers may conflict when the engineer's goal of innovation (which tends to be expensive) conflicts with the project manager's goal to keep costs down. Finally, those innovations created by the design engineers will cause the manufacturing engineers fits as they try to maintain product quality and keep costs down. Note that simple goal differences do not always result in conflicts. For example, design engineering has a budget goal, while marketing and sales does not. There is no conflict regarding this because engineering's budget goal is not inconsistent with any of marketing and sales' goals.

Vertical Goal Consistency: Vertical goal consistency is that between different levels of the organization. A manager's goals should be consistent with those of the people he/she supervises; goals at the top of the company should be consistent with and supported by goals at lower levels. This is assessed using the Vertical Goal Consistency Matrix, which is actually a linked series of matrices (Figure 11.8), in which the rows of one matrix become the columns in the next "lower" matrix, much as is done in Quality Function Deployment (QFD). In fact, this process is often called Policy Deployment, since it is concerned with the deployment of company policy (goals) down the chain of command. In each matrix we compare the two different levels' goals for consistency. You needn't actually do the entire organization chart this way (although it's an interesting exercise); you just need to work with the levels that affect the parts of the company with which you're working.

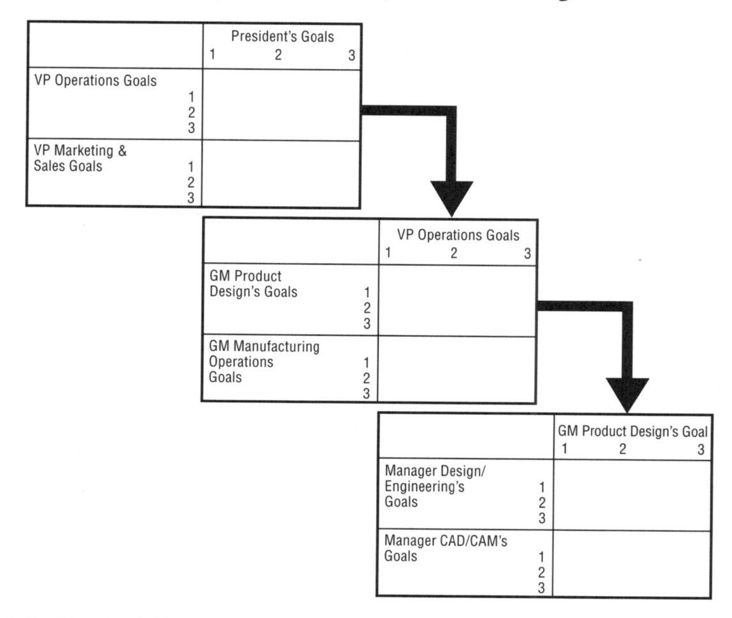

Figure 11.8 • Vertical Goal Consistency Linkages

In Exhibit 11.25 we show one example of a Vertical Goal Consistency Matrix that Coseat Inc. used to compare the General Manager of Product Design's goals with those of his direct reports. This was done by asking each individual what his/her goals were as assigned or implied by their management. Often one's goals are not explicit, but most people can identify what it is that their supervisor wants them to do. We can see that there is quite a bit of consistency, but there are also some serious inconsistencies. In particular, none of the people at the director level has any goals associated with the General Manager's goal of increased innovation. We suspect that under these circumstances innovation will not be a high priority in the work that gets done. Also, some of the staff reduction goals are in at least partial conflict with the goal to get products out sooner. This is a bit less problematic, since staff hours are not the only determinant of product development speed. It may well be possible to reduce staff and get products out sooner if other changes are made, such as technical or work process improvements.

Key:
+ = Goals Are Consistent
- = Goals Are Inconsistent

		GM Product Design's Goals		
		Reduce Product Cost	Get Products Out Sooner	Increase Innovation In Products
Director Design Engineering's Goals	Reduce Engineering Changes	+	+	
	Reduce Engineering Staff Hours	+	-	-
	Reduce Product Cost	+		-
	Meet Customer Schedules		+	-
Director CAD/CAM's Goals	Support Design Engineering	+	+	
	Support Manufacturing	+	+	
	Support Tooling	+	+	
	Support Prototyping	+	+	
	Reduce Staff Hours		-	
	Reduce Equipment Costs			

Exhibit 11.25 • As-Is Vertical Goal Consistency Matrix for Coseat Inc.

Consistency Among Goals and Metrics: Consistency among goals and metrics in the hierarchy is the third issue to consider. This assumes that there are indeed both goals and metrics. However, the absence of either goals or metrics should be considered as data for the assessment. Most companies will in fact have both, even if one or the other is implicit. Assuming you have both, it is a relatively simple exercise to compare the two for consistency as we have done for the Design Engineers at Coseat Inc., in Exhibit 11.26. As you can see, the design engineers have several metrics that work against goal accomplishment, a potentially serious problem. Also, one goal (keep a neat work area) is not associated with any metrics.

	Metrics				
Goals	Fewer Engineering Changes	Fewer Design Staff-Hours	Keep Within Design Phase Cycle Time	Reduce Product Cost	Meet Customer Schedule
Keep Within Project Budget	+	+	0	+	+
Develop Innovative Products	-	-	-	-	-
Assist Other Teams When Requested	0	0	-	0	-
Keep A Neat Work Area	0	0	0	0	0
Run Efficient Meetings	0	+	0	0	+

Key:
+ = Metric Helps Indicate Goal Accomplishment
0 = Metric And Goal Are Unrelated
- = Metric Works Against Goal Accomplishment

Exhibit 11.26 • Goals and Metrics Worksheet for Coseat Inc., Design Engineers

Appraisal Systems. As we indicated in Chapter 6, the appraisal system needs to be perceived as being fair and accurate. Different parts of the system may be viewed as being more or less fair or accurate. In Exhibit 11.27, Coseat Inc. uses the Appraisal System Assessment Worksheet to record views about the system. The Worksheet provides indicators of good practice. The Coseat Inc. team's comments are in italics. From the comments, it is clear that Coseat Inc. does appraisals annually, but that the process is inconsistent and highly dependent on the supervisor applying it.

Reward Systems. The critical issues for reward systems are that they must be perceived as fair and that they must not provide any disincentives to better performance. Coseat Inc. used the Reward System Assessment Worksheet in Exhibit 11.28 for its assessment. At Coseat Inc., Ed Cazotzky and his team did an informal survey of their colleagues

Performance Appraisal System Assessment Worksheet

This worksheet is used to assess appraisal practices for either a single job or a class of jobs. For example, management jobs may be appraised differently from hourly jobs, but all management jobs may be assessed with the same system. The same worksheet can be used for the whole class of jobs which is assessed using the same system.

Job Class Assessed - *all classes below manager*

1. **Are there regular established times for formal appraisals?** *Yes - annual*
2. **Are regular appraisals supplemented with assessment and feedback as needed in day-to-day job activities?** *depends on supervisor - inconsistent*
3. **Does the measurement system use a stable and important set of performance domains?** *Yes, although it appears the domain set is incomplete*
4. **Are performance judgments behaviorally anchored? That is, can concrete examples be given for appraisers' judgments?** *not at all*
5. **Are appraisals tied to business results and employees contributions?** *partially - link to schedule and cost*
6. **Does the appraisal reflect the acquisition and use of important knowledge and skills?** *no*
7. **Does the appraisal get inputs from multiple levels and functions (supervisors, peers, subordinates, self, departments, teams, support staff, customers)?** *no, supervisor only*
8. **Are appraisals followed by feedback, development, counseling, and training?** *depends on supervisor, most do not*

Exhibit 11.27 • Coseat Inc., As-Is Performance Appraisal System Assessment Worksheet

about how the reward system is perceived. If your team is cross-functional and cross-level enough, it may include enough diversity of opinion that you wouldn't need to bother with this. Alternatively, you could do a formal survey to get at more objective opinions about fairness and what is really being rewarded. It really all depends on the atmosphere in your company. If there is a climate of fear and intimidation, then people will need to be assured of anonymity before they'll give you their real opinions. In most companies, the Guidance Team or a group they appoint should be able to get enough information to complete this worksheet reasonably well. Note that in Coseat Inc.'s case, several of the rewards were perceived as being distributed unfairly. Even professional development opportunities were perceived as being distributed unfairly due to favoritism by managers, rather than by need. The several blank rows in the worksheet indicate that those rewards are not used by Coseat Inc.

Job Title: Design Engineer	Check If You Use	Frequency	Perceived As Fair?	Basis	What Is Really Rewarded?
Hourly Wage Or Salary	X	Monthly	Yes	Individual Background	Education And Experience For Base Pay
Pay Raises	X	Annual	No	Individual Performance As Determined By Manager	Meeting Project Deliverables For Time And Cost
Bonuses	X	Annual	No	Individual Performance As Determined By Manager	Meeting Project Deliverables For Time And Cost, But Higher Or Lower Depending On Company Performance
Profit-Sharing Or Gain-Sharing					
Time-Off					
Stock Or Other Ownership					
Gift Equivalents Of Cash					
Recognition Awards (Plaques, Certificates)	X	Irregular	No	Individual Performance As Determined By Manager	Extraordinary Performance On A Project And Making Sure Manager Knows About It
Professional Development, Education	X	Annual	No	As Determined By Manager	Initiative To Seek PD Opportunities
Social Activities (Parties, Event Tickets, Etc.)	X	Annual	Yes	Company-Wide Party	Being Alive And Working For Coseat
Fringe Benefits (Pension, Insurance, Etc.)	X	W/Pay	Yes	Part Of Pay Package	Being Alive And Working For Coseat
Other Rewards					

Exhibit 11.28 • Reward System Assessment Worksheet for Coseat Inc.

TECHNOLOGY DESCRIPTION

We now move to the final element in the assessment before we tie all the pieces together — the technology description (Figure 11.9). We are primarily interested in the technology used by your company and key suppliers for product development as it affects integration across boundaries, such as functions, organization hierarchy, and supply chains. See Chapter 7 in Part I for a discussion of the integration issues. We call this a technology description rather than an assessment because it is usually not worth your time to do a detailed assessment of how your current technology fits with your current work flow, organization, and people systems. The detail can come later when you consider what new technologies are needed to support your new vision in the to-be design phase. At this point, we just want an organized listing of what you are using and what your key suppliers are using. The description is done in two steps. First, describe the internal technical systems used by each job classification related to product development within your company. Next, describe the technical systems used by each of your key suppliers.

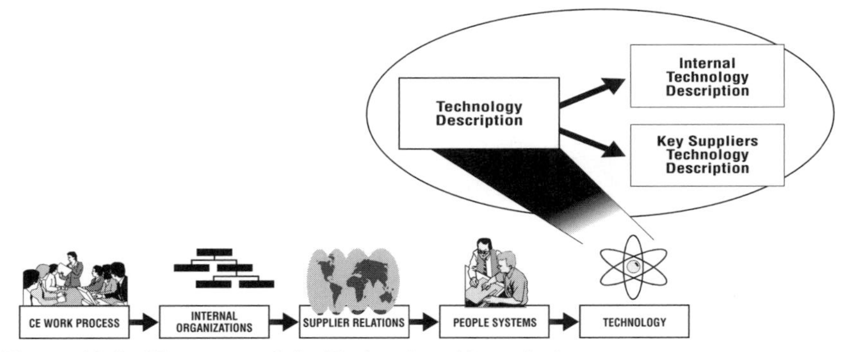

Figure 11.9 • Elements of the Technology Description

Describe Internal Technical Systems

For each job related to product development (the same ones you used to assess the people systems), describe the technical system used by the people in that job. The technical system is all of the equipment and software they use. It does not include any "methods" which are part of the skill base of the individual using the technical system. In other words,

Job Classification	Technology Description	Comments
Design Engineer	Personal computer with word processing, spreadsheet, project management software; LAN with fileserver and internal e-mail. Also use QFD, design of experiments (DOE), and structural analysis.	Engineers don't use CAD - believe it is a waste of their time. They can view CAD files on PCs and find that valuable. Internal e-mail used extensively; would like this extended to suppliers. QFD & DOE used 1 or 2 times, too cumbersome to use on most projects. Structural analysis done occasionally on seat structure and then outside consultants are used - seems to work fine.
Designer	Workstation with CAD software; LAN with fileserver; files transferred to suppliers by IGES and phone lines or tapes; have access to a pool of PCs.	Designers spend most of their time on CAD system. They complain about poor lighting. System response time is slow. CAD database is poorly organized; designers generally start each project fresh without using older related designs. File transfer over the phone is slow with many data errors. IGES translates only partial geometry and much rework is done by suppliers. Designers seldom use PCs; check email once or twice a day.
Drafters	Workstation with CAD software; LAN with fileserver.	Drafters take instruction primarily from designers. They spend their time on CAD; have same complaints as designers.
Purchasing	PC with word processing, spreadsheet, MRP; LAN with fileserver and internal e-mail.	Outside of e-mail, purchasing has no technology linked to design engineers or product development.
Sales/Marketing	PC with word processing, spreadsheet, database; LAN with fileserver and internal e-mail.	Limited e-mail with design engineering. Marketing contributes market data to QFD but has little direct involvement in the process.
Tool Engineers	PC with word processing and spreadsheet; LAN with fileserver and internal e-mail. Tool design is done on a drawing board.	Tool engineers rarely use the PCs or e-mail. No electronic links to CAD.
Tool Makers	NC equipment to cut tools programmed manually.	Do not use e-mail; NC programming is not linked electronically to CAD.
Test Engineers	PC with word processing, specialized data analysis and spreadsheet; LAN with fileserver and e-mail.	Use PCs and e-mail extensively. E-mail is a main connection to design engineers. No access to CAD database.
Manufacturing Engineers	PC with word processing & spreadsheet; LAN with fileserver and e-mail.	Use e-mail occasionally. No access to CAD database. Mostly spend time on shopfloor reacting to equipment problems.
Prototype Makers	Access to one PC with word processing and spreadsheet; LAN with fileserver and internal e-mail. Build prototypes with manual tools.	Occasionally use e-mail. No access to CAD database. No rapid prototyping technologies.

Exhibit 11.29 • Coseat Inc. Internal Technology Worksheet

if you can see it and feel it, it may be part of the technical system. If it is inside someone's head, then it is not. We are often asked if something is part of the technical system if it is included in a book or some kind of worksheet used on a white board, for example. If it is a major systematic methodology like Quality Function Deployment, you probably want to include it. But it is usually easier to not include every detailed table and equation in use. In the Coseat Inc. example (Exhibit 11.29) you'll notice we've confined ourselves to hardware and software and a few major design methodologies. We have focused on "integrative technologies," that is, technologies that help information and data cross functional and organizational boundaries. Also included in the technology description worksheet is a comments column for capturing any thoughts at this point about usefulness or problems with particular technologies.

The assessment team at Coseat Inc. collected information about the technical system by interviewing representatives of each function. A number of points came through quite clearly:

1. There is relatively little CAD data exchange and designers do not even use old designs as starting points. CAD is also not used to transfer data to manufacturing (i.e., no CAD/CAM). Thus, CAD is serving little integrative purpose.
2. CAD is not used by design engineers, though they find it useful to be able to call up the CAD designs on their PC.
3. Data transfer with suppliers is problematic.
4. No functions outside of design are technically integrated with design engineering except through e-mail which is used to a limited degree for communication across functions. Design engineers use e-mail extensively and find it very useful, as do test engineers.

Describe Supplier Technical Systems

For each supplier with some involvement in product development (based on Exhibit 11.15) describe the technical system used for design or prototyping of components provided to your company. Even contractual suppliers who do not design components should be considered if they are involved in testing, prototyping, or manufacturing that might benefit from CAD/CAM linkages. In the case of Coseat Inc. (Exhibit 11.30), they decided to focus attention on the four suppliers identified in the "Parental" role. The rest of the suppliers were "Contractual" and they all have a similar level of integrative technology, namely, none except telephones and faxes.

Supplier	Technology Description	Comments
Ann Arbor Spring & Wire	Internal CAD system not directly compatible with Coseat's. Some IGES transfer by phone and tape.	IGES transfer is not satisfactory; often files need to be extensively revised once transferred. It is used mainly for the outside geometric shape. They can dial-in to Coseat's e-mail system.
Springmasters	Internal CAD system not directly compatible with Coseat's. No data exchange takes place.	CAD system is old and needs upgrading. Hard copy blue prints are the common mode of communication.
Burlington	Internal CAD system not directly compatible with Coseat's. No data exchange takes place.	Data exchange typically not needed.
Mitch's Switches	Internal CAD system not directly compatible with Coseat's. No data exchange takes place.	CAD system is old and needs upgrading. Hard copy blue prints are the common mode of communication.
Contractual Role Suppliers	No integrative technology except phone and fax.	Some discussions of adding e-mail access by setting up one terminal with modem access to Coseat's has been underway.

Exhibit 11.30 • Coseat Inc., Supplier Technology Worksheet

The assessment team at Coseat collected information about these technological systems by calling supplier representatives as well as from what they had learned from their own design engineers. It became very clear that there was almost no use of technology to integrate with suppliers. The one exception was Ann Arbor Spring & Wire, their most technologically advanced supplier. Even in this case, data transfer was not working because of buggy phone lines and the limitations of IGES. Ann Arbor Spring & Wire used e-mail connections regularly with Coseat's design engineers and purchasing, and reports this is an invaluable resource. There have been discussions about extending e-mail access to other suppliers, but this has yet to lead to real action.

ANALYSIS AND CONSOLIDATION

So far in this chapter we've assembled a large amount of data about processes, organization, supplier relations, people, and technology. This is where we try to make sense of it all. Our basic approach is called the *Consolidation Analysis*, and the tool we use for it is called the *Consolidation Matrix*.

CONSOLIDATION ANALYSIS

The ultimate purpose of the Consolidation Analysis is to help you plan your *CE Elements*. The CE Elements are the things you have been learning how to assess in this chapter: work process, internal organization, technology, people systems, and supply chain relations.

While you initially do a Consolidation Analysis as the last task in the As-Is Assessment, it should become an ongoing activity for you to perform as you proceed through the design process (Chapter 12). If you use the Consolidation Analysis in the way we suggest, you will have a way to *continually assess the impact of changes as you plan them*, rather than having to wait for a complete plan to be done before you know what you have. In a sense we are applying the principles of CE to the process of planning CE.

The Consolidation Analysis brings three sets of data together:

- ***CE Elements***, which are the assessment data we have discussed in this chapter,
- ***Process Benchmarking*** data which describe best practices in CE, and
- ***Performance Benchmarking*** data which describe best performance in your industry.

The Consolidation Analysis looks at the data in light of your own company's objectives. More specifically, we attempt to look at three types of relationships (Figure 11.10).

- ***Objectives versus Performance Benchmarks.*** How do your performance objectives compare to the performance of your competitors? Are you aiming too low? Or, are you aiming for a difficult goal which your competitors aren't even close to?
- ***CE Elements versus Objectives.*** How well are your current work process, organization, people, and technology (the CE Elements) helping you to meet your objectives? Elements which have a strong impact on your ability to meet your objectives, but which are getting in the way, will need to be changed.

- ***CE Elements versus Process Benchmarks.*** How are your current CE Elements similar to *or* different from those of your benchmarks? Looking at benchmarks in specific areas where you know you need to change can give you suggestions about *how* to change.

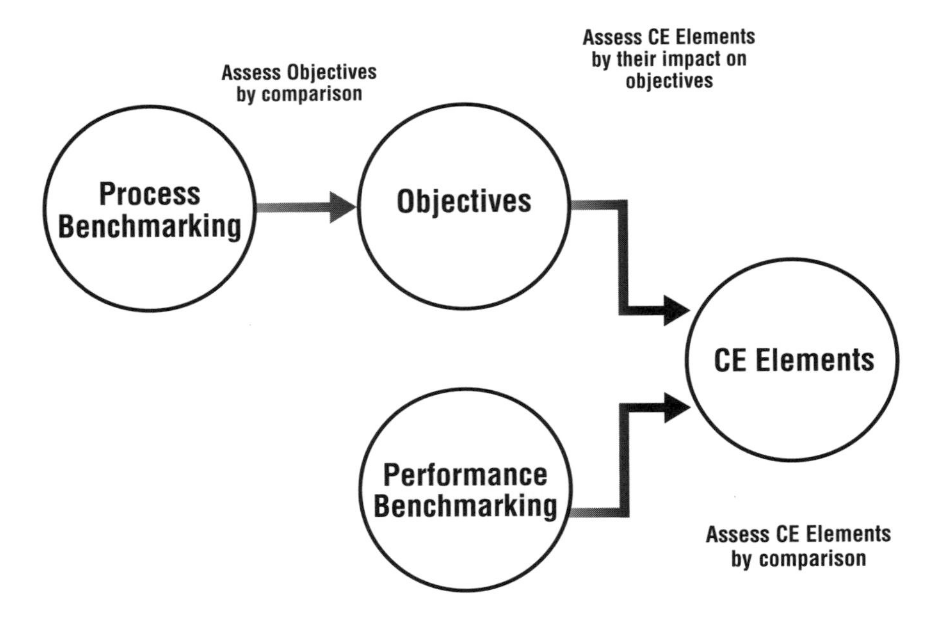

Figure 11.10 • Consolidation Analysis

As you can see in Figure 11.10, the Consolidation Analysis brings the pieces together with the ultimate purpose of providing input to the assessment and design of the CE Elements. The Consolidation Matrix is a tool that you can use to perform the Consolidation Analysis.

Consolidation Matrix

The Consolidation Matrix has its origins in Quality Function Deployment (QFD). If you are familiar with QFD (Akao, 1990; Hauser and Clausing, 1988), you will recognize many similarities between the two methods. There are some differences, however, which we will note as we go along. The Consolidation Matrix helps you in two ways.

1. ***It makes the data visible.*** One of the problems with collecting lots of data is finding a way to look at a lot of it all at once. In our experience, being able to look at more data helps you understand the overall situation. But that's very difficult when you have everything separated into different matrices and forms. The Consolida-

tion Matrix will permit you to see a large amount of summarized data all at once to help you find trends and in general make sense of it all.

2. ***It helps to find relationships.*** As we have suggested several times already, no single piece of data really tells you very much. When you look at relationships among data elements, you are in a much better position to identify inconsistencies which translate into opportunities. The Consolidation Matrix enables you to look at many relationships.

The Consolidation Matrix has six components (Figure 11.11).

1. ***Objectives.*** These are your objectives for doing CE which you developed during the Scoping the Change Effort activity in Chapter 10. These should not be your objectives for a specific design project, but for your overall CE effort.
2. ***CE Elements.*** These are the elements which you have been learning how to assess in this chapter: work process, internal organization, technology, people systems, and supply chain relations. You can see those as column headings in Figure 11.11. In a real Consolidation Matrix (Exhibit 11.31) there would be more detailed subelements under each of those headings.

			CE Elements					Performance Benchmarks
			Work Process	Internal Organization	Supply Chain	People	Technology	
		Weight 1 = Low 5 = High	(Specific CE Sub-Elements Go In This Row)					
Objective	Objective 1	Rating						
	Objective 2	Rating	Design Relationships					
	Objective 3	Rating	(Rate How Well Each CE Sub-Element Helps To Meet Each Objective)					
	Objective 4	Rating						
	Objective 5	Rating						
Summaries	Importance		(Summarize The Importance Of Each CE Sub-Element In This Row)					
	Describe The Sub-Elements		(Summarize Key Description Of Each CE Sub-Element In This Row)					
	Comments		(Additional Comments About Each CE Sub-Element)					
	Ease Of Achieving (To-Be Only)		(Indicate How Hard It Will Be To Change Each CE Sub-Element)					
Practice Benchmarks			(Describe Benchmarks Of Best Practices In Each CE Sub-Element From Other Companies)					

Figure 11.11 • Skeleton Consolidation Matrix

3. ***Practice Benchmarks.*** These are benchmarks for CE processes — for example, best practices in project management. You can find these in the lower rows of the matrix in Figure 11.11. In a real Consolidation Matrix (Exhibit 11.31), we place information about any benchmark cases we know about, including cases in this book.
4. ***Design Relationships.*** This is the center of the matrix at the intersection between Objectives and CE Elements. Here you place ratings of how well each CE Element affects each objective.
5. ***Performance Benchmarks.*** In the far right columns, you place information about how well your competitors fare on the substance area of your objectives. For example, if one of your objectives concerns time to market, you would place information in these columns about your competitors' time to market for similar products.
6. ***Summaries.*** In the middle rows, entitled Summaries, we place summarized information about the CE Elements.

Consolidation Matrix Example

The Consolidation Matrix looks complicated, and indeed it is — after all, we are putting a lot of information in a relatively small space. On the other hand, it is really quite easy to fill out because you have already collected the information needed to do so as part of the As-Is Assessment. The most serious problem with the matrix is that it is so large (in particular, wide). In the example matrix in Exhibit 11.31, there are 25 columns. Such a matrix is quite easily created on a computer spreadsheet such as Microsoft Excel® or Lotus 123®. The spreadsheet can then be printed onto several pages which are taped together for easy viewing. Alternatively, the spreadsheet can be projected onto a screen, although you may find it impossible to view the whole thing at once unless you have a very large screen.

The following sections discuss the Coseat Inc. example Consolidation Matrix (Exhibit 11.31). This can be viewed in the following pages.

Objectives. The objectives are taken directly from Coseat Inc.'s Scoping effort (see Exhibit 10.5). In the column just to the right of the objectives, they have rated the weight (i.e., how important it is) of each objective on a 1–5 scale, with 5 being the highest importance. For Coseat Inc., the most heavily weighted objectives are to reduce time to market by 50 percent and to reduce late deliveries by 90 percent.

CE Elements. The CE Elements are listed in the column headings across the top, first in categories, then more specifically as sub-elements. These come straight out of the As-Is Assessment framework described in this chapter. You can, if you choose, decide to drop or add sub-elements.

CE Elements								
		Work Process				Internal Organization		
		weight 1=low 5=high	Activities	Interfaces	Sequencing & Coordination	Org. Architecture	Project Management	Responsibilities
Objectives	Reduce time to market by 50%	5	-9	-9	-9	-3	-9	-9
	Improve quality 10X	3		-3	-3	-9		-3
	Reduce engineering changes 25%	2	-3	-9	-3	-3	-3	-9
	Reduce warranty costs by 30%	2		-3				
	Reduce scrap rate by 25%	1		-3				
	Reduce late deliveries by 90%	4		-3			-3	
	Importance		-51	-93	-60	-48	-63	-72
	Characterize the Sub-elements				No formal process	Functional	Lightweight PM	
	Comments		Get market info too late; tool & die build process not defined	Too much throw it over the wall; info flow not well defined	Prototype should wait for market info; tooling design starts late	Difficult coordination across functions, goals tend not to be customer driven; design and tooling too separated	PM lacks authority; technically strong w/in disciplines	Design not involved in market rsch, marketing not involved after concept; test engin. not involved to support design & build; tool engin. and mfg. engin. have input late; purch. & suppliers involved too late
Practice Benchmarks	Motorola				Structured development process	Product line organization with functional managers, councils at all levels	Heavy weight PM	
	Chrysler				No formal dev. process - focus on schedule based on need to meet intermediate schedule deadlines	Platform teams in eng., purch., matrixed, mfg separated as a functional group	Autonomous PM	
	Eaton/Ford						Lightweight PM	
	Newport News Shipyard				Highly structured development process	Hybrid with product lines that have a very strong functional structure	Space Kings - lightweight PM	
	Whirlpool				C2C program management process	Hybrid primarily product line, functional divisions within product lines		Cross-functional involvement in early stages
	Texas Instruments				Very detailed structured workflow process	Strong functional units	Lightweight PM	
	Ford Alpha				World Class Timing - 8 stage structured development process	Functional	Lightweight PM	Use SOIA to assure agreement about responsibilities

Exhibit 11.31 • Example Consolidation Matrix for Coseat Inc.

(This exhibit is reproduced on the fold-out chart at the front of the book)

CE Elements								
		Internal Organization			Supply Chain			
		Communication	Cross-Functional Coordination	Culture	Supplier Responsibilities	Supplier Comm.	Supplier Management Mechanisms	Supplier Roles
Objectives	Reduce time to market by 50%	-9	-9	-9	-3	-3	-9	-9
	Improve quality 10X	-3	-9	-9	-3	-9	-3	-3
	Reduce engineering changes 25%	-9	-9	-3	-3	-3	-3	-3
	Reduce warranty costs by 30%		-9	-3		-3		
	Reduce scrap rate by 25%						-3	-3
	Reduce late deliveries by 90%	-3	-3		-9	-9	-3	-3
	Importance	-84	-120	-84	-66	-90	-75	-75
	Characterize the Sub-elements		Primarily informal meetings, outputs in the form of requirements					Mostly contractual, a few parental
	Comments	Most comm. is 1-way and low frequency; exceptions are tool engin w/ tool making, purchasing with suppliers	Most coordination with project mgrs and design engin thru meetings; standardization of output primary mechanism otherwise	Authoritarian, resist formal processes, not participative, teams may be a hard sell; can-do attitude; major status differences	Poor relations with tooling suppliers, most suppliers involved late	Low frequency one way with parts & some component makers before production	Qualification is only mechanisms used consistently, early involvement with 2 suppliers	Even highly involved supliers are at best parental
Practice Benchmarks	Motorola	Teams promote communication	Emphasis on cross-functional teams	Strong team culture for design, production & improvement, heavy emphasis on measurement	Strong supplier involvement early		Supplier with early involvement usually gets the contract, price agreements	
	Chrysler	Teams promote communication; cross-team communication promoted by physical facility	Cross-functional teams, technology reviews and technology clubs for within-function coordination	Very cooperative, participative	Strong supplier involvement throughout design, more design responsibility at supplier	Generally high levels of communication; much face to face at Chrysler Tech Center	Early price commitments based on target pricing	Many at mature, moving toward partner
	Eaton/Ford		Cross-functional teams		Full service supplier-develop complete valve system; now work w/their suppliers to validate process		Early commitment from customer, but no contract until late-"strategic alliance" not "partnership"	
	Newport News Shipyard		Cross-functional teams; Space Czars to resolve problems it Kings can't do it	Major culture change being attempted - moving away from military model				
	Whirlpool		Strong emphasis on cross-functional teams		Strong emphasis on early involvement in design, suppliers responsible for product test		Strategic suppliers member of team early, pre-source to them	Many in mature roles
	Texas Instruments	Strengthened communication with customer	Cross-functional teams of part-time staff, QFD, DFM, design-to-cost model	Still trying to deal with military contractor mindset	Battery supplier involved early in new technology development			Battery supplier in mature or partner role
	Ford Alpha		QFD, cross-functional teams					Developing full-service suppliers such as Eaton-much like mature or partner roles

CE Elements											
		People			Technology		Competitive Benchmarking			
		Job Designs	Skills	Motivation Systems	Internal	Suppliers	Coseat Now	Coseat Future	Competitor A	Competitor B
Objectives	Reduce time to market by 50%		3	-3	3	-3	16 months	8 months	14 months	10 months
	Improve quality 10X		9	-3	3	-3	230 ppm	23 ppm	300 ppm	200 ppm
	Reduce engineering changes 25%		3	-3	3	-3	250	190	200	150
	Reduce warranty costs by 30%		3				$75	$53	$100	$100
	Reduce scrap rate by 25%						8%	6%	5-10%	5-10%
	Reduce late deliveries by 90%					-3	10%	7%	8%	7%
	Importance	NA	54	-30	30	-42				
	Characterize the Sub-elements									
	Comments	NA in assessment	Highly skilled workforce; engineers up to date on technical issues; willing, energetic staff; training not systematic	Rewards tend to focus on functional organization at expense of product goals; some goal conflicts; appraisal inconsisent; some rewards viewed as unfair	Slow CAD system, IGES for translation, no links from CAD to CAM	Most have primitive technology				
Practice Benchmarks	Motorola		Heavy commitment to training and skill development	Strong PDA and QSR processes, measure everything	Wide array of tools					
	Chrysler	Traditional roles maintained			Central CAD database using single CAD system (CATIA)	All required to use CATIA for CAD - some do but many use translation internally or thru service				
	Eaton/Ford		40 hours training per year for all employees							
	Newport News Shipyard				VIVID system used as single solid model - major advance in visualization, comm. across functions	Exchange data with GD Electric Boat using IGES				
	Whirlpool			Objectives focused in team goals						
	Texas Instruments				ProE very effective	IGES effectively used to link with other systems				
	Ford Alpha				Ford uses a single CAD system within major functions	Attempts to require suppliers to use same CAD system, but has not been totally successful - some IGES used				

Design Relationships. In the center of the matrix, at the intersection between objectives and CE Elements, are the design relationship cells. The numbers in these cells[12] are recorded as follows:

9 = the CE element has a strong *positive* effect on the objective
3 = the CE element has a moderate *positive* effect on the objective
blank = little or no relationship between the CE element and the objective
-3 = the CE element has a moderate *negative* effect on the objective
-9 = the CE element has a strong *negative* effect on the objective

As an example, at the intersection of the objective "reduce time to market by 50 percent" and the CE element "project management," we see that Coseat Inc.'s current project management structure (lightweight) has a strongly negative effect on the ability to reduce time to market. This is because Coseat typically runs into trouble through lack of control over the different functions. The project manager lacks enough authority to bring everyone in line and keep them to schedule. It is important to recognize that *not every company with lightweight project managers will find this exact relationship*. For example, in some companies the functional managers may be strongly reinforced for keeping to schedule, making the project manager's job much easier and requiring less authority.

Summaries. There are four summary rows below the design relationships matrix. Each serves a different purpose.

- ***Importance.*** This row summarizes the importance of each CE sub-element. The Importance score is calculated by multiplying each design relationship score by the weight of its objective and adding the sum of those products. Thus, the Importance score for Project Management is the sum of $(-9 \times 5) + (-3 \times 2) + (-3 \times 4) = (-63)$.
- ***Characterize the Sub-Elements.*** This row is a place to provide a summary characterization of the CE sub-elements. In general, we only use this row when we have a specific way to categorize the sub-element — e.g., we know that project management has five types, or we know that there are four different supplier roles in

[12] Note that while it is possible to use intermediate numbers between the ones shown (e.g., a score of 2 or 5), it tends not to be a good idea. The reason is that you are in fact using a scale of small, moderate, and strong for the effect size. The numbers are a simple translation of those adjectives. We have further simplified things by not recording any scores in which the effect is small. Those familiar with Quality Function Deployment (QFD) will recognize the numbering scheme as those attached to the symbols used in the center of a QFD matrix. The major difference (other than dropping the symbols) is the use of negative numbers. In QFD, negatives are never used since we would not want to design a product which had negative impact on customer satisfaction. Hence, if a negative were to appear you would simply work to remove it. In this analysis we are looking for negatives as much as positives, so we can later work to change them.

product development. As an example, Coseat Inc. shows that their project management type is lightweight.

- ***Comments.*** This is a place to put other comments which will remind you about the sub-element, but which are not included in the categorization. For example, Coseat Inc. describes their project managers as technically strong but lacking in authority.
- ***Ease of Achieving.*** This row is not used in the As-Is Assessment. It will be used later on when you make changes in one or more CE Elements. Then you will indicate how easy or difficult it will be to make the desired change.

Practice Benchmarks. The Practice Benchmarks in the bottom set of rows are examples of how other companies used their CE Elements to achieve CE. In our sample matrix (Exhibit 11.31) we have included the information available about our case studies in the Appendix. We do not have enough information about each case to be able to assess all the sub-elements. You should add your own benchmarks to ours by using data that you collect, either from reviewing the literature or making company visits. In most cases you will also not gather enough information to know about all CE Elements, but get what you can. The real utility is that you can see how a number of other excellent companies rated on the CE Elements. For example, Coseat Inc. is interested in changing how they do project management. They note that several of the benchmark companies use heavyweight project management, but some have succeeded with lightweight project managers. In this way they can see that there are different approaches to success. Looking across this section, they can also get a sense of what some tradeoffs might be. For example, one company (e.g., Newport News Shipyard) might use lightweight project management, but compensate by using other coordination methods.

Performance Benchmarks. In the rightmost columns, we have provided a place to put information about how your competitors perform relative to your objectives. In the first column you would put your own current performance, in the second column your desired performance, and in the others your competitors' performance. You can see that, while Coseat Inc.'s current time to market is worst in class, its desired performance will place it first.

Interdependencies of CE Elements

The CE Elements are not totally independent. This means they tend to affect one another. For example, communication and responsibilities are often closely linked — if two functions share responsibility in some way, then their mutual communication is affected. What this means is that we can't change one without either changing the other, or at least recognizing the impact the change will have on the other. Potentially, any one of the CE Elements may be interdependent with any other, *depending on the circumstances*. If we take the example of responsibilities and communication, depending on just how my responsibilities change, our communication may or may not be affected. This means we can't provide general guidelines for the interdependencies — they will or will not interact, depending on the specific circumstances.

In Table 11.1 we give several examples of interdependencies, just to give you a sense of the logic. In the next section we will show how we believe you should address interdependencies.

Interdependency	Justification
Organization Architecture and Project Management	Formal structure and project management structure often involve trade-offs between the two; e.g., a heavyweight project management structure might be used to make up for weaknesses apparent in a functional formal structure.
Organization Architecture and Cross-functional Coordination	Formal structure and coordination methods also may involve trade-offs, e.g., a company might use a large number of cross-functional teams to make up for the lack of project-level coordination inherent in a functional structure.
Project Management and Cross-functional Coordination	As with the formal structure, there may be trade-offs among the project management structure and the coordination methods.
Cross-functional Coordination and Organization Responsibilities	The distribution of responsibilities will affect how coordination needs to take place, hence coordination methods.
Organization Responsibilities and Motivation	Changes in the distribution of responsibility is dependent on those with new responsibilities having the goals and incentives needed for them to want to carry them out. Equally, changes in reward systems often change the perception of what one's responsibilities are, hence a careful examination of responsibilities and how they are affected may be needed if an independent change in goals and rewards is made.

Table 11.1 • Examples of Interdependencies

USING THE CONSOLIDATION MATRIX

There are two main uses for the Consolidation Matrix. The first is to help you make your recommendations for change that comes out of the As-Is Assessment. The second, which we will discuss in more detail in Chapter 12, is to help you keep track of the impact of the changes you design.

For purposes of the As-Is Assessment, we suggest that you go through four steps in your analysis of the Consolidation Matrix.

1. ***Look at the importance ratings.*** Sub-elements with high negative numbers are probably always important to work on for change. After all, they represent significant barriers to reaching your objectives. In the Coseat Inc. example, Cross-functional Coordination, Interfaces, and Supplier Communication are very strong negatives and consequently high change priorities.
2. ***Use the process benchmarks.*** The process benchmarks may give you some hints about what seems to work well together. For example, if you look at Motorola, they have both heavyweight project managers and a very strong emphasis on teams. In contrast, several other benchmarked companies have lightweight project managers and teams. This suggests that teams will work regardless of project management type.
3. ***Rank the sub-elements.*** Based on the importance ratings, the process benchmarks, and your own sense of what is important, rank order the first five to ten sub-elements you think you might want to work on to change. We have not suggested that you go about this by some formula, although you obviously could do that. There are a lot of informed judgments to be made here. You are the one who (if you've done all this) has collected a ton of information, and you are the one who is attempting to synthesize it. The Consolidation Analysis is just one means to help you do that.
4. ***Consider the interdependencies.*** Decide what other elements are interdependent with the ones you've selected. Determine the impact of the interdependence, and decide whether to change priorities or to add sub-elements to your list.

You can use the Consolidation Worksheet (Exhibit 11.32) to record your ranks, your initial ideas about what kinds of changes should be made in each sub-element, and your thoughts about the impact of interdependencies.

Coseat Inc.'s Assessment

So, what conclusions does Coseat Inc. draw from all this analysis? Before we get carried away, remember that we created this example to demonstrate the use of all the tools; so Coseat Inc. looks like a real mess — they have problems in almost every area! On the other hand, they're not all that different from some companies we have seen; companies that are not doing all that badly in the marketplace. The difficulty is that the Coseats of the world did well in an older environment. In today's environment, they begin to lag behind; and unless they change over the course of the next few years, their decline will accelerate. Thus, we can have a company which appears to be relatively successful, but which still shows up poorly in this kind of assessment.

In any case, the assessment is obviously not complete until we draw some conclusions about the company. We will structure our conclusions in the same framework as the analysis. We will then use the information from the Consolidation Matrix to suggest priorities for change. The details of change will be discussed in the next chapter. Coseat Inc., of course would put this assessment in the form of a report and most likely a set of briefing slides for presentation purposes.

Work Processes

The most obvious conclusion is that concept design begins without a good understanding of customer needs. This is partly because the company lacks data about the customer's needs, and partly because, in the absence of a structured product development process, the design engineers simply jump into their job of designing what they think is a better seat. The lack of a formal product development process also stands out quite strongly in comparison to the benchmark companies, most (but not all) of which have such processes. Given that a structured process is not absolutely essential (witness Chrysler's success without one, and Toyota's success with only a minimal process), it might be argued that Coseat does not need to do this. However, the problems with getting and using information about the customer could be solved by creating a formal process that required the use of such information. Furthermore, until the concept of CE is deeply ingrained in the practices of engineers and others in the company, a structured process is needed to teach them how to go about it.

Internal Organization

Coseat has a strongly functional structure, with only weak mechanisms for coordination across functions. This includes lightweight project management and the use of relatively weak, impersonal methods for communication and coordination. Resolving this would be one of the most important ways to move Coseat in the direction of CE. There are a number of approaches which might be used, including moving to a product-line structure, strengthening the form of project management, and developing stronger cross-functional coordination methods such as teams. There are also a number of cultural barriers to changing to a more flexible form of organization.

Supply Chain

Coseat has serious problems with its supply chain management. Relations are distant and often adversarial. Suppliers are involved late in the process, and little or no advantage is taken of their expertise in product development. Closer, longer-term relations with suppliers, and a stronger involvement of suppliers in product development, would seem to be the direction to move.

People Systems

While the individuals who work for Coseat Inc. appear to be one of the company's strongest assets, the company is failing to take full advantage of this by having poor job designs and weak motivational systems.

Technology

Coseat is using CAD for all its designs, and has adequate traditional prototyping and tooling capabilities. But technology is not being used to help integrate functions. CAD is used as a stand alone system with designs developed by designers and drafters. Design engineers do not use CAD to develop designs, though they can call up designs and find this feature useful. There is very little exchange of CAD data internally or with suppliers, so the potential benefits of CAD for integration are not realized. Data must be generated separately by each supplier, by tool engineers, and by manufacturing to program CNC equipment. E-mail is used by some groups and with one supplier; wherever it is used, it receives rave reviews. There is great potential for using technology to help facilitate integration across functions and with suppliers.

In Exhibit 11.32, Ed Cazotzky and his team have recorded their priorities in the Consolidation Worksheet. We have only shown their top five sub-elements. Chances are they would really look at a few more than that.

CE Subelements	Initial Priority	Interdependencies	Possible Changes	Effects On Interdependencies
Interfaces	2	Communication, Coordination, Supplier Communication, Internal And Supplier Technology	Develop Structured Process That Shows Interfaces	Make Sure Supplier Communication And Technology Support Changes
Communication	4	Interfaces, Sequencing, Responsibilities, Coordination, Culture, Supplier Communication, Job Designs	Increase Frequency And Make More 2-Way Comm	As Interfaces Become Apparent; Coord, Culture, Job Designs Needs To Support Changes
Cross-Functional Coordination	1	Interfaces, Sequencing, Org Architecture, Project Management, Responsibilities, Communication, Culture, Job Designs	Make Coordination Less Centralized, More Team Based	Need Work Process Changes; Make Sure Chain Of Command Supports This; Clarify Responsibilities, Job Designs To Support Team Based Work
Culture	4	Management, Responsibilities, Communication, Coordination, Supplier Responsibilities, Supplier Communication, Supplier Roles, Job Designs, Motivation		Support Other Elements; Need Top Management Support For Change
Supplier Communication	3	Activities, Interfaces, Communication, Culture, Supplier Responsibilities, Supplier Mechanisms, Supplier Technology	Improve Communication On Both Social And Technical Levels	Work Process Must Support Greater Supplier Involvement; Mechanisms Needed To Get It Done; Technology Must Be Improved At Supplier End

Exhibit 11.32 • Coseat Inc., As-Is Consolidation Worksheet

CONCLUSIONS

OK! You have gotten through a lengthy, complex, and perhaps tedious process. In fact, reading about it is significantly more tedious than doing it. But there were two purposes for doing this. First, the As-Is Assessment provides a baseline against which you can compare future performance after you have introduced CE. Second, by looking in detail at existing practices and structures, you have a rich database of opportunities for improvement. These will provide a strong foundation for design—and in this way, the new design will be grounded in the reality of your design practices rather than just being a set of blue-sky notions. The As-Is Assessment is clearly a complex process, but we have two suggestions to help keep it manageable.

1. Use informed judgments wherever possible. The purpose of assembling a cross-functional team is to get a rich set of informed judgments. Some assessments will be obvious, and all will rapidly agree; but in other cases, there will be disagreements, and these are where you should expend your resources on investigating further. Just keep in mind that we are looking for rough judgments, not precise numerical measurements.
2. Keep the analysis at a high level. Going into a great deal of detail will not be useful for the design. We are looking at designing the new CE system at a relatively high level (e.g., heavyweight versus lightweight project manager). Details to the design can be added in later by operating personnel as they are working on projects.

A good as-is assessment will make the later design stage go much more quickly and smoothly. And it will provide the basis for developing a business case to get management's attention. Now we can move on to the fun, creative part—Design!

CHAPTER 12

•

DESIGN THE NEW CE SYSTEM

[Life in a] New England town is like jazz. Admittedly, it's very constrained jazz, pianist Bill Evans say, not Fats Waller. But like jazz, it involves improvisation, and as in jazz, this does not mean that the result is accidental or that there are no rules. • Witold Rybcznski

In this third and perhaps most critical phase of the change effort (Figure 12.1) you will design the new, concurrent engineering-based system for product development. Although the arrows in Figure 12.1 show this as taking place after the assessment, in true CE fashion you can begin design work while you are doing the assessment. Indeed, implementation can (and should) start before the design is complete, but we'll discuss that in Chapter 13.

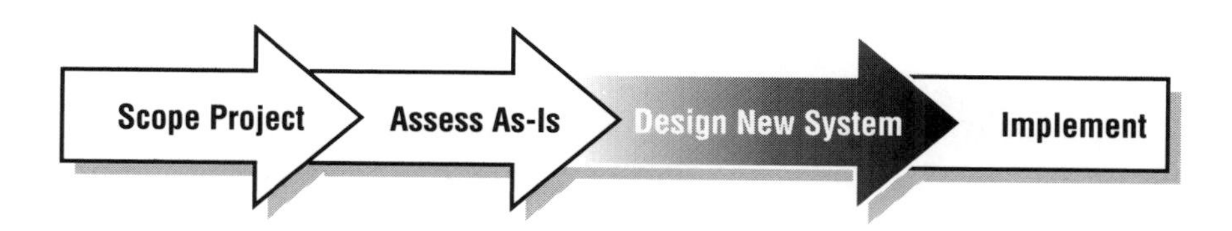

Figure 12.1 • Four Phases of the DAO Change Process

Ed Cazotzky has completed about 2/3 of his As-Is Assessment and his boss is getting concerned that nothing has happened yet. Ed's team has been working for over a month with nothing to show for their efforts yet. Ed thinks it would be nice if he could finish the assessment before he started designing the new CE system. On the other hand, his team has a pretty good idea of how things stand and where they should get started, so they agree that they can certainly begin their design work.

Purpose of the Design Phase

Design the New System means just that — design how your new CE system will look and operate before you implement it. If we may make the analogy with product development, most people would say it was perfectly obvious that you should design your product before you start to build it, rather than the other way around. Yet many companies will make significant organizational changes without taking the time to conduct a serious, systemic design effort for those changes.

This is not to suggest that these companies just go off one day and start changing the organization (although that does seem to happen from time to time!). No, they undertake some design effort, but it tends to be piecemeal, a little bit at a time, without considering the changes that might be needed in other parts of the company. The most common example of this is when a company introduces technological changes without changing other aspects of the organization which are needed to support the technology and take advantage of it. Consider the following (mostly true) story.

Magnificent Motors Inc. decided to introduce Electronic Data Interchange (EDI) technology between its automobile assembly plants and the plants of its supplier for seats, Comfy Seating Inc. The idea was that Comfy Seating would receive schedule data using EDI, and would easily and automatically send the data to its suppliers so the entire supply chain would always be up to date with Magnificent Motor's build schedule. Data for the next week were to be sent each Tuesday afternoon as Magnificent Motors updated its schedule. Prior to EDI, schedules were sent by fax. Someone at Comfy Seats would have to interpret the (sometimes blurry) fax, enter the data into Comfy's computer, then generate a report which would be faxed to Comfy's suppliers. With EDI, all of the interpretation and data entry were eliminated. Six months later, as Magnificent Motors evaluated the benefits of EDI, the conclusion was that it was a failure. Schedules at the second-tier suppliers (i.e., suppliers to Comfy Seats) were as far behind as ever. Further investigation revealed the problem: the individual at Comfy who was responsible for sending weekly schedules to Comfy's suppliers continued to send the schedule for the week out on Tuesday morning, the day after the week had already started. He had originally done this because it took about a week to deal with the faxed data from Magnificent Motors. Although he was receiving data by EDI and not fax, his work process hadn't changed. Suppliers still got data a week after it was generated by Magnificent Motors, when it was too late to really be very useful.

The problem in this story is that, although one company (Magnificent Motors) changed its process, its supplier (Comfy) didn't make an equivalent change. Moreover, the individual responsible for using the data clearly didn't understand what the data were being used for, nor was he motivated to think about it and change his process. If more care had been taken to design the system in which the EDI technology was being embedded, the system would not have been perceived as a failure, and substantial improvements would have been realized much more quickly. In this case, the problem was one person who, through training, could have easily changed his or her behavior. In the design of a complex product, there are often hundreds or thousands of people involved whose activities must be coordinated.

A Model for Designing New Systems

The process we recommend to design the new CE system follows a logical sequence which at least partially parallels the sequence we recommended for the As-Is Assessment (Figure 12.2).

1. ***Characterize Environment.*** In Chapter 2 we discussed the need to characterize your environment so that you could match your design elements to the environment. Here we revisit this idea and provide methods for developing the characterization.
2. ***Develop Design Principles.*** Here we suggest you develop a set of "principles" that will be simple enough to guide much of your design work without constant referral to elaborate benchmarks and literature reviews.
3. ***Design the CE Work Process.*** This means to design the tasks, work and information flow, and timing of tasks. We suggest doing this early (at least at a high level) since getting product development work done is your primary task, and you will need this for the rest of the process.
4. ***Design the Internal Organization.*** The organization design is the embodiment of how the organization will structure itself to perform the tasks described in the work process. This should follow the design of the work process, but could be concurrent with design of the people systems, technology, and supply chain context.
5. ***Design the People Systems.*** These are the supporting structures for the people who must perform the work and function within the organization. This should follow the design of the work process,

but could be concurrent with design of the internal organization, technology, and supply chain context.

6. ***Design the Supplier Relations System.*** When you design the supply chain context, you are making critical decisions about the environment in which everything else will operate. This should follow the design of the work process, but could be concurrent with design of the internal organization, technology, and people systems.
7. ***Design the Technology.*** Technology needs to support the work people do and must fit within the internal and supply chain context. On the other hand, the design of those elements must be informed by the technical possibilities. Hence this design will take place concurrently with the designs of the organization, people systems, and supply chain, rather than being treated as a separate step.
8. ***Consolidation and Feedback.*** Here you consider the interactions among the design elements, and provide feedback to each so they become mutually supportive.

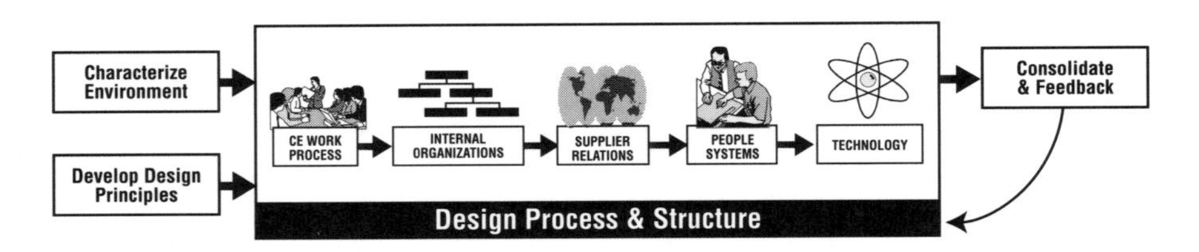

Figure 12.2 • Model for Designing the New CE Systems

Remember that in the As-Is Assessment, you developed some priorities for what you want to change. We aren't throwing those priorities out the window here. However, we do need to show you how to change any and all of the elements. You can (and probably should) choose to keep some elements the same. But you need to *know* how to change any of them if that's what you choose. Consequently, the structure of this chapter will be to assume you want to change all of the CE elements, since that provides us with the opportunity to show you how to design the necessary changes in each.

Characterize Your Environment (Measuring Environmental Uncertainty)

As we described it in Chapter 2, Environmental Uncertainty is the extent to which the environment around your company is unpredictable. There are three components to environmental uncertainty: *predictability of change*, *rate of change*, and *complexity*. We will provide you with a relatively simple method to quickly assess your company's environment in each of the three components.

MEASURING PREDICTABILITY OF CHANGE

Following are three vignettes that represent high, medium, and low on Predictability of Change. Score yourself 1, 2, or 3, depending on which comes closest to your own situation.

Predictability of Change Vignettes

1. ***High Predictability Example.*** The market for most traditional foods (e.g., frozen peas, canned beans) is relatively easy to predict. It doesn't change much from year to year, and seasonal changes are quite predictable (e.g., ice cream sales rise in the summer). The rate of change in computer chip technology is another example of highly predictable change.
2. ***Medium Predictability Example.*** The current automobile industry has medium predictability. While overall sales are relatively stable (mostly tracking economic conditions), the mix of vehicles (e.g., luxury cars versus sport-utility vehicles, price concerns,) can change significantly over a few years. Given the long lead time to develop a vehicle, this instability can result in automakers being significantly out of synch with the market. Consider Ford's introduction of a new Taurus in 1996, with a relatively high price caused by the inclusion of too many non-essential features. Ford was forced to quickly develop a more basic version of the Taurus and cut prices sharply.
3. ***Low Predictability Example.*** Both the market and the technology in the biotechnology industry are low in predictability. Some amazing new technology might be right around the corner that would revolutionize the way in which biotech products are developed or made, or it might not — the industry is clearly in tremendous ferment, but both the direction and pace of change seem to be totally unpredictable.

Unless your company is in one of the industries we have just described, you may feel your company (or industry) doesn't quite fall right into one of these categories. Rather, you may feel you fall between categories or fall into several categories. Here is a tie-breaker.

Predictability of Change Tie-Breaker

How well are you able to predict the kinds of changes that your company will face over the next few years?

a. We usually can predict what's happening short- to medium-term in most environmental sectors fairly well.
b. We can predict what's happening short to medium-term in most sectors fairly well, but not in others; *or* we can predict most sectors short-term, but not medium-term at all.
c. We can't predict much of what's going to happen in the short-term, much less medium-term.

If you were uncertain about your category (i.e., you fell between two categories) *and* you answered "a," then move yourself to the next higher category; if you answered "c," then move yourself to the next lower category; if you answered "b," move to the middle predictability category. Be careful with this one, low predictability is associated with higher uncertainty, and hence gets a "higher" score for this.

Coseat Inc. rates at medium predictability. This is partly because it is primarily in the automotive industry. In seats, there is some new technology being developed such as new fabric, new cushioning, and, most important, new types of power controls. However, none of this is very radical and it is not too hard to forecast the changes which will take place.

Measuring Rate of Change

Following are three vignettes that represent high, medium, and low on Rate of Change. Score yourself 1, 2, or 3, depending on which comes closest to your own situation.

Rate of Change Vignettes

1. ***Low Rate of Change Example.*** The industries which produce dinnerware and china (e.g., plates and bowls) haven't changed much for many years. While the specific shape and colors of the product changes regularly, the fundamentals have changed very slowly, with only the occasional introduction of a new material (e.g., various plastics).
2. ***Medium Rate of Change Example.*** The office furniture industry has a medium rate of change. Every ten years or so, there is a new trend in office design which results in radical changes in the kinds of furniture made. For example, the widespread introduction of computers resulted in significant changes in furniture beginning in the mid-1980s. Recently we have seen the beginnings of a more mobile workforce which may only have temporary quarters available to them — the industry has been developing furniture suitable for them.
3. ***High Rate of Change Example.*** The personal computer industry has a high rate of change. This is the result of the rapid change in processing power of the chips used, and the steadily increasing memory and disk space demands of more increasingly powerful software. Thus, computer makers must cope with developing new systems that take maximum advantage of increasing power and still keep prices low in a highly competitive market.

As with predictability, you may feel your company (or industry) doesn't quite fall right into one of these categories, but rather falls between categories or falls into several categories. Here's a tie-breaker.

Rate of Change Tie-Breaker

What would you say is the rate of change that your company will face over the next few years?

a. There will be very few changes in most environmental sectors.

b. There will be a moderate amount of change in most sectors; *or,* there will be a high rate of change in a few sectors, but not in others.

c. There will be a lot of changes in most environmental sectors.

If you were uncertain about your category (i.e., you fell between two categories) and you answered "a," then move yourself to the next higher category; if you answered "c," then move yourself to the next lower category; if you answered "b," move to the middle Rate of Change category.

At Coseat Inc., the rate of change is medium. There are some changes in seat technology, but they only happen every few years. The market could be said to evolve more than rapidly change.

Complexity

Complexity refers to the number of dissimilar environmental sectors your company needs to deal with, in this case, for product development. Following are three vignettes that represent high, medium, and low on Complexity. Score yourself 1, 2, or 3, depending on which comes closest to your own situation.

Complexity Vignettes

1. ***Low Complexity Example.*** The small appliance industry deals with a relatively low complexity environment. It sells mostly to distributors and large retailers. While there is some government regulation, it is minor as compared to that in aerospace or automotive. Most products are sold on a national or at most regional basis. The level of technological complexity is also relatively low.
2. ***Medium Complexity Example.*** The construction industry has a medium complexity environment. It is highly regulated, although the regulation changes little from year to year. The technology varies a great deal —some of it will be very complex and high tech (HVAC, for example), while others will be quite simple (carpentry tools), as will the work techniques used. With a few exceptions, almost all of the work of the typical construction company is done within a single national (or more likely state or regional) boundary.
3. ***High Complexity Example.*** The commercial aerospace industry has an environment that is high in complexity. It sells worldwide to a wide variety of airline customers and is governed by an incredibly complex array of different government regulations. It includes an enormous supplier infrastructure, and it needs to provide support over decades, rather than years as in most industries.

Again, you may feel your company (or industry) doesn't quite fall right into one of these categories. Use this Complexity tie-breaker to make your final decision.

Complexity Tie-Breaker

How many of the following environmental sectors[1] are relevant to product development in your company?

1. ***Industry*** — competitors, competitiveness
2. ***Suppliers*** — raw material and parts suppliers, service providers, real estate market
3. ***Human Resources*** — labor market, unions, availability of education and training
4. ***Financial Resources*** — stock market, banks, private investors
5. ***Market*** — customers, potential customers
6. ***Technology*** — suppliers of production equipment, developers of new technology and materials
7. ***Economic Conditions*** — unemployment rate, rate of growth in the economy, inflation rate, interest rates
8. ***Government*** — laws and regulations, taxes, political concerns
9. ***Socio-cultural*** — values, beliefs, education, work ethic in both the community and the work force
10. ***International*** — participation in international markets, exchange rates, foreign customs, international competition

If you said that six or fewer sectors applied to your product development efforts in your company, then move down to the next lower category. If you said seven or more sectors applied to you, then move up to the next higher category.

Coseat Inc. scores high for the complexity of its environment. Its industry as a whole (the automotive industry) is relatively high in complexity. Coseat serves a global customer base, and its product is regulated to a significant degree through safety regulations. Since Coseat builds seat assembly plants near each of its customer's plants, it has a large number of locations in an unusually large number of countries and regions. This leads to a complex set of relations with a highly diverse workforce.

Combined Environmental Uncertainty Scoring

Once you have scores for the three components of Environmental Uncertainty, you should *average* the scores for your Combined Environmental Uncertainty score. We will use this score later in this chapter and in Chapter 13.

[1] The discussion of environmental sectors is based on a taxonomy developed by Daft (1992).

Coseat Inc.'s Combined Environmental Score is about 2.3 on a 3-point scale — fairly high in complexity. This was calculated by adding the two medium scores (2 points each) and the one high score (3 points) and then dividing by 3.

Develop Design Principles

There is no cookbook for designing your CE process and supporting context. But we believe the design will go much more smoothly if your design team can agree on a set of principles that define what a good CE process should look like.

What do we mean by principles? Principles are the essential beliefs we hold about the way things ought to be and on which we base our actions. They should be sufficiently *general* so they can apply to many different situations, and *fundamental* so they can provide powerful guides to action. For example, the constitutional principle of freedom of speech is both general and fundamental. We can find many applications in everyday life, though how the principle applies in any given situation is not always obvious and is the subject of great debate. In fact, the full-time job of the U.S. Supreme Court is to interpret general Constitutional principles, such as freedom of speech, as they apply to specific cases.

Your principles have to be developed by you. We will show you what Coseat Inc.'s principles are (Exhibit 12.1), but theirs may not be best for you. We suggest you develop your principles using the following process.

1. ***Assemble a cross-functional team representing key actors in the product development process.*** This probably should be your guidance team.
2. ***Review the Coseat Inc. principles*** (Exhibit 12.1). This can provide a basis for discussing what principles are and what kinds of principles you need for your organization. You might decide some of Coseat's principles are not what you consider to be principles. You might decide that some of their principles are very appropriate for you while others are not.
3. ***Identify the categories of principles needed.*** Coseat's principles were organized into three categories: work process, organizational design, and the product development process. You may have a different set of categories that make sense for your organization.
4. ***Brainstorm principles in each category.*** If you've read Chapters 1–8, you have a lot of background to add to your own

experiences. Encourage creativity. Don't judge or criticize. At this point, don't worry about specific wording.

5. ***Refine your principles.*** This might involve combining some, dropping others, and adding ones that did not come up in the original brainstorm. You may want to go back to Coseat's principles to see whether a concept that should be captured in your principles is missing.
6. ***Revise the principles based on stakeholder input.*** Review the principles with management and groups of stakeholders to get their input and buy-in, and revise the principles as needed. If you redesign things based on principles that only your small group believes in, then implementation will be very difficult.

What will you do with these principles once you have developed them? They will be useful throughout the design and implementation phases. They can suggest ways to redesign your organization and provide a set of standards for evaluating organizational options. They can also help guide the implementation process. For example, you can judge whether adjustments to the plan are consistent with the principles. Finally, they can provide guidance for continuous improvement of your new CE system.

Ed Cazotzky and his Guidance Team decide to develop a set of design principles to guide the work they are going to do next. Lacking a model, they just go ahead and review all the written material they have and then engage in a brainstorming session. The brainstorming session results in about 50 "principles," or at least statements that look vaguely like principles. They next undertake an exercise to refine these into a small set of categories. The result is 15 principles divided into three categories. These are then presented to three stakeholder groups – upper management, middle management at each of the affected functions, and design staff in the affected functions. Minor changes are made as a result of these meetings. The results of their efforts are in Exhibit 12.1.

Design the CE Work Process

Revising the Work Process

As we suggested in Chapter 3, there is a lot of disagreement about what it takes to design a work process. We are not going to attempt to resolve those differences here. Nor are we sure it matters very much, so long as you do not take an inordinate amount of time and effort to design your process. Our minimum recommendation is to build a high-level phases-and-gates process combined with a schedule. We also rec-

Exhibit 12.1 • Coseat Inc. CE System Design Principles

Work Process Principles

1. ***Non-value-added data and data flows should be eliminated to minimize inaccuracy and redundancy.*** While information is at the core of the design process, too much information can create overload and lead to waste, while inaccurate information can lead to bad decisions. Examine the information and work flow to minimize redundancy and opportunities for information to get distorted in the process of communication. A useful question is: does having this information add value to the core design process?
2. ***Necessary information and other outputs of work should go directly to the point of action in a timely manner.*** Too often, information travels everywhere in the organization except where it is needed most. For example, information on warranty claims might stop at quality control, then go to management, and never get to design engineers who can make the revisions which would eliminate the defects in the future. Data should go directly to people who are in a position to act on that information immediately without wasteful side trips through staff departments and layers of management.
3. ***Decisions should be traceable to the source.*** We should be able to identify where in the process a decision is first made so that feedback on the outcome of the decisions can be provided to the source. The source may be an individual or a group.
4. ***All parties to the development process should use a common vocabulary.*** Communication is unmanageable if terms used by the sending party mean different things to the receiving party. A common vocabulary ensures sender and receiver share the same understanding of the intent of the message and reduces waste due to misunderstanding and inappropriate actions.

Organizational Principles

1. ***Interrelated tasks should be assigned to intact work groups.*** When possible, assign relatively complete, interrelated tasks to intact work groups. These groups may be collocated, or provided with means for regular, effective communication. Avoid assigning pieces of highly interrelated tasks to different work groups separated geographically and organizationally.
2. ***Product-focused leaders should be appointed for all important development projects.*** Organizations based purely on functional leadership (e.g., managers of electrical engineering, information systems, manufacturing engineering, etc.) are not effective at bringing different specialists together as a team to focus on developing world class products. The organization should include influential leaders whose job is to focus on the design of particular product lines drawing on all appropriate functional specialties.
3. ***Appropriate coordinating mechanisms should be developed to manage interfaces.*** In any complex product development process that involves large numbers of people, it is impossible to get everyone into the same room dedicated full time to product development. For instance, suppliers will be located in different places, functions will report up different chains of command, etc. It is important that all key interfaces between different parties who must coordinate their tasks and decisions be managed with appropriate linking mechanisms (e.g., liaisons, project managers, etc.).
4. ***All stakeholders affected by a decision should be represented in the decision.*** If we fail to represent a stakeholder, two things can happen. First, we

(continued)

may miss critical input and make a poor quality decision. Second, the stakeholder may not be willing to support the decision imposed on them by others. Unfortunately, in any complex process, too many individuals are affected to involve every single one of them in any reasonable time frame. But it can be effective to represent groups of stakeholders in the decision, particularly if the representative is carefully selected.

5. ***Hierarchy should be minimized with decisions made at the point of action.*** Decisions should be made at the point of action, not several layers or steps removed. Multiple levels of management reduce the likelihood that strategic decisions made at the top will be accurately communicated to people doing the work at the bottom. Each additional level increases the likelihood that managers with superficial knowledge of specific design decisions will micro-manage and ultimately make bad decisions. If people doing the core work involved in product development are empowered to make decisions in their area of expertise, decisions will generally be of higher quality and will be made faster, thereby reducing overhead.
6. ***Single points of primary responsibility and authority are required for each major task.*** This is the organizational corollary to the work process principle that decisions should be traceable to the source. There should be one party or group that has clear responsibility and the authority to carry out each major task. Others may share in the task and even in the responsibility, but one party should have primary responsibility. Without this, it is likely that important decisions will not be made on time, finger pointing will occur, and deadlines will slip.
7. ***Motivation systems should reinforce organizational objectives.*** Motivation systems in established organizations have typically evolved for historical reasons and may have little relationship to the current mission and challenges in today's environment. When organizations are designed with a product focus and objectives are clear, the motivation systems should act to reinforce those objectives. Objectives to be reinforced can include product characteristics (e.g., reduced weight) as well as service characteristics (e.g., improving the quality of information provided to those who depend on your product and services to do their jobs).

Product Development Process Principles

1. ***A structured design process should be developed and used.*** A structured design process is a key coordinating mechanism. It tells all parties involved what inputs are needed from them and when. It should specify major activities, key decision points, and timing for product development programs. It need not be volumes long, but might be as simple as a small number of flow charts.
2. ***The voice of the customer should be the focus throughout the process.*** This means accurately identifying customer wants and needs, and keeping these in front of everyone involved throughout the development process. Tools such as Quality Function Deployment provide a useful framework for capturing the voice of the customer, but these are of limited value if this information is not utilized throughout concept formation, prototype development and revision, and preparation for production. All key stakeholders must hear the voice of the customer — preferably directly from the customers themselves.
3. ***Stable strategic targets should be developed early in any design project.*** These targets should reflect critical specifications, and should be determined strategically in the concept stage so that they do not have to be revised later in the development process. These targets should be based on the voice of the

(continued)

customer and used to review progress at critical checkpoints in the structured development process. By stable we mean they should not be a moving target changing from month to month. Stable strategic targets are important mechanisms for coordinating the activities of different parties in the development process.

4. ***Design history should be captured and re-used. Reinventing the wheel is obviously wasteful.*** One repository of design history is the human brain, but memory is less than perfect, and inevitable personnel movement limits the effectiveness of this mechanism. Critical design decisions and their rationale should be captured in a way that is usable and useful for future product generations. This might include use of QFD, continually updated design rules, engineering checklists, and a database of key test results.

[end of Exhibit 12.1]

ommend that each level should build *its own* phases, gates, and schedule designed to meet the requirements of the level above it. These approaches were discussed in Chapter 3.

In discussions about their work process and about various problems they were experiencing, the Coseat Inc. team realized that a major issue for them was that they didn't know enough about customer needs and wants. At least part of their analysis of this issue was that they didn't really start thinking about what a customer might want in a seat until they got an order for a seat, and then they simply did what their automaker customer asked of them. They began to realize that if they got more pro-active by explicitly thinking about customer needs before an order came in, they would be in a better position to sell and they would be much more

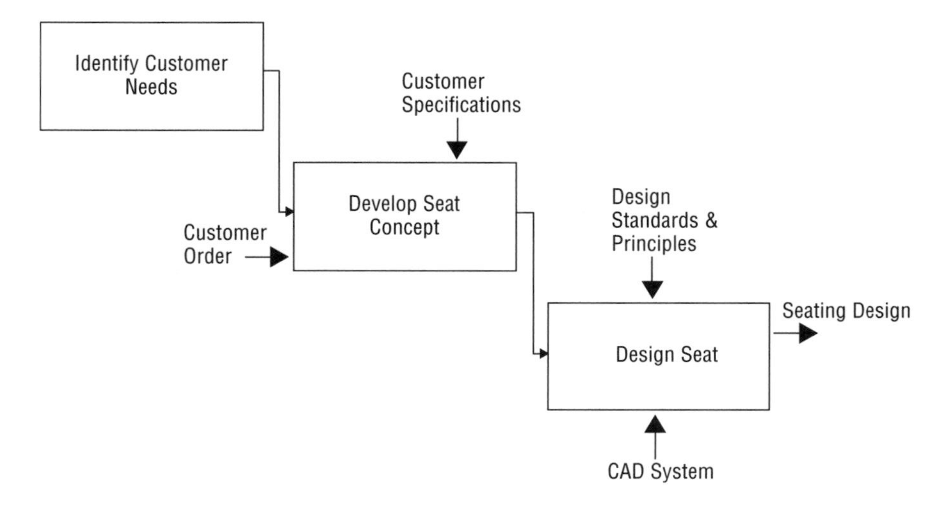

Exhibit 12.2 • Coseat Inc. To-Be Design Process (Design Seat System Phase)

responsive to what the automaker customer really wanted. They also thought about their principle that read, "The voice of the customer should be the focus throughout the process." *This led to a significant change in their design process (Exhibit 12.2). First, they introduced a new phase in their process, "Identify Customer Needs." This involved a number of activities that would test out various seating ideas on potential autobuyer customers. Second, they removed the step called "Get Market Reaction" as a high level step in the process. With their newfound early knowledge about the market, the team believed that this step could become a subordinate activity under developing the seating concept.*

This IDEF0 diagram in Exhibit 12.2 only shows the revised portion of the process — everything else remained the same. Note that Exhibit 12.2 is a modification of Exhibit 11.3 which shows the old process. Naturally, the team also develops the lower-level activities and detail in the new box, and makes revisions in the "Develop Seat Concept" box to reflect the addition of activities related to getting customer feedback. We won't show you all that detail.

The Coseat team also worked on revising the timing of these activities. Working closely with individuals from the various functions, they conclude that significant overlap could be introduced into the timing. They developed the Gantt Chart in Exhibit 12.3 to show the approximate timing.

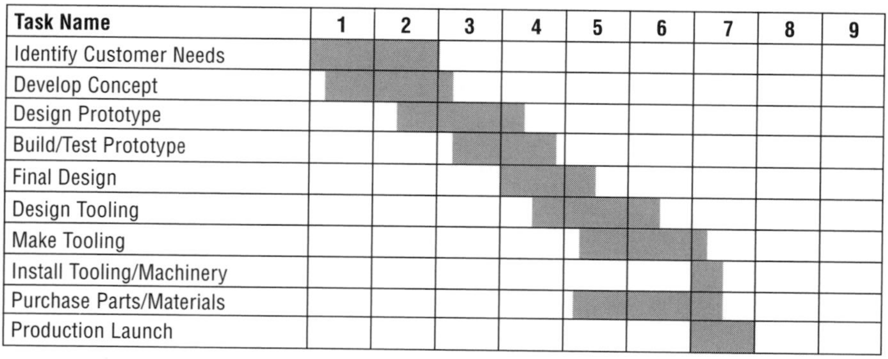

Task Name	1	2	3	4	5	6	7	8	9
Identify Customer Needs									
Develop Concept									
Design Prototype									
Build/Test Prototype									
Final Design									
Design Tooling									
Make Tooling									
Install Tooling/Machinery									
Purchase Parts/Materials									
Production Launch									

Exhibit 12.3 • Coseat Inc. Gantt Chart for New Process

You can compare Exhibit 12.3 to Exhibit 11.4 to see the difference in timing between the two processes.

We haven't shown an example in which the process is totally transformed since our emphasis in this book is not on the details of process redesign. Moreover, we believe the incremental type of change described in this section is more typical of the kinds of changes needed by most companies. The changes described by Coseat Inc. (in combination with other changes described below) will result in significant improvements in Coseat's customer responsiveness, even if it is not a totally revolutionary change.

DESIGN THE INTERNAL ORGANIZATION

Design of the internal organization includes eight interconnected steps:

1. ***Determine To-Be Task Responsibilities***. Decide which functions should be performing which tasks.
2. ***Determine Core Design Technology Needs by Function.***
3. ***Determine To-Be Communication Patterns***. Decide which functions need to communicate with each other, along with the approximate frequency, direction, and synchronicity.
4. ***Determine Coordination Needs***. Decide which functions need to coordinate, and at what level of frequency.
5. ***Design Coordination Mechanisms***. Decide which coordination mechanisms should be used for coordination between functions, given the to-be roles and responsibilities, communication patterns, and coordination needs.
6. ***Design Project Management Approach***. Decide which project management approach should be used, given the above decisions.
7. ***Design Organization Architecture***. Design the organization architecture to support all of this.
8. ***Consider Organization Culture***. Consider what elements of the organization's culture will need to change in order to support the changes you are planning.

DETERMINE TO-BE TASK RESPONSIBILITIES

Given a work process, you next need to decide which functions will have different levels of involvement and responsibility for each activity in the process. We can classify the involvement and responsibilities of the functional groups by modifying the Functional Involvement Matrix (Exhibit 12.4). The To-Be Functional Involvement Matrix shows the new "high-level" activities in the To-Be Work Process. If you compare Exhibit 12.4 with Exhibit 11.11a, you see that Coseat Inc. has added an activity "Identify Customer Needs." They have also eliminated "Get Market Reaction" as an activity at this level. You will also see that each cell in Exhibit 12.4 has a top and bottom half. In the top half, they have placed the old, As-Is responsibilities. In the bottom half, they have placed the new, To-Be responsibilities.

The easiest way to fill out the To-Be Functional Involvement Matrix is to start with the existing responsibilities. How do you want them to change, based on your As-Is Assessment, your benchmarking, and your design principles? Recognize that these are only first-cut decisions, which you will probably have to revisit later when you look at the downstream implications of them.

At Coseat Inc., Ed Cazotzky and his team modified their Functional Involvement Matrix to reflect the revised work process in Exhibit

Phase	Activity	As-Is (Top) / To-Be (Bottom)	Project Management	Marketing & Sales	Design Engineering	Prototyping	Test Engineering	Tool Engineering	Tool Production	Manuf. Engineering	Purchasing	Suppliers
	ID Customer Needs	As-Is										
		To-Be	A	R	R	I	I			I	I/S	S
DESIGN SEATING SYSTEM	Develop Concept	As-Is	A	A	R							
		To-Be	A	R	R			S		S		S
	Design Prototype	As-Is	A		R	I						
		To-Be	I	A	R	S	S	S	S	S		S/R
	Build/Test Prototype	As-Is	A		A	R						
		To-Be	I		A	R	R			S		S/R
	Final Design	As-Is	A		R			I	I	I	I	I
		To-Be	A	A	R		I	S	S	S		S/R
DESIGN & BUILD TOOLING	Design Tooling	As-Is						R	I	I	I	I
		To-Be	I					R	S	S	I	S
	Make Tooling	As-Is							R	I		
		To-Be	I					A	R	S		S
INITIATE PRODUCTION	Install Tooling/ Equipment	As-Is	I							R	S	R
		To-Be	I		S			S	S	R	S	R
	Purchase Parts & Materials	As-Is	I								R	S
		To-Be	A		S		S	S			R	S
	Production Launch	As-Is	I		I			I		R	S	S
		To-Be	S		S			S		R	S	S

KEY: R Responsible; A Approval; S Support; I Informed

Exhibit 12.4 • Coseat Inc. To-Be Functional Involvement Matrix

12.2. The team then discussed the level of involvement they thought would be best for each function in each activity in the process. In most cases, they were able to consider the current level of involvement and then decide whether that should change. In a very large number of cases, they wanted a higher level of involvement. The team took about three hours to complete this exercise. No formal decisions were made yet about any team assignments or responsibilities.

Exhibit 12.4 is really quite rich with evidence that Coseat learned a lot in its As-Is Assessment. As an example, look at the role of account managers before and after. In the As-Is analysis, the account managers (the Marketing and Sales function) drop out after they approve the concept. In the new work process, the account managers retain approval authority all the way through final design. This means the people who are closest to the customer retain authority over the design until it is completed. As another example, look at the changing roles of most functions in the early "Design the Prototype" phase. In the As-Is process, no one from tooling, manufacturing, or suppliers was involved in the design or test of the prototype. Now they play a variety of supporting roles which encourages their input and early planning.

It is important to realize that the Functional Involvement Matrix provides no information about how to implement these responsibilities. But it does serve as a set of requirements for other design elements. For example, it can suggest who should be on teams for a given task. It can also suggest that some communication media is necessary to keep all those who need to be informed up to date on the progress of the task. And it is very useful for deciding who should have access to basic design technologies.

Determine Core Design Technology Needs by Function

This is a good point at which to think about what functions need access to core technologies. By core technologies we mean the basic design technologies needed for each function to do their work, such as a CAD system for a designer, or rapid prototyping technologies for prototype makers. Later we will think about technologies that link different functions like communication technologies, data transfer, and standardized tools and methods.

The new work process should lead naturally to the selection of technologies. The analysis of task responsibilities will suggest what functional groups should have access to these technologies. In Exhibit 12.5, Coseat Inc. has described the core technologies needed within each

Function	Technology Description	Comments
Project Management	PC with word processing, spreadsheet, database, project management software; LAN with fileserver, internal e-mail and Internet access.	
Marketing & Sales	PC with word processing, spreadsheet, database software; LAN with fileserver, internal e-mail and Internet access.	
Design Engineering	CAD workstation + PC. Workstation includes 3D CAD with solid modeling capability, STEP translation capability, engineering analysis modules + ability to exchange data with the PC which contains design of experiments software. PC includes word processing, spreadsheet, project management software. QFD software available on selected systems. LAN with fileserver, internal e-mail and Internet access.	Not all design engineers will use CAD though all should at least learn how to access files and read them on the screen.
Prototyping	Access to several PCs; Stereo Lithography rapid prototyping system which is programmed from the PC; PC based CAD system which can accept STEP files from designer's workstation over the LAN. Supplemental manual tools with NC programming to cut clay models. LAN with fileserver and internal e-mail.	
Test Engineering	Personal computer with word processing, spreadsheet, and specialized data analysis programs; LAN with fileserver and internal e-mail.	
Tool Engineering	CAD workstation - includes 3D CAD with solid modeling capability and STEP translation capability. LAN with fileserver and internal e-mail.	Tool design will be done on CAD.
Tool Production	CNC equipment to cut tools programmed from CAM system; CAM can use CAD files accessed over LAN. LAN with fileserver and internal e-mail.	
Manufacturing Engineering	PC with CAD, word processing and spreadsheet; LAN with fileserver and internal e-mail.	
Purchasing	PC with word processing, spreadsheet, MRP system; LAN with fileserver, internal e-mail and access to Internet.	

Exhibit 12.5 • To-Be Core Internal Technology Description Worksheet

function. They have not yet attached technologies to specific jobs; that will come later on.

DETERMINE TO-BE COMMUNICATION PATTERNS

We next need to think about how different functions should communicate. Just as you did with the Functional Involvement Matrix, you can start with your As-Is Cross-Functional Communication Matrix and convert it into a To-Be version by splitting the cells in half. The top half of each cell represents the As-Is, while the lower half of each cell represents the To-Be.

At Coseat Inc., Ed Cazotzky and his team discussed the direction and frequency of communication which they thought would be best for each pair of functions. In most cases, they were able to consider the current level of communication and then decide how that should change. In a very large number of cases, they wanted a higher level of communication frequency and greater synchronicity. The team took about two hours to complete this exercise.

Looking at the To-Be Cross-Functional Communication Matrix (Exhibit 12.6), we can see a continuation of themes started in the Functional Involvement Matrix. Look at the intersection of design engineering with Marketing and Sales. In order to support the account manager's new role as authority over design activities, there needs to be much more intense communication between the account managers and the design engineers. This is reflected in the To-Be communication as being high intensity and two-way synchronous communication. Some changes are much less dramatic. For example, in the As-Is situation, test engineers did not communicate with tooling engineers or toolmakers. In the To-Be, they will be expected to send their test reports to those functions. This is represented by a one-way arrow with low intensity.

There are probably several ways to achieve any given level of communication intensity and directionality. These will be discussed later when we talk about Designing Coordination Mechanisms.

DETERMINE COORDINATION NEEDS

The next step is to revise the Cross-Functional Coordination Matrix to reflect your new needs for coordination between the functions. You should do this after you've done the Functional Involvement and Communication matrices, since those will frame your thinking about who should do what, and who needs to talk to whom.

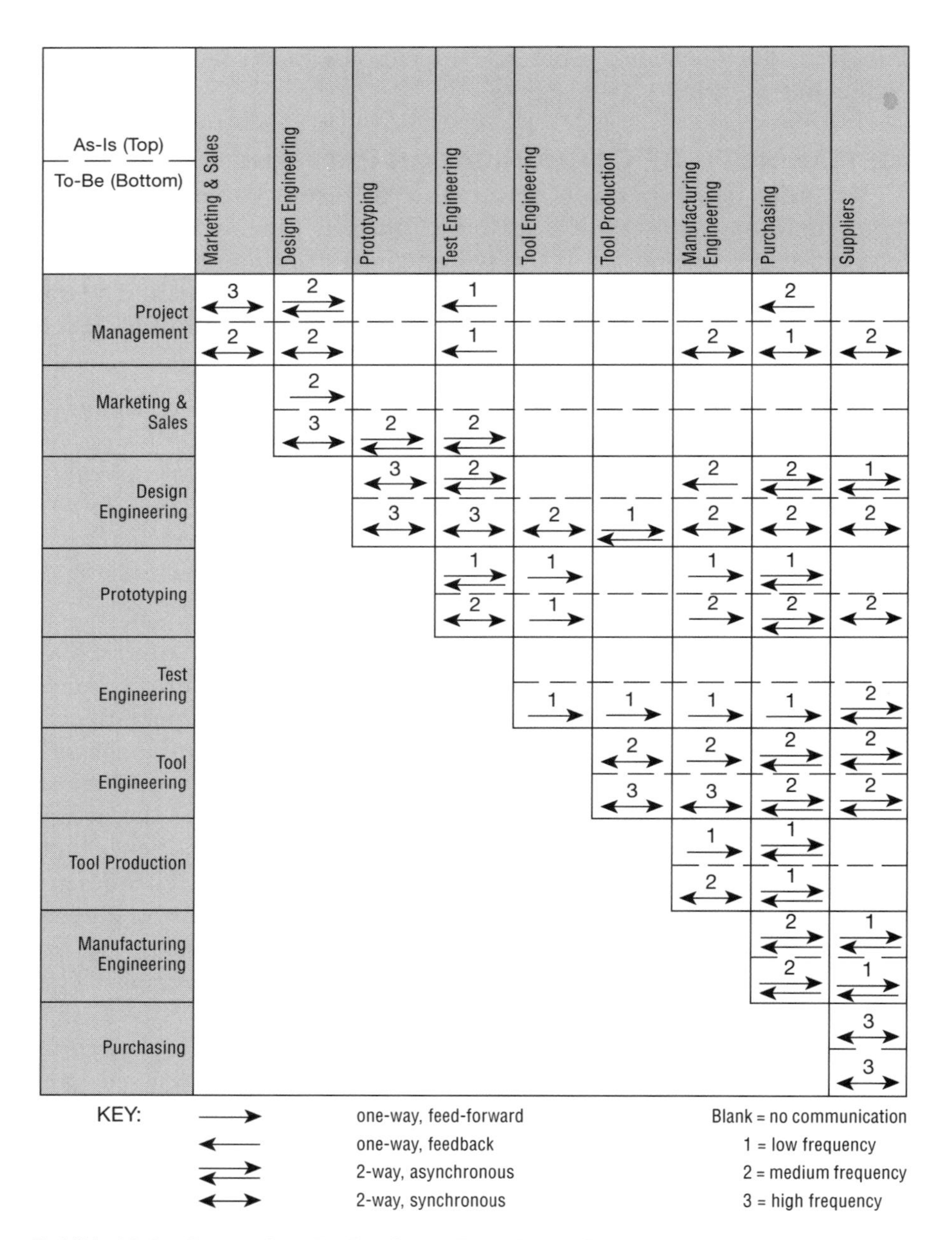

Exhibit 12.6 • Coseat Inc. To-Be Cross-Functional Communication Matrix

Ed Cazotzky and his team at Coseat Inc. move on to their To-Be Cross-Functional Coordination Matrix (Exhibit 12.7). First they look at the To-Be Cross-Functional Communication Matrix. They only work on cells for which there is communication in that matrix — if there's communi-

Cross-Functional Coordination Matrix	As-Is (Top) / To-Be (Bottom)	Project Management	Marketing & Sales	Design Engineering	Prototyping	Test Engineering	Tool Engineering	Tool Production	Manufacturing Engineering	Purchasing	Suppliers
Project Management	As-Is	2 DS, WP	3 MA, O	3 MA, DS, S	3 DS, S	2 DS, S	3 DS, S	3 DS	3 DS	3 WP	2 O, MA
	To-Be	MA, O	MA, O	MA, S	MA, S, O	MA, S, O	MA, S, O	MA, S, O	MA, S	MA	MA, O
Marketing & Sales	As-Is			1 O							
	To-Be			MA							
Design Engineering	As-Is			2 O, S, MA	3 MA, WP	3 MA, WP, O	2 MA, WP, O	1 MA, WP, O	2 MA	2 MA, WP	2 MA, O
	To-Be			MA, O	MA, WP, S	MA, WP, O	MA, O	S, O	MA	MA, WP	MA, O
Prototyping	As-Is					3 MA, WP			2 MA		
	To-Be					MA, WP	O		O	MA, O	MA, O
Test Engineering	As-Is						2 O		1 O		
	To-Be						O		O		
Tool Engineering	As-Is							3 MA, WP	2 MA		
	To-Be							MA, O, S	MA, O		
Tool Production	As-Is								2 O		
	To-Be								MA, O		
Manufacturing Engineering	As-Is										2 MA, O
	To-Be								MA		MA, O
Purchasing	As-Is										2 MA, O, WP
	To-Be										MA, WP, O
Suppliers	As-Is										
	To-Be										

KEY:
MA = Mutual Adjustment
DS = Direct Supervision
WP = Standardization of Work Process
O = Standardization of Outputs
S = Standardization of Worker Skills

1 = Inadequate Coordination
2 = Barely Adequate Coordination
3 = Very Active And Useful Coordination

Exhibit 12.7 • Coseat Inc. To-Be Cross-Functional Coordination Matrix

cation there's likely to be a need for coordination. In each of those cells they try to decide if the more impersonal forms of coordination (standards) fit. Then they determine if direct supervision is needed, and finally consider needs for mutual adjustment, such as teaming. After they've done the cells which had communication, they go back to see if any other cells need to be filled in with the impersonal methods such as standards. The logic is that if there's no communication needed, then neither direct supervision nor mutual adjustment makes sense. The entire exercise takes them about three hours.

The biggest As-Is versus To-Be difference you'll note in Coseat Inc.'s Cross-Functional Coordination Matrix is that project managers need to use mutual adjustment much more. This suggests they will need to become much more personal and flexible in their approach to working with the people they manage on their projects. In general there's much more expectation that coordination will involve standardization of outputs (i.e., I know what to expect from you) than there was in the As-Is.

DESIGN COORDINATION MECHANISMS

So far, the analysis in this section has been relatively abstract. We have decided who needs to communicate, and how much; we have decided who should have various kinds of responsibilities; we have decided in general what coordination mechanisms should be used; but we have not done anything really specific. For example, what specific teams should be formed, and who should be on them? Here we will design three specific types of coordination mechanisms: teams, communication media, and standardization methods.

Teams. One of the most critical coordination mechanisms is the team. As we discussed in Chapter 4, there are many different kinds of teams, and indeed most companies use a wide variety of these forms. In Table 4.5 (reproduced here as Table 12.1) we summarized the circumstances under which you would use the four different types of teams.

	Low Permanence	**High Permanence**
Low Task Scope	Task Force	Standing Committee
High Task Scope	Temporary Team	Semi-permanent Team

Table 12.1 • Types of Cross-Functional Teams

Ed Cazotzky's Coseat Inc. guidance team completed the Cross-Functional Team Matrix (Exhibit 12.8) in order to map out their proposed participation on various teams. They first decided that they would create a somewhat different team for each phase in their design process, and that these would be classified as temporary teams since they only exist for the single phase in the life cycle of an individual product. Some functions (e.g., project management, account management) stay on all of these teams. Other functions (e.g., test engineering, tool and die shop) are only members for one or two phases. In addition to the temporary teams asso-

ciated with the design process phases, the group decided there should also be a more permanent product based team which is focused on specific customers. Ed included several existing committees like the Quality Standards and the Design Standards Committees on the matrix. However, the group expanded the composition of these committees from their present membership. Finally, the group made note of task forces which Coseat establishes on a temporary basis for specific problem solving activities. The work of completing this matrix takes the group half a day since there is so much discussion about which functions should be represented on which teams.

Team Categories	Specific Teams	Project Management	Marketing & Sales	Design Engineering	Prototyping	Test Engineering	Tool Engineering	Tool Production	Manufacturing Engineering	Purchasing	Suppliers
Task Forces	Problem-Solving Teams	As Needed									
Standing Committees	Quality Standards Committee		X	X	X	X	X		X	X	
	Design Standards Committee			X	X	X	X		X	X	
Temporary Teams	Design Seat System Team	X	X	X	X				X	X	X
	Design And Build Tooling Team	X	X	X	X	X	X	X	X	X	X
	Prototype Team	X	X	X	X	X	X		X	X	X
	Manufacturing Team	X	X	X					X	X	X
Semi-Permanent Teams	Customer A Seat Team	X	X	X					X	X	
	Customer B Seat Team	X	X	X					X	X	

Exhibit 12.8 • Coseat Inc. To-Be Cross-Functional Team Matrix

Exhibit 12.8 doesn't show all of the possible teams that might be set up by a company like Coseat Inc. Ed and his group chose a fairly formalized way of evolving the membership of the product development teams over time. They decided to change the name of the team and its membership with the passing of each phase. For example, they started with a "Design Seat Team" and then changed its membership and name to the "Design and Build Tooling Team." They might well have decided to include everyone on a single team which stayed constant over time, recognizing the reality that people would vote with their bodies to participate or not as they were needed. Equally, they might have broken things down into subphases, with many more teams forming and disappearing over time. A more detailed version of the Cross-Functional Team Matrix could be developed to show individual participants rather than their functions.

Communication Technology. A second coordination mechanism is Communication Technology. In Table 12.2, which is adapted from Table

7.1, we define eight categories of communication and suggest the form of communication for which each level of technology would be best suited. For example, formal written messages (such as impersonal memos) are best suited for communication that goes one way and is relatively infrequent. In contrast, video conferences are best suited for two-way synchronous communication that takes place infrequently (given the relatively high cost of the medium). This is not to suggest that individuals engaged in two-way synchronous communication will never use shared databases (for example); indeed, they will typically use all of the low- or medium-rich forms of media when those are sufficient.

Communication Technology	Richness	Best for
Formal written messages (paper or electronic mail)	Low	← One-way, low frequency
Shared databases	Medium	⇄ Two-way asynchronous, low frequency
Computer Conferences	Medium	⇄ Two-way asynchronous, low frequency
Personal written messages (paper or electronic mail)	Medium	⇄ Two-way asynchronous, high frequency
Voice mail	Medium	⇄ Two-way asynchronous, high frequency
Telephone	High	↔ Two-way synchronous, high frequency
Video Conferences	High	↔ Two-way synchronous, low frequency
Face-to-face meetings (coming together from distant places)	Very High	↔ Two-way synchronous, low frequency
Face-to-face meetings (collocation)	Very High	↔ Two-way synchronous, high frequency

Table 12.2 • Prime Uses for Communication Technology

Table 12.3 takes the information in Table 12.2 and translates it into a form that will enable you to determine the best form of communication media for a given communication need that you might derive from the Cross-Functional Communication Matrix. Thus, if the Cross-Functional Communication Matrix (Exhibit 12.6) suggests having two-way asynchronous communication at a high (3) level of frequency, then you know you need personal written messages and voice mail.

Frequency of Communication	Direction of Communication		
	←	⇄	↔
1	Formal messages & shared databases	Databases and computer conferences	Telephone to meetings — depends on content
2	Formal messages & shared databases	Personal messages and voice mail	Telephone to meetings — depends on content
3	Formal messages & shared databases	Personal messages and voice mail	Collocation

Table 12.3 • Interpretation of Cross-Functional Communication Matrix

Combining the information in Table 12.3 with what you know about communication needs from the Cross-Functional Communication Matrix (Exhibit 12.6), you can now decide what communication media to make available to different functions, and you can educate the different functions about new ways they might communicate.

The Coseat Inc. Guidance Team next completes the Communication Media Matrix (Exhibit 12.9) in order to make sure everyone has the media available that they need. In each of the six categories of media, they list existing technologies and possible technologies they might add. They consider the Cross-Functional Communication Matrix and the Functional Coordination Matrix. In cases where a high level of communication or mutual adjustment is required, some serious consideration is given to ensuring these parties have the same media required for communication. Many decisions are obvious. For example, worldwide web access, e-mail, and conference calls are to be available to everyone. On the other hand, there are databases that the engineering-oriented staff (e.g., design engineering, test engineering) need to share with each other, but not with the marketing and sales staff. Equally, there is a sales-oriented database used by the project mangers and account managers which is needed occasionally by design engineers; no one else needs access to this database.

This work is tempered by financial reality as well. Although manufacturing engineering could make some use of in-building video conferencing capabilities to meet with the design engineers, the fact that the engineers are located in various plant facilities around the country would make installing video conference facilities for them prohibitively expensive. In the case of suppliers, the team can only recommend that suppliers have certain technology. For example, the "CuCme" technology (a form of video groupware attached to an individual workstation) will be used by design engineering and the prototype shop. Coseat will then recommend that suppliers involved with Coseat's design process get such technology, though whether or not they do so is out of Coseat's hands. Other Coseat suppliers will not need it.

Communication Media Categories	Specific Media Provided	Project Management	Marketing & Sales	Design Engineering	Prototyping	Test Engineering	Tool Engineering	Tool Production	Manufacturing Engineering	Purchasing	Suppliers
Formal Written Messages	Internet Email	X	X	X	X	X	X	X	X	X	X
	Library Access	X	X	X	X	X	X	X	X	X	
	Guidebooks	X	X	X	X	X	X	X	X	X	
Shared Databases	Common File Server Access	X		X	X	X	X	X	X	X	X
	Common File Server Access	X	X	X							
	Internet Web Site Access	X	X	X	X	X	X	X	X	X	
Personal Written Messages	Internet Email	X	X	X						X	
Computer Conferences	Conference Line On Phone	X	X	X	X	X	X	X	X	X	X
	AT&T Conference System	X	X	X	X	X	X	X	X	X	X
	WWW Site Conferences	X	X	X	X	X	X	X	X	X	X
	Bulletin Boards	X	X	X	X	X	X	X	X	X	X
Video Conferences	In-Building Video Conference Suite	X	X	X						X	
	CuCme			X	X						X
Face-To-Face Meetings	Conference Rooms	X	X	X	X	X	X	X	X	X	X

Exhibit 12.9 • Coseat Inc. To-Be Communication Media Matrix

The Communication Media Matrix in Exhibit 12.9 includes only six of the eight categories of media. This is because the two face-to-face categories have been combined; and since everyone already has a telephone, the telephone category is left out. Exhibit 12.9 also does not show all the communication media in use at Coseat Inc. Consider these to be examples of technologies which need to change.

Standardization Methods. As we discussed in Chapter 4, there is a wide variety of standards and tools which are used as coordination mechanisms. These were divided into three categories: work processes, outputs, and worker skills. We will address the issue of standardization of worker skills later under People Systems. The Standards and Tools Matrix (Exhibit 12.10) shows which standards need to be available for use within various functions in the company. This is not intended to be a complete list of all the standards and tools that any company might want to use, but it represents a range of examples.

Ed's team next moves on to discuss which standards and tools should be available for use by the different functions. He uses the Standards and Tools Matrix (Exhibit 12.10) to record the results of their discussions. Some of the standards and tools are already in use by these functions, so many decisions are very easy. In other cases (e.g., QFD) where a technique is not currently in use, a long argument ensues about the utility of the technique and who can gain the most benefit from it. They add engineering checklists as a tool to guide design based on their benchmarking of Toyota. The matrix takes about three hours to complete in total, although Ed's team is unable to do this in a single day because they have to investigate a number of different options before they can make final decisions.

	Standardization Categories	Specific Standards Or Tools Provided	Project Management	Marketing & Sales	Design Engineering	Prototyping	Test Engineering	Tool Engineering	Tool Production	Manuf. Engineering	Purchasing	Suppliers
Standardization Of Work Processes	Standard Operating Procedures	ECN Procedures	x		x					x	x	x
		Database Maintenance Procedure			x		x	x	x	x		
	Planning & Scheduling Systems	Project Schedule	x	x	x	x	x	x	x	x	x	x
	Monitoring Systems	Design Reviews	x	x	x	x	x	x	x	x	x	x
		Management Reports	x									
		Fiscal Reviews	x									x
	Structured Dev. Processes	Coseat Dev. Process	x	x	x	x	x	x	x	x	x	x
	Tools And Techniques	QFD	x	x	x					x		x
		DFMA			x							x
		FMEA			x		x			x		x
		FEA			x							x
		Robust Design			x							x
Standardization Of Outputs	Design Standards	Drafting Standards			x			x		x		
		Engineering Checklists			x			x		x		
		Design For Downstream Data Use Standard			x							
	Performance Metrics	Project Targets	x	x	x							x
		Corporate Goals	x	x	x							
		Quality Improvement Goals	x		x							

Exhibit 12.10 • Coseat Inc., To-Be Standards and Tools Matrix

Like any large company, Coseat Inc. has a large number of standards and standard operating procedures. We have not attempted to represent the full range of such things. In any case, it is probably not necessary to be exhaustive in this kind of exercise. Many of the standards and tools are not very applicable to product development and can be safely ignored for our purposes.

Design Project Management Approach

You should now be ready to make decisions about the project management approach you want to take for CE in your company. You can use the Project Management Worksheet (Exhibit 12.11) for this purpose.

Ed Cazotzky and his team complete the Project Management Worksheet (Exhibit 12.11) for the To-Be situation at Coseat Inc. They decide that they want their project managers to have more control over staff. As a consequence, they decide to allow project managers to select their own staff, rather than having them just given to them by the functional managers. Project managers will also evaluate the performance of staff who work for them on a project. This results in a "classic" heavyweight project management approach, as opposed to the previous lightweight approach. Ed updates his Consolidation Matrix to reflect the new design choices.

Project Management Responsibilities	Absent	Liaison	Lightweight	Heavyweight	Autonomous	Your Company
Distribute And Share Technical Information Among Project Members And Facilitate Problem Solving		X	X	X	X	*Yes*
Distribute Reports, Minutes Of Meetings		X	X	X	X	*Yes*
Set Project Goals			X	X	X	*Yes*
Schedule And Coordinate Project Activity			X	X	X	*Yes*
Allocate Funds And Equipment For Project			X	X	X	*Yes*
Select Staff For Project (Or Have Significant Influence)				X	X	*Yes*
Evaluate Performance Of Project Members (Or Have Significant Influence)				X	X	*Yes*
Evaluate Overall Performance Of Project Members					X	*No*
Long Term Professional Development Of Project Members					X	*No*

Exhibit 12.11 • To-Be Project Management Worksheet for Coseat Inc.

Design Organization Architecture

Let's say right up front that in many of the cases in which a company wants to do CE, the organization's architecture is not up for grabs. However, in a surprising number of cases, it is changed. You will note that we do not suggest that changing the organization architecture is essential. Remember, the architecture is just one coordination mechanism among many, though it is an especially powerful one. If your organization's architecture works against coordination but cannot be changed, then you will simply have to strengthen other coordination mechanisms.

The Coseat Inc. team was unsure if they were empowered to propose changes in organization architecture. However, they felt sufficiently concerned about how dysfunctional the current architecture was that they decided to go ahead and propose some changes. They felt quite strongly that the company was not customer-focused enough, and that the strongly functional organization was a major source of the problem. Therefore, they proposed a product line organization, based on administrative services and the three major product lines in the company: automotive, heavy equipment, and aerospace (Exhibit 12.12).

Note the *iterative* nature of this work. Changing the architecture has implications for other coordination requirements, in this case for cross-product line coordination. Upon discovering the problem of cross-product line coordination they had created, the Coseat team went back to look at previously reviewed coordination mechanisms that might be used to correct the problem.

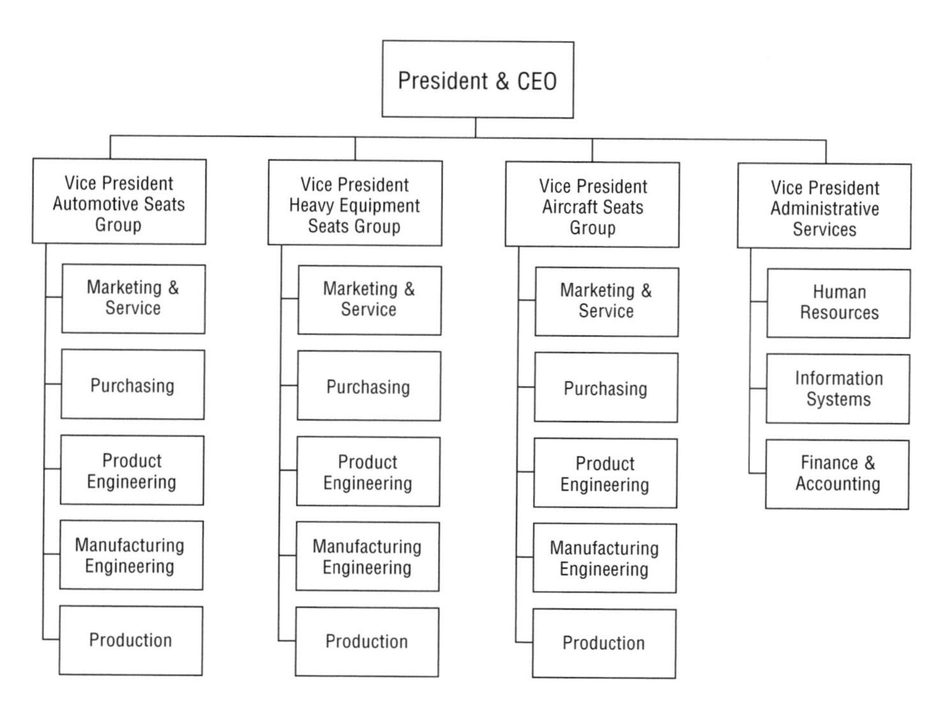

Exhibit 12.12 • Coseat Inc., To-Be Organization Architecture

However, given the broad spread of each of the product lines, it was clear that an architectural shift would not solve all of the company's problems regarding customer focus. Furthermore, there would be some loss of learning across product lines. Upon noticing this, the team decided to go back and create several semi-permanent teams which will serve the purpose of coordinating across product lines. For example, they plan to create several commodity teams which will coordinate purchasing in specific commodity areas, such as cushioning or seat frames. These new teams were then added to the list shown previously in Exhibit 12.8.

Consider Organization Culture

The cultural issues discovered in the As-Is Assessment could present serious barriers to the success of the new CE system you are designing. Since changing the culture is a difficult, long-term process, you need to address those issues in the beginning, rather than waiting while the cultural problems lie in wait to wreck your change efforts. We suggest that you consider each of the cultural problems you discovered in the Assessment, and decide how you would like the culture to be, and then derive a set of actions for each change you would like to introduce. You can see how Coseat Inc. did this in Exhibit 12.13.

Ed and his team at Coseat Inc. go through the list of cultural conclusions they came up with in the As-Is Assessment (Exhibit 11.14). For each one that they identified as a problem, they describe how they would like the changed culture to be. In a way, they are coming up with a vision for their culture. Finally, they consider actions that would need to take place if those changes were to actually be implemented. In many cases, the actions include top management making serious changes in the way they behave. Other changes are easier to implement. They record their deliberations in the To-Be Cultural Change Matrix (Exhibit 12.13).

Design the People Systems

There are four elements to be designed as part of the People Systems in the company: job designs, the technology associated with the jobs, skills acquisition systems, and motivation systems.

Job Designs

The task here is to change job responsibilities so that individual jobs are a better "fit" with doing CE and with the other parts of the organization that have already been designed, such as the project management approach and coordination mechanisms. In Chapter 11 we described

CE Elements	As-Is Cultural Conclusions	To-Be Cultural Changes	Actions Needed
Work Process	*People may not believe that teams are a good idea or that team work is better than individual efforts.* *May resist efforts to introduce formal processes.*	Widespread belief in the value of teams. Acceptance of the need for formal processes.	Top management must participate in teams. Reward team behavior in reward and appraisal systems.
Internal Organization	*Very authoritarian, may resist participative, team based activity, despite statement from management.*	Less authoritarian. More participative.	All levels of management must demonstrate this, especially the top.
Supplier Relations	*Supplier relations are very top-down; resistance should be expected from buyers to idea of suppliers participating closely with design teams.*	View suppliers as members of our enterprise, as partners where appropriate.	Formal changes in policy and procedure. Training of purchasing staff. Top management commitment.
People	*"Can-do" attitude should help with change in the long run. Emphasis on training will help. Status differences will need to be broken down before team based culture can be adopted.*	Reduced emphasis on status differences to confirm identity.	Formal systems need to break down status barriers - no more special parking, dining rooms. Mfg engineers pay needs to rise to be near design engineers; offices and other perks need to be equivalent.
Technology	*Technology per se will not be a problem, but engineers' attitudes toward soft issues will be a source of resistance to change.*	no changes	

Exhibit 12.13[2] • Coseat Inc. To-Be Cultural Change Matrix

how to assess current jobs. Based on that assessment, you need to change jobs in one or more of the following ways:

[2] Note that the column labeled "As-Is Cultural Conclusions" comes right out of the As-Is Cultural Assessment Matrix (Exhibit 11.14).

- ***Combine Tasks.*** Either enhance variety or establish identity (doing a "whole" piece of work).
- ***Form Natural Work Groups.*** Some form of team is the most common way to do this.
- ***Empowerment.*** Provide additional authority to the job.
- ***Establish Relations with the Environment.*** Add responsibility for relations with customers, vendors, and other elements of the environment.
- ***Provide Feedback.*** Arrange the job so the individual gets feedback from doing the work itself.

In Chapter 6 we discussed a set of contingencies for which we would use these strategies (Table 6.1, reproduced here as Table 12.4).

	Task Interdependence	Task Uncertainty	Task Skills Needs	Environmental Uncertainty
Combine Tasks	HIGH	LOW	LOW	
Form Natural Work Units	HIGH	HIGH +/OR HIGH SKILL	HIGH +/OR HIGH TASK UNCERTAINTY	
Empower	LOW	HIGH	HIGH	
Establish Relations with Environment				HIGH
Give Feedback on Work		HIGH		

Table 12.4 • Contingencies for Job Design Approaches

You should revise your Job Characteristics Matrix based on your new understanding of CE work processes, communication, and responsibilities. That will provide the basis for making decisions about whether any jobs should be changed. You should also add an assessment of environmental uncertainty as it applies to each job class.

Applying Environmental Uncertainty to Jobs. We have defined environmental uncertainty as a company-wide characteristic, i.e., it's something that applies to the whole company. Some definitions of environmental uncertainty are at the job or department level. For our purposes, we think it is better *defined* at the company level. However, jobs do differ in terms of how much they are *exposed* to that environment. For example, a purchasing agent is more exposed to the environment than a draftsman. We have therefore defined another contingency characteristic for jobs designs — *Job Exposure to Uncertainty*. This is rated as shown in Table 12.5.

		Level Of Environmental Uncertainty		
		Low Uncertainty	Medium Uncertainty	High Uncertainty
Exposure To Relevant Elements Of The Environment	Rare Exposure	Low	Medium	High
	Occasional Exposure	Low	Medium	High/Medium
	Regular Exposure	Low	Low	Medium

Table 12.5 • Job Exposure to Uncertainty Scoring

Add a column to the To-Be Job Characteristics Matrix (Exhibit 12.14) to reflect this additional assessment characteristic.

The Coseat Inc. team took their As-Is Job Characteristics Matrix (Exhibit 11.20) and revised it based on their new understanding of the jobs which resulted from their analysis and design work. The result of their efforts can be seen in Exhibit 12.14. The most significant change was the addition of the Job Exposure to Uncertainty scores for all positions. They also listed the possible job design approaches that can be derived from these characteristics. For example, they noted that the design engineer job is high on all four characteristics. Looking back at Table 12.4 we can see that this means they might form natural work groups, establish environmental relations, and give feedback on the work to the design engineers. For the designer's job, on the other hand, the only recommendation is to combine tasks. This might mean combining his tasks with those of the draftsman, or it might mean combining the draftsman's tasks with those of the design engineer.

Note that the approaches listed in the "Possible Job Design Approaches" column in Exhibit 12.14 are just *possible* approaches. The next step is to consider which approaches are currently being applied (e.g., the account managers and design engineers already have customer relations as part of their jobs), and which make sense to change. Actual determination of which "natural" work groups should be formed also needs to be made. You can see Coseat's initial resolution of this in the "Actual

	Task Interdependence	Task Uncertainty	Task Skill Needs	Job Exposure To Uncertainty	Possible Job Design Approaches	Actual Job Design Changes
Project Manager	H	H	H	H	Form Work Units, Env. Relations, Feedback	Form Leadership Team With Design Engineer And Accout Manager
Account Manager	L	H	H	H	Empower, Env. Relations, Feedback	Form Leadership Team With Design Engineer And Accout Manager
Design Engineer	H	H	H	H	Form Work Units, Env. Relations, Feedback	Form Leadership Team With Design Engineer And Accout Manager; Form Design Team With Designer And Prototype Builder
Designer	H	M	M	L	Combine Tasks	Combine Jobs Of Designer And Draftsman; Form Design Team With Designer And Prototype Builder
Draftsman	H	L	L	L	Combine Tasks	Combine Jobs Of Designer And Draftsman
Prototype Builder	H	M	H	L	Form Work Units	Form Design Team With Designer And Prototype Builder
Test Engineer	H	M	M	L	Form Work Units	Form Tooling Team With Tool Engineer, Tool Maker And Manufacturing Engineer
Tool Engineer	H	M	H	L	Form Work Units	Form Tooling Team With Tool Engineer, Tool Maker And Manufacturing Engineer
Tool Maker	H	M	H	L	Form Work Units	Form Tooling Team With Tool Engineer, Tool Maker And Manufacturing Engineer
Manuf. Engineer	H	H	H	M	Form Work Units	Form Tooling Team With Tool Engineer, Tool Maker And Manufacturing Engineer
Purchasing Agent	L	M	M	H	Environmental Relations	No Change

Exhibit 12.14 • Coseat Inc. To-Be Job Characteristics Matrix

Design Changes" column. Note that you don't see quite this same arrangement of teams in Coseat's initial cut at teaming (Exhibit 12.8). Having decided here that they might need some additional teams (a leadership team, and separate design and tooling teams), they should go back and revise the Cross-Functional Team Matrix. In doing so, they will consider if these choices are in fact the right ones.

As we have noted several times, there is no single best "solution" to these problems. As with almost any design problem, the payoff you achieve is the result of working through the various pieces iteratively until you reach a satisfactory solution that meets all your requirements. Then you need to try it.

Technology for Job Classes. So far we have discussed technology by functions (Exhibit 12.5) and by communication needs (Exhibit

12.9). We now want to take that into slightly greater detail and allocate technology by job classification. All you need to do here is to take the technology you listed under the functions and spread that out among the different job classifications within that function. In Exhibit 12.15 we have done that for the Design Engineering function at Coseat Inc. There are two primary job classes within Design Engineering: design engineer and designer. There is also a design manager position, but from a technology perspective we can ignore that job.

Job Class	Technology Description	Comments
Design Engineer	CAD workstation + PC. Workstation includes 3D CAD with solid modeling capability, engineering analysis modules + ability to exchange data with the PC which contains design of experiments software. PC includes work processing, spreadsheet, project management software. QFD software available on selected systems. LAN with fileserver, internal e-mail and Internet access.	Not all design engineers will use CAD through all should at least learn how to access files and read them on the screen.
Designer	CAD workstation - includes 3D CAD with solid modeling capability and STEP translation capability. PC includes word processing, and spreadsheet software. LAN with fileserver, internal e-mail and Internet access. LAN provides ability to exchange product data over the Internet.	This is for the combined role of designer and draftsman.

Exhibit 12.15 • Coseat Inc. To-Be Technology by Job Classification

Note the differences between the two positions. The design engineer has some capabilities that the designer lacks, including engineering analysis and design of experiments software, as well as QFD. The designer has STEP translation capability, which is lacking in the engineer's system.

Skills Acquisition

There are two basic issues to address regarding skills. The first is how to fill skill gaps that were identified in the As-Is Assessment. It is essential to fill the most critical of these quickly or else CE may be impossible to execute. The second issue is to change the systems by which skills are acquired and enhanced (if necessary) so that skill gaps do not reappear in the future.

Filling the Skill Gaps. The approach to filling the skill gaps is quite straightforward. Based on the Skill Profiles you completed in the As-Is Assessment (Exhibit 11.21), identify the gaps, prioritize them, and plan for how to fix them. You can use the Skills Gaps Matrix (Exhibit 12.16) to help work through this.

Position	Gaps	# With Gap	How Critical For CE?	How To Fill?
Project Manager	Project Management Skills	3	Very Important	5-Day PM Course
	Engineering Skills	1	Very Important	Probably Can't - Individual Should Not Take Role Of PM - Return To Purchasing Role
Account Manager	None			
Design Engineer	Design Troubleshooting	3	Very Important	1-Day Troubleshooting Workshop; Mentoring By Senior Engineers
	Basic CAD Skills	8	Mildly Important	3-Day Training Course Provided By Vendor Or Local Community College
	Interpersonal Communication	4	Very Important	1/2-Day Communication Workshop; Mentoring By Senior Managers
Designer	Interpersonal Communication	2	Important	1/2-Day Communication Workshop; Mentoring By Design Supervisor

Exhibit 12.16 • Skill Gaps Matrix

The Coseat Inc. team next works on their skill gaps. For each Skill Profile, the team identified gaps by looking at important skills (e.g., above a score of 3 on the importance rating) for which the job holders lack a high level of the skill (below a skill level of 3). They highlighted these gaps and then went to the experts — the people involved and their managers. They discussed the gaps along with representatives from Coseat's HR department. From these discussions, they came up with the numbers of people with significant skill gaps, the importance of the gap, and the proposed solutions. While funding was not available to fill all the gaps, the importance ratings suggested which ones should be filled first. Since the affected managers were part of the process of completing the Skill Gaps Matrix, their support for increased levels of training was relatively easy to achieve. As an example, look at the Project Manager changes suggested in Exhibit 12.16. Three of the Project Managers lacked project management skills that could be filled with a 5-day project management course. One of the current Project Managers came out of purchasing and lacks engineering skills. The recommendation is for that person to return to his original purchasing job, since he can't easily obtain the needed engineering skills.

The Skill Gaps Matrix records decisions made by the Guidance Team with input from the right people. Note that Exhibit 12.16 only includes a few job classes as an example.

Changing the Skills System. The skill gaps at Coseat Inc. that we just discussed came about because the company wasn't paying enough attention to the skills of its employees. In response, the Coseat team worked with the various managers to upgrade training and education programs where necessary, and revised the Training and Education Assessment Worksheet (Exhibit 12.17) to record the updates about these systems. As an example, designers now get training in design and customer contact skills in addition to CAD system training.

Job Classes	Type Of Training	Training Sources Used	Hours/ Year	Education Opportunities
Account Manager	1 Workshop/Year + 1 Professional Conference	Professional Organizations	40	Tuition For Grad Degree
Design Engineer	1 Workshop/Year + 1 Professional Conference	Professional Organizations	40	Tuition For Grad Degree
Designer	CAD System Updates	CAD Vendor	15-30	Tuition For BS In English
	Design Skill Upgrades & Customer Contact Skills	Various	10-15	
Tool Engineer	1 Workshop/Year + 1 Professional Conference	Professional Organizations	40	Tuition For Grad Degree
Tool Maker	CAD System Updates	CAD Vendor	15-30	Tuition For AA Or BS
	Design Skill Upgrades	Local Community College	10-15	
Test Engineer	1 Workshop/Year + 1 Professional Conference	Professional Organizations	40	Tuition For Grad Degree
Manufacturing Engineer	1 Workshop/Year + 1 Professional Conference	Professional Organizations	40	Tuition For Grad Degree
Prototype Maker	CAD System Updates	CAD Vendor	15-30	Tuition For AA Or BS
	Design Skill Upgrades	Local Community College	10-15	
Purchasing Agent	1 Workshop/Year + 1 Professional Conference	Professional Organizations	40	Tuition For Grad Degree

Exhibit 12.17 • To-Be Coseat Inc. Training Assessment Worksheet

MOTIVATION SYSTEMS

In Chapter 11 we assessed three issues concerning Motivation Systems: goal consistency, appraisal systems, and reward systems.

Goal Consistency. In the case of Coseat Inc., both the Horizontal Goals Consistency Matrix (Exhibit 11.24) and the Vertical Goals Consistency Matrix (Exhibit 11.25) showed some inconsistency of goals which could disrupt the efforts to develop a CE system.

Ed and his team at Coseat Inc. realized right off that they would never eliminate all such inconsistencies in their systems, so they decided to prioritize. Initially they focused on the potential conflict between design engineers and manufacturing engineers. They could think of three strategies for resolving the inconsistency:

1. *One function can change its goals to be consistent with the other.*
2. *Both functions could change to become consistent.*
3. *The goals could be reinterpreted so as to redefine the inconsistency out of existence. If they do this, they will need to be careful to not just treat it as a semantic exercise.*

Unfortunately, the resolution to this problem is complicated by the changes made in organization architecture — specifically, the move to a product line architecture means that the inconsistency found between the goals of the functions may no longer be there once the new product line executives are in place. The Guidance Team assumes, for the sake of argument, that the architecture will be unchanged for the foreseeable future.

The team brings the various design engineering and manufacturing engineering managers together along with a representative from HR to discuss their concerns. These are reflected in the To-Be Horizontal Goals Consistency Matrix (Exhibit 12.18). It quickly becomes apparent that quality and cost are vitally important from the customer perspective, as well as from the perspective of top management, and that design engineering needs to assume some responsibility for meeting those objectives. While no changes appear needed in manufacturing's goals, all involved agree that manufacturing will need to change their practices substantially in order to accommodate design's continuing needs for innovation. This will get reflected in changes in the skills profile for manufacturing engineers, as well as changing patterns of communication between the two functions (already reflected in the To-Be Communication Matrix in Exhibit 12.6).

The team also noticed the vertical conflict between the GM for Product Design's goals for innovation and the Director of Design Engineering's

	Goals Of Each Function	Project Manager	Marketing & Sales	Design Engineering	Manuf. Engineering	Purchasing
Project Management	Time; Budget; Cost			Budget	Cost	Cost
Marketing & Sales	Accurate Data	ACCURACY VS. TIME				
Design Engineering	Innovation; Budget; Quality Cost	INNOVATION VS. COST				
Manuf. Engineering	Quality; Cost					Cost
Purchasing	Cost					

KEY:
TOP RIGHT = Common Goals (Small Letters)
BOTTOM LEFT = CONFLICTING GOALS (Capital Letters)

Exhibit 12.18 • To-Be Horizontal Goals Consistency Matrix for Coseat Inc.

goals to cut costs and labor hours (Exhibit 11.25). After some discussion with the individuals involved, they come to the conclusion that while there is some conflict between innovation and product cost and schedule, they're going to have to live with that conflict because the market is demanding it. On the other hand, the conflict with reducing total engineering staff hours is another story. No one on the team is empowered to deal with this issue. The best they can do at this time is to make sure that Robert Ferry, the GM for Product Design (Ed's boss) is aware of the issue. Ferry agrees to take it up with his boss to see if the conflict can be resolved. No changes are made in the Vertical Goals Consistency Matrix at this time, but if the conflict is resolved, it will be reflected in changes to that matrix.

Another issue is the consistency of goals and metrics. Here you change either goals or metrics to ensure consistency.

The Coseat Inc. team now revises its Goals and Metrics Worksheet for each function. In the example of Design Engineering (Exhibit 12.19), two new goals are added based on the discussion which resulted in the To-Be Horizontal Goals Consistency Matrix (Exhibit 12.18) — meet product quality targets and help meet manufacturing cost targets. Two goals were removed: assist other teams and keep a neat workspace. The goal to assist other teams was removed for being too nonspecific and because the two new goals were a specific version of it. The neat workspace goal was removed since it wasn't being measured, nor was it likely to be. Note that all metrics are now associated with at least one goal, and all goals have at least one measure.

Function: *Design Engineering*	**Metrics**				
Goals	Fewer Engineering Changes	Fewer Design Staff-Hours	Keep Within Design Phase Cycle Time	Reduce Product Cost	Meet Customer Schedule
Keep Within Project Budget	+	+	0	+	+
Develop Innovative Products	-	-	-	-	-
Assist Other Teams When Requested	+	0	+	0	0
Keep A Neat Work Area	+	0	+	+	+
Run Efficient Meetings	0	+	0	0	+

Key:
+ = Metric Helps Indicate Goal Accomplishment
0 = Metric And Goal Are Unrelated
- = Metric Works Against Goal Accomplishment

Exhibit 12.19 • To-Be Goals and Metrics Worksheet for Coseat Inc.

Appraisal System. The Appraisal System revision should look back to the As-Is Assessment in order to make the system fairer and more accurate.

Ed Cazotzky and his team at Coseat Inc. worked with the HR department to revise the performance appraisal system. While they couldn't develop a perfect system, they were able to agree to some substantive changes that should improve the system quite a bit. The system will be more open, more complete, and more behaviorally based. The changes in the system are shown as the underlined items in Exhibit 12.20 (compare to Exhibit 11.27).

Performance Appraisal System Assessment Worksheet

This worksheet is used to assess appraisal practices for either a single job or a class of jobs. For example, management jobs may be appraised differently from hourly jobs, but all management jobs may be appraised with the same system. The same worksheet can be used for the whole class of jobs which is assessed using the same system.

Job Classes Assessed - *all classes below manager*

1. **Are there regular established times for formal appraisals?** *Yes - annual*
2. **Are regular appraisals supplemented with assessment and feedback as needed in day-to-day job activities?** *depends on supervisor - inconsistent*
3. **Is the process open to continuous input from employees, colleagues, subordinates and supervisors?** *<u>Yes</u>*
4. **Does the measurement system use a stable and important set of performance domains?** *<u>Yes, domain set is revised to insure completeness</u>*
5. **Are performance judgments behaviorally anchored? That is, can concrete examples be given for appraisers' judgments?** *<u>Yes, required</u>*
6. **Are appraisals tied to business results and employees contributions?** *partially - link to schedule and cost*
7. **Does the appraisal allow for inclusion of suggestions for improvement?** *depends on supervisor, most do this*
8. **Does the appraisal reflect the acquisition and use of important knowledge and skills?** *<u>Yes</u>*
9. **Does the appraisal get inputs from multiple levels and functions (supervisors, peers, subordinates, self, departments, teams, support staff, customers)?** *<u>Yes</u>*
10. **Are appraisals followed by feedback, development, counseling, and training?** *depends on supervisor, <u>supervisors will be trained to do this</u>*

Exhibit 12.20 • Coseat Inc. To-Be Performance Appraisal System Assessment Worksheet

Reward System. As with the Appraisal System, the Reward System review should look back to the As-Is Assessment to fix problems.

Ed and his team also worked with HR to revise some elements of the Reward System at Coseat Inc. (Exhibit 12.21). They managed to introduce a slight change in the way raises and bonuses were awarded, by making sure that project managers for whom employees worked were part of the rating system. Professional development also ceased to be considered a "reward," but rather is now just like any other work activity — everyone does it, and for a specific reason. This is now consistent with the Education and Training system changes. Finally, Coseat introduced a profit sharing package, although Ed and his team had nothing to do with that change.

Job Title: Design Engineer	Check If You Use	Frequency	Perceived As Fair?	Basis	What Is Really Rewarded?
Hourly Wage Or Salary	X	Monthly	TBD	Individual Background	Education And Experience For Base Pay
Pay Raises	X	Annual	TBD	Individual Performance As Rated By Project Managers	Quality Of Project Work
Bonuses	X	Annual	TBD	Individual Performance As Rated By Functional Manager	Meeting Project Deliverables For Time And Cost, But Higher Or Lower Depending On Company Performance
Profit-Sharing Or Gain-Sharing	X	Annual	TBD	Company Performance	Working Toward Company Goals
Time-Off					
Stock Or Other Ownership					
Gift Equivalents Of Cash					
Recognition Awards (Plaques, Certificates)					
Professional Development, Education	X	Annual	TBD	All Employees Based On Discussion Between Manager And Employee	No Longer A Reward, Simply A Work Activity
Social Activities (Parties, Event Tickets, Etc.)	X	Annual	TBD	Company-Wide Party	Being Alive And Working For Coseat
Fringe Benefits (Pension, Insurance, Etc.)	X	W/Pay	TBD	Part Of Pay Package	Being Alive And Working For Coseat
Other Rewards					

Exhibit 12.21 • To-Be Reward System Assessment for Coseat Inc.

DESIGN THE SUPPLY CHAIN CONTEXT

Designing the Supply Chain Context involves five steps which are revisions of those taken as part of the As-Is Assessment.

1. ***Determine Supplier Roles in Product Development.*** Decide at a high level the roles that key suppliers will play. This might include dropping some suppliers.
2. ***Determine Supplier Responsibilities in Product Development.*** Based on their revised roles, what responsibilities will each supplier take on?
3. ***Determine Supplier Communication in Product Development.*** Based on their new responsibilities, how will suppliers communicate with you?

4. ***Suggest Supplier Integration Technologies***. Based on each supplier's roles, responsibilities, and communication, determine what integration related technologies will need to be in place at the supplier.
5. ***Design Supplier Management Mechanisms***. Design specific mechanisms to implement the new roles and responsibilities.

DETERMINE SUPPLIER ROLES IN PRODUCT DEVELOPMENT

Determining the new roles you want your suppliers to take in product development is a two step process (Figure 12.3). First, you decide what the *ideal* supplier in each category would be like. Second, you face the reality of your *actual* suppliers and decide how closely you want them to come to your ideal. While you obviously want all of your suppliers to be ideal, known constraints may make this impossible, at least in the near term. You have to face those facts, and in this section we provide a means to do that.

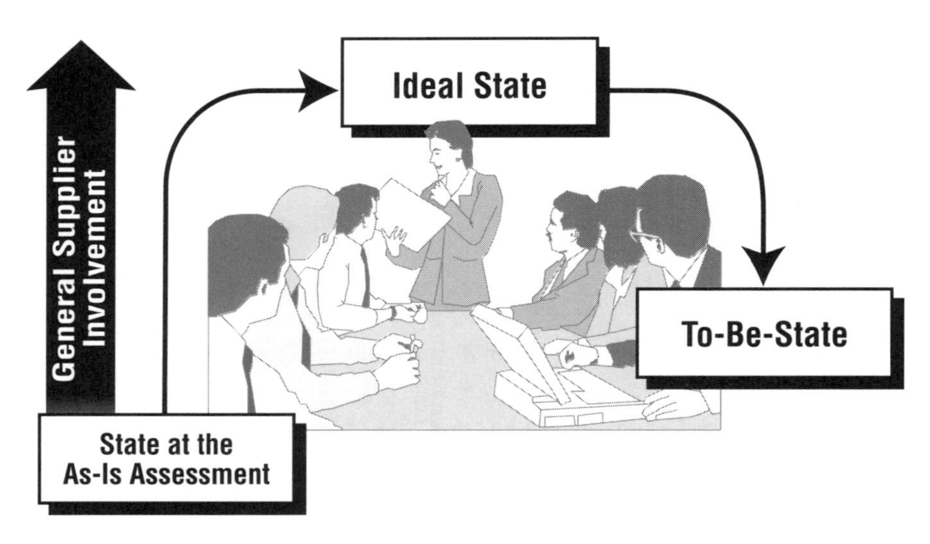

Figure 12.3 • Determining Supplier Roles

Determine Ideal Supplier Roles in Product Development. In the As-Is Assessment you divided suppliers into categories based on what products and services they provide for you (Exhibit 11.15). You also rated your existing suppliers on the roles they currently take in product development (Exhibit 11.18) — so you know what's currently happening. Now, in the best of all possible worlds, how would you like suppliers in each category to be involved in product development? Interestingly, the best of all possible worlds does not mean that all suppliers would be involved at the highest levels. First of all, there are many things you want to continue to do in-house. Second, there may be little or no need for heavy involvement depending on the specific nature of the product or service being supplied. For example, makers of commodity parts will typically have little or no involvement. Finally, some types of industries are typically so lacking in certain capabilities that you may find it hard to imagine that they will ever get involved in some ways — so why try to change the very nature of the way they do business?

The Coseat Inc. team now goes to work on the ideal roles they'd like Coseat's suppliers to play in product development (Exhibit 12.22). In general, they want to significantly increase the role of most suppliers. Tooling suppliers will continue to have a relatively limited role, based on their limited capability. The primary change for them will be that they will now be responsible for designing their own tooling. In contrast, the spring, fabric, and component makers can and should have a very large influence on the direction of the product. The simple parts makers, on the other hand, have less need to be involved.

Key: Assign each characteristic a score on a 1-4 scale. The anchor points are provided in parentheses. For example, assign 1 if the customer is solely responsible and 4 if the supplier is responsible and 2 or 3 if it is somewhere between.	**Tooling**	**Specialty Springs**	**Fabric**	**Make Parts**	**Components**
Component Design Responsibility (1 = Customer Only—4 = Supplier Only)	4	4	4	3	4
Product Complexity (1 = Simple Parts—4 = Complete Subsystems)	2	2	2	2	3
Level Of Specifications Provided (1 = Complete Designs—4 = General Concepts)	1	4	4	2	4
Supplier Influence On Specifications (1 = None—4 = Collaborative)	2	4	4	3	4
Earliness Of Supplier Involvement (1 = Late Prototype State—4 = Pre-Concept)	2	4	4	3	4
Supplier's Component Testing Responsibility (1 = None—4 = Complete)	3	4	4	4	4
Supplier Development Capabilities (1 = None—4 = Autonomous)	1	4	4	4	4
Average =	2.1	3.7	3.7	3.0	3.9
Overall Relationship (C = Contractual, P = Parental, M = Mature, PP = Partnership)	P	PP	PP	M	PP

Exhibit 12.22 • Coseat Inc. Ideal Supplier Roles in Product Development

Determine To-Be Supplier Roles in Product Development. With this ideal in mind, you can now look at your existing suppliers (and new suppliers as well) to see what roles you would like them to play in your product development process. Many companies have chosen to reduce the number of suppliers they deal with. Looking at a supplier's ability to meet your ideal along several dimensions of product development may provide a mechanism for making these decisions.

Within each category of supplier, the Coseat Inc. team inserts the names of their current suppliers (Exhibit 12.23). They do this for their expectations of what the supplier will be able to do over the next two to three years, rather than what might happen over any longer period of time. Almost all of the suppliers fall short of the ideal. Some fall quite short, but are retained because of special skills or special relationships. For example, Joe's Tool and Die is obviously less capable than Fred's Molds, but Joe does certain kinds of relatively simple dies very well and at a very low price. Marshall Wire gets dropped as a supplier because of their low level of capabilities and the ability of the other two spring suppliers to make their products.

	Supplier's Primary Product Service or Function									
	Tooling		Specialty Springs		Fabric	Make Parts			Components	
Key: Assign each characteristic a score on a 1-4 scale. The anchor points are provided in parentheses. For example, assign 1 if the customer is solely responsible and 4 if the supplier is responsible and 2 or 3 if it is somewhere between. **NOTE:** Cases Where TO-BE Has Changed From AS IS Shown In Bold-Face Italics	Joe's Tool and Die	Fred's Molds	Ann Arbor Spring & Wire	Springmasters	Burlington	Franklin Stamping	Lenawee Stamping	Henry's Molding	Mitch's Switches	Acme Frame
Component Design Characteristic (1 = Customer Only—4 = Supplier Only)	1	4	***4***	***4***	***4***	***3***	***2***	***3***	***4***	***2***
Product Complexity (1 = Simple Parts—4 = Complete Subsystems)	1	2	***2***	***2***	***2***	***2***	***1***	***2***	***3***	***2***
Level Of Specifications Provided (1 = Complete Designs—4 = General Concepts)	1	1	***4***	***3***	***3***	***2***	***2***	***3***	***3***	***2***
Supplier Influence On Specifications (1 = None—4 = Collaborative)	***2***	***2***	***4***	***3***	***3***	***2***	***2***	***3***	***3***	***2***
Earliness Of Supplier Involvement (1 = Late Prototype State—4 = Pre-Concept)	***2***	***2***	***4***	***3***	***3***	***3***	***3***	***3***	***3***	***3***
Supplier's Component Testing Responsibility (1 = None—4 = Complete)	2	3	***4***	***3***	***3***	***3***	***2***	***3***	***4***	***3***
Supplier Development Capabilities (1 = None—Complete, 4 = Autonomous)	1	1	***3***	***3***	***1***	***3***	***2***	***3***	***4***	***3***
Average =	1.4	2.1	3.6	3.0	2.7	2.7	2.0	2.9	3.4	2.4
Overall Relationship (C = Contractual, P = Parental, M = Mature, PP = Partnership)	**C**	**P**	**PP**	**M**	**M**	**M**	**P**	**M**	**M**	**P**

Exhibit 12.23 • Coseat Inc. To-Be Supplier Roles Matrix

DETERMINE TO-BE SUPPLIER RESPONSIBILITIES IN PRODUCT DEVELOPMENT

Having decided what roles you want individual suppliers to take, you can now translate that into specific responsibilities in the product development process. This can be done using the Supplier Responsibility Matrix as we did in the As-Is Assessment (Exhibit 11.16).

Based on the roles they developed for each supplier (Exhibit 12.23), the Coseat team determines levels of responsibility for each supplier in each major step in the product development process (Exhibit 12.24). In keeping with the overall higher roles played by most suppliers, those same suppliers are also taking on higher levels of responsibility. For example, consider the responsibilities of Ann Arbor Spring and Wire. In the As-Is situation (Exhibit 11.16), they had supporting responsibilities in the early stages of product development; in the To-Be situation, they take responsibility for early activities such as identifying part requirements. In addition to this expanded responsibility by many suppliers, there is much more information being spread around about system level requirements and concepts. Also, many more suppliers are taking supporting responsibilities in activities they once had no role in at all.

	Supplier's Primary Product Service or Function									
	Tooling		Specialty Springs		Fabric	Make Parts			Components	
KEY R = Takes Responsibility A = Authority S = Provides A Support or Resources I = Is Informed	Joe's Tool and Die	Fred's Molds	Ann Arbor Spring & Wire	Marshall Wire	Burlington	Franklin Stamping	Lenawee Stamping	Henry's Molding	Mitch's Switches	Acme Frame
Identify Seat Requirements			I	I	I	I	I	I	I	I
Develop Seat Concept			I	I	I	I	I	I	I	I
Identify Part/Component Requirements	I	S	R	S	S	S	I	S	S	I
Develop Part/Component Concept			R	S	R	R	I	S	R	I
Design Seat Prototype			S	S	S	S		S	S	
Design Part/Component Prototype	I	S	R	R	R	R	S	R	R	S
Build and Test Part/Component Prototype	S	S	R	R	R	R	S	R	R	S
Build and Test Seat Prototype			S	S	S	S		S	S	
Final Seat Design			S	S	S	S		S	S	S
Final Part/Component Design	I	S	R	R	R	R	S	R	R	R
Design Tooling/Equipment	S	R								
Make Tooling/Equipment	R	R								
Install Tooling/Equipment										
Purchase Parts and Material (Select Suppliers)										
Production Launch										

Exhibit 12.24 • Coseat Inc. To-Be Supplier Responsibilities Matrix

Determine To-Be Supplier Communication in Product Development

Once you know the responsibilities of individual suppliers, you can decide what levels of communication are necessary for them to be able to perform as needed. This is done using the Supplier Communication Matrix (Exhibit 12.25).

KEY

1 = Infrequent Communication
2 = Occasional Communication
3 = Frequent Communication

↔ 2-WAY SYNCHRONOUS
→ ONE WAY FROM COSEAT
⇄ 2-WAY ASYNCHRONOUS
← ONE WAY TO COSEAT

	Supplier's Primary Product Service or Function									
	Tooling		Specialty Springs		Fabric	Make Parts			Components	
	Joe's Tool and Die	Fred's Molds	Ann Arbor Spring & Wire	Marshall Wire	Burlington	Franklin Stamping	Lenawee Stamping	Henry's Molding	Mitch's Switches	Acme Frame
Identify Seat Requirements			1 →	1 →	1 →	1 →	1 →	1 →	1 →	1 →
Develop Seat Concept			1 →	1 →	1 →	1 →	1 →	1 →	1 →	1 →
Identify Part/Component Requirements	1 →	1 ⇄	2 ⇄	1 ⇄	1 ⇄	1 ⇄	1 →	1 ⇄	1 ⇄	1 →
Develop Part/Component Concept			2 ↔	1 ↔	2 ↔	2 ↔	1 →	1 ↔	2 ↔	1 →
Design Seat Prototype			1 ⇄	1 ⇄	1 ⇄	1 ⇄		1 ⇄	1 ⇄	
Design Part/Component Prototype	1 →		2 ↔	2 ↔	2 ⇄	2 ↔	2 ⇄	2 ↔	2 ↔	2 ⇄
Build and Test Part/Component Prototype	1 ↔	1 ↔	1 ⇄	1 ⇄	1 ⇄	1 ⇄	1 ⇄	1 ⇄	1 ⇄	1 ⇄
Build and Test Seat Prototype			1 ⇄	1 ⇄	1 ⇄	1 ⇄		1 ⇄	1 ⇄	
Final Seat Design			1 ⇄	1 ⇄	1 ⇄	1 ⇄		1 ⇄	1 ⇄	
Final Part/Component Design	1 →	1 ↔	2 ↔	2 ↔	2 ⇄	2 ↔	2 ⇄	2 ↔	2 ↔	2 ↔
Design Tooling/Equipment	2 ↔	2 ↔								
Make Tooling/Equipment	1 ↔	1 ↔								
Install Tooling/Equipment										
Purchase Parts and Material (Select Suppliers)										
Production Launch										

Exhibit 12.25 • Coseat Inc. To-Be Supplier Communication Matrix

SUGGEST TO-BE SUPPLIER INTEGRATION TECHNOLOGIES

Based on the combination of roles, responsibilities, and communication requirements, you are in a position to suggest specific kinds of supplier integration technologies that each supplier should use. If we look at the Coseat Inc. example (Exhibit 12.26) we can see that Ann Arbor Spring and Wire is expected to play a partner role in Coseat's product development and will have a high level of responsibilities and communication. This is reflected in the technology they will need: STEP CAD data exchange, a dedicated line to Coseat Inc. for transmission of the CAD data, EDI, and e-mail. In contrast, Joe's Tool and Die has a much lower level of responsibility (and capability) and will not be expected to exchange CAD data at all. They will, however, be expected to use EDI and e-mail.

	Supplier's Primary Product Service Or Function									
	Tooling		Specialty Springs		Fabric	Make Parts			Components	
	Joe's Tool and Die	Fred's Molds	Ann Arbor Spring & Wire	Springmasters	Burlington	Franklin Stamping	Lenawee Stamping	Henry's Molding	Mitch's Switches	Acme Frame
CAD Data Exchange Using STEP		X	X	X		X		X	X	
CAD Data Exchange Using IGES					X		X			X
CAD Data Exchange Using Native Format										
File Exchange Over Dedicated Line			X	X				X	X	
File Exchange Over ISDN					X	X				X
File Exchange Over Modem										
File Exchange Using Disk/Tape							X			
Electronic Data Interchange	X	X	X	X	X	X	X	X	X	X
Internet E-Mail	X	X	X	X	X	X	X	X	X	X

Exhibit 12.26 • Coseat Inc. To-Be Supplier Integration Technologies

DESIGN TO-BE SUPPLIER MANAGEMENT MECHANISMS

Given the roles and responsibilities you have developed for each supplier, you now need to decide which supplier management mechanisms to use in order to implement those roles and responsibilities. In Chapter 5 we defined a number of mechanisms along with their benefits and potential problems. Choose from that list (Figure 5.4) or develop others as needed. The mechanisms you choose for each supplier can be recorded in the Supplier Management Matrix (Exhibit 12.27).

The Coseat Inc. team assembled all of the information about their suppliers and moved to the final step of the process — the specific management mechanisms they would use with each supplier. Considering each supplier's role and responsibilities, they moved down the columns of the Supplier Management Mechanisms Matrix to decide which mechanisms would be used. For example, Ann Arbor Spring and Wire has a very strong Partner role in Coseat's product development (Exhibit 12.23), and they have responsibility for many tasks in the process (Exhibit 12.24). Coseat will therefore need to exchange product data with them. They will also be treated as a partner in planning, be involved early in development projects, have a long-term contract, and sit as a member of the spring commodity team. All components except tooling will use target costing, and all suppliers must be qualified. Given Ann Arbor Spring and Wire's level of development, no supplier development is needed.

Note that Coseat has kept its existing mechanisms (e.g., supplier qualification for all suppliers) and added relevant ones for the suppliers it intends to get much closer to.

	Joe's Tool and Die	Fred's Molds	Ann Arbor Spring & Wire	Springmasters	Burlington	Franklin Stamping	Lenawee Stamping	Henry's Molding	Mitch's Switches	Acme Frame
Supplier Data Integration		X	X	X	X	X		X	X	X
Supplier Partnership			X		X				X	
Preferred Supplier		X		X		X		X		X
Target Costing			X	X	X	X	X	X	X	X
Supplier Rating										
Early Involvement			X	X	X	X			X	X
Supplier Qualification	X	X	X	X	X	X	X	X	X	X
Supplier Development	X	X		X		X	X	X		X
Long Term Contracting			X	X	X				X	
Commodity Team		X	X		X	X			X	X

Exhibit 12.27 • Coseat Inc. To-Be Supplier Management Mechanisms Matrix

Consolidation and Feedback

Consolidation in Design is very different from consolidation in the As-Is Assessment (Chapter 11). The idea is to use the original Consolidation Matrix (Exhibit 11.31) as the basis for *modification* as you design new CE Elements. For example, suppose Coseat Inc. were to design changes in Cross-Functional Coordination (to include a strong emphasis on teams, the use of tools, etc.) and Communication (altering communication patterns). Ed Cazotzky would take the As-Is Consolidation Matrix and change the wording under those two elements. He would then change the ratings given for Cross-Functional Coordination and Communication for each objective, as befits the new forms of coordination and communication. Consider the example in Exhibit 12.28. The only difference between it and the As-Is Consolidation Matrix (Exhibit 11.31) is the words and ratings in the Cross-Functional Coordination and Communication columns. Note the change in the Importance scores for Cross-Functional Coordination and Communication. Cross-Functional Coordination has gone from a very strong negative (-120) to a fairly strong positive (+78). Communication has gone from a strong negative (-84) to a strong positive (+90).

		CE Elements					
		Work Process			Internal Organization		
Objectives	weight 1=low 5=high	Activities	Interfaces	Sequencing & Coordination	Organization Architecture	Project Management	Responsibilities
Reduce time to market by 50%	5	-9	-9	-9	-3	-9	-9
Improve quality 10X	3		-3	-3	-9		-3
Reduce engineering changes 25%	2	-3	-9	-3	-3	-3	-9
Reduce warranty costs by 30%	2		-3				
Reduce scrap rate by 25%	1		-3				
Reduce late deliveries by 90%	4		-3			-3	
Importance		-51	-93	-60	-48	-63	-72
Characterize the Sub-elements					Functional	Lightweight PM	
Comments		Get market info too late; tool & die build process not defined	Too much throw it over the wall; info flow not well defined	Prototype should wait for market info; tooling design starts late	Difficult coordination across functions, goals tend not to be customer driven; design and tooling too separated	PM lacks authority; technically strong w/in disciplines	Design not involved in market rsch, marketing not involved after concept; test engin. not involved to support design & build; tool engin. and mfg. engin. have input late; purch. & suppliers involved too late
Ease of Achieving							
Process Benchmarks							
Motorola				Structured development process	Product line organization with functional managers, councils at all levels	Heavyweight PM	
Chrysler				No formal dev. process - focus on schedule based on need to meet intermediate schedule deadlines	Platform teams in eng., purch., matrixed, mfg separated as a functional group	Autonomous PM	
Eaton/Ford						Lightweight PM	
Newport News Shipyard				Highly structured development process	Hybrid with product lines that have a very strong functional structure	Space Kings - lightweight PM	
Whirlpool				C2C program management process	Hybrid primarily product line, functional divisions within product lines		Cross-functional involvement in early stages
Texas Instruments				Very detailed structured workflow process	Strong functional units	Lightweight PM	
Ford Alpha				World Class Timing - 8 stage structured development process	Functional	Lightweight PM	Use SOIA to assure agreement about responsibilities

Exhibit 12.28 • Slightly Changed Consolidation Matrix

CE Elements								
		Internal Organization			Supply Chain			
		Communication	Cross-Functional Coordination	Culture	Supplier Responsibilities	Supplier Communication	Supplier Management Mechanisms	Supplier Roles
Objectives	Reduce time to market by 50%	9	9	-9	-3	-3	-9	-9
	Improve quality 10X	3	3	-9	-3	-9	-3	-3
	Reduce engineering changes 25%	9	3	-3	-3	-3	-3	-3
	Reduce warranty costs by 30%	3	3	-3		-3		
	Reduce scrap rate by 25%						-3	-3
	Reduce late deliveries by 90%	3	3		-9	-9	-3	-3
	Importance	90	78	-84	-66	-90	-75	-75
	Characterize the Sub-elements		Planning, QFD, cross-functional teams, e-mail, shared database					Mostly contractual, a few parental
	Comments	Much more communication expected, clarified	New methods should result in much stronger coordination	Authoritarian, resist formal processes, not participative, teams may be a hard sell; can-do attitude; major status differences	Poor relations with tooling suppliers, most suppliers involved late	Low frequency one way with parts & some component makers before production	Qualification is only mechanisms used consistently, early involvement with 2 suppliers	Even highly involved supliers are at best parental
	Ease of Achieving		Fairly easy -req. training and mngt committment					
Practice Benchmarks	Motorola	Teams promote communication	Emphasis on cross-functional teams	Strong team culture for design, production & improvement, heavy emphasis on measurement	Strong supplier involvement early		Supplier with early involvement usually gets the contract, price agreements	
	Chrysler	Teams promote communication; cross-team communication promoted by physical facility	Cross-functional teams, technology reviews and technology clubs for within-function coordination	Very cooperative, participative	Strong supplier involvement throughout design, more design responsibility at supplier	Generally high levels of communication; much face to face at Chrysler Tech Center	Early price commitments based on target pricing	Many at mature, moving toward partner
	Eaton/Ford		Cross-functional teams		Full service supplier-develop complete valve system; now work w/their suppliers to validate process		Early commitment from customer, but no contract until late-"strategic alliance" not "partnership"	
	Newport News Shipyard		Cross-functional teams; Space Czars to resolve problems it Kings can't do it	Major culture change being attempted - moving away from military model				
	Whirlpool		Strong emphasis on cross-functional teams		Strong emphasis on early involvement in design, suppliers responsible for product test		Strategic suppliers member of team early, pre-source to them	Many in mature roles
	Texas Instruments	Strengthened communication with customer	Cross-functional teams of part-time staff, QFD, DFM, design-to-cost model	Still trying to deal with military contractor mindset	Battery supplier involved early in new technology development			Battery supplier in mature or partner role
	Ford Alpha		QFD, cross-functional teams					Developing full-service suppliers such as Eaton-much like mature or partner roles

		CE Elements								
		People			Technology		Competitive Benchmarking			
		Job Designs	Skills	Motivation Systems	Internal	Suppliers	Coseat Now	Coseat Future	Competitor A	Competitor B
Objectives	Reduce time to market by 50%		3	-3	3	-3	16 months	8 months	14 months	10 months
	Improve quality 10X		9	-3	3	-3	230 ppm	23 ppm	300 ppm	200 ppm
	Reduce engineering changes 25%		3	-3	3	-3	250	190	200	150
	Reduce warranty costs by 30%		3				$75	$53	$100	$100
	Reduce scrap rate by 25%						8%	6%	5-10%	5-10%
	Reduce late deliveries by 90%					-3	10%	7%	8%	7%
	Importance		54	-30	30	-42				
	Characterize the Sub-elements									
	Comments		Highly skilled workforce; engineers up to date on technical issues; willing, energetic staff; training not systematic	Rewards tend to focus on functional organization at expense of product goals; some goal conflicts; appraisal inconsisent; some rewards viewed as unfair	Slow CAD system, IGES for translation, no links from CAD to CAM	Most have primitive technology				
	Ease of Achieving									
Practice Benchmarks	Motorola		Heavy commitment to training and skill development	Strong PDA and QSR processes, measure everything	Wide array of tools					
	Chrysler	Traditional roles maintained			Central CAD database using single CAD system (CATIA)	All required to use CATIA for CAD - some do but many use translation internally or thru service				
	Eaton/Ford		40 hours training per year for all employees							
	Newport News Shipyard				VIVID system used as single solid model - major advance in visualization, comm. across functions	Exchange data with GD Electric Boat using IGES				
	Whirlpool			Objectives focused in team goals						
	Texas Instruments				ProE very effective	IGES effectively used to link with other systems				
	Ford Alpha				Ford uses a single CAD system within major functions	Attempts to require suppliers to use same CAD system, but has not been totally successful - some IGES used				

The advantage of using the Consolidation Matrix in this way is that it enables you to keep track of the changes you plan and their expected impacts. It provides an opportunity to do a series of *what-if* scenarios without going through the trouble of planning deep changes before you understand what their effects would be.

In Exhibit 12.29, Coseat uses the Consolidation Worksheet to consider interdependencies with the suggested changes in Cross-Functional Coordination and Communication. This worksheet would normally be longer, but since we have only changed two subelements for purposes of our example, it has only two rows.

CE Subelements	Priority	Interdependencies	Possible Changes	Effects On Interdependencies
Cross-Functional Coordination	1	Interfaces, Sequencing, Org Architecture, Project Management, Responsibilities, Communication, Culture, Job Designs	Make Coordination Less Centralized, More Team Based	Need Work Process Changes; Make Sure Chain Of Command Supports This; Clarify Responsibilities, Job Designs To Support Team Based Work
Communication	2	Interfaces, Sequencing, Responsibilities, Culture, Technology, Coordination	Increase Communication As Indicated In Comm Matrix	Need Work Process Changes; Communication Should Follow Changes In Joint Responsibilities, Technology Needed To Support Communication

Exhibit 12.29 • Coseat Inc. To-Be Consolidation Worksheet

In Exhibit 12.30 we show the final To-Be Consolidation Matrix for Coseat Inc. This is what it would look like if Coseat Inc. planned to change *every* CE Element, as we have discussed in this chapter. As we have noted several times, it would be a rare company that has the nerve to change everything. Nonetheless, the planning team might well plan the design in this way, just to get a sense of priorities. In Chapter 13 we discuss the need to roll things out in a gradual manner, using pilots to get people used to the ideas we've been formulating.

CE Elements								
			Work Process			**Internal Organization**		
		Importace 1=low 5=high	Activities	Interfaces	Sequencing & Coordination	Organization Architecture	Project Management	Responsibilities
Objectives	Reduce time to market by 50%	5	3	3	3	3	9	9
	Improve quality 10X	3			3	3	3	3
	Reduce engineering changes 25%	2		3	3		3	9
	Reduce warranty costs by 30%	2						3
	Reduce scrap rate by 25%	1						
	Reduce late deliveries by 90%	4	3		3			
	Importance		27	21	42	24	60	78
	Characterize the Sub-elements					Product line with functional managers inside product lines	Heavyweight PM	
	Comments		Customer needs new first step; remove customer feedback as high level step	Flow explicit; interfaces clarified	Structured process with gates and phases, making concurrency explicit	Much tighter focus on customer, stronger cross-functional links within products lines	PM has authority over project staff, very senior people with deep technical expertise	Cross-functional responsibilities established, responsibility clarified for all roles
	Ease of Achieving (to-be only)		Not too hard - most staff willing & interested	Easy - most staff have asked fo this kind of clarification	Moderately difficult - full development will require much additional work; engineers likely to resist since it will be seen as restricting	Difficult - significate org. shuffle, expensive	Moderate difficulty - people available, but requires cultural change in addition	Moderate difficulty - both training and culture change required to implement
Process Benchmarks	Motorola				Structured development process	Product line organization with functional managers, councils at all levels	Heavyweight PM	
	Chrysler				No formal dev. process - focus on schedule based on need to meet intermediate schedule dealines	Platform teams in engineering matrixed, mfg. separated as a functional group	Autonomous PM	
	Eaton/Ford						Lightweight PM	
	Newport News Shipyard				Highly structured development process	Hybrid with product lines that have a very strong functional structure	Space Kings - lightweight PM	
	Whirlpool				C2C program management process	Hybrid primarily product line, functional divisions within product lines		Cross-functional involvement in early stages

Exhibit 12.30 • Coseat Inc. To-Be Consolidation Matrix

CE Elements								
		Internal Organization			**Supply Chain**			
		Communication	Cross-Functional Coordination	Culture	Supplier Responsibilities	Supplier Communication	Supplier Management Mechanisms	Supplier Roles
Objectives	Reduce time to market by 50%	9	9	3	3	3	3	3
	Improve quality 10X	3	3	3	9	3	3	9
	Reduce engineering changes 25%	9	3	3	3	3		3
	Reduce warranty costs by 30%	3	3	3	3			3
	Reduce scrap rate by 25%			3	3			3
	Reduce late deliveries by 90%	3	3	3	9	3	3	3
	Importance	90	78	51	93	42	36	69
	Characterize the Sub-elements		Planning, QFD, cross-functional teams, e-mail, shared database					
	Comments	Much more communication expected, clarified	New methods should result in much stronger coordination		Focus on improving relations with suppliers, involving them when possible from start	Higher intensity, more two-way, build teams that include suppliers	Multiple machanisms to be put in place	Higher level roles for many suppliers
	Ease of Achieving	Moderate difficulty-both training and culture change required to implement	Fairly easy-requires training and management commitment	Difficult-requires top management commitment to change	Moderate difficulty-each company will require a change effort, but made easier by our leverage	Easy-most suppliers say they want much more	Moderate difficulty-we have to learn and they have to accept-both will be hard	Difficult for some
Practice Benchmarks	Motorola	Teams promote communication	Emphasis on cross-functional teams	Strong team culture for design, production & improvement, heavy emphasis on measurement	Strong supplier involvement early		Supplier with early involvement usually gets the contract, price agreements	
	Chrysler	Teams promote communication; cross-team communication promoted by physical facility	Cross-functional teams, technology reviews and technology clubs for within-function coordination	Very cooperative, participative	Strong supplier involvement throughout design, more design responsibility at supplier	Generally high levels of communication; much face to face at Chrysler Tech Center	Early price commitments based on target pricing	Many at mature, moving toward partner
	Eaton/Ford		Cross-functional teams		Full service supplier-develop complete valve system; now work w/their suppliers to validate process		Early commitment from customer, but no contract until late-"strategic alliance" not "partnership"	
	Newport News Shipyard		Cross-functional teams; Space Czars to resolve problems it Kings can't do it	Major culture change being attempted - moving away from military model				
	Whirlpool		Strong emphasis on cross-functional teams		Strong emphasis on early involvement in design, suppliers responsible for product test		Strategic suppliers member of team early, pre-source to them	Many in mature roles

CE Elements									
	People			Technology		Competitive Benchmarking			
	Job Designs	Skills	Motivation Systems	Internal	Suppliers	Coseat Now	Coseat Future	Competitor A	Competitor B
Objectives									
Reduce time to market by 50%	9	3	3	3	3	16 months	8 months	14 months	10 months
Improve quality 10X	3	3	9		3	230 ppm	23 ppm	300 ppm	200 ppm
Reduce engineering changes 25%	3	3	3	3		250	190	200	150
Reduce warranty costs by 30%	3					$75	$53	$100	$100
Reduce scrap rate by 25%					3	8%	6%	5-10%	5-10%
Reduce late deliveries by 90%			3	3	3	10%	7%	8%	7%
Importance	66	30	60	33	63				
Characterize the Sub-elements		Highly skilled workforce; engineers up to date							
Comments	Combine designer + draftsman; form more teams	Provide more training opportunities	Focus goals on organizational objectives, more rewards away from non-productive areas	Upgrade technology for each functional area; improve communication media	Improve communication integration for most suppliers				
Ease of Achieving									
Practice Benchmarks									
Motorola		Heavy commitment to training and skill development	Strong PDA and QSR processes, measure everything	Wide array of tools					
Chrysler	Traditional roles maintained			Central CAD database using single CAD system (CATIA)	All required to use CATIA for CAD - some do but many use translation internally or thru service				
Eaton/Ford		40 hours training per year for all employees							
Newport News Shipyard				VIVID system used as single solid model - major advance in visualization, comm. across functions	Exchange data with GD Electric Boat using IGES				
Whirlpool			Objectives focused in team goals						

Conclusion

In this chapter we've described the many things you could and should do in order to design a new concurrent engineering system for your company. We've described a large number of tools and methods which cover a wide range of possibilities. But you don't have to use them all — you just need to pick and choose those you need, according to your own priorities. In the next chapter we'll discuss how to take your design and turn it into reality by implementing it.

CHAPTER 13

IMPLEMENTING THE NEW CE SYSTEM

If you can't ride two horses at once, you shouldn't be in the bloody circus. • British Labour politician Jimmy Thomas

In this book we have expended a lot of effort to describe the processes you should undertake to think about and plan your CE system. But the really hard part is implementing that system. Implementation isn't what we might call "doing CE;" in fact, that represents "turning the crank" on the CE system once you've set it up. No, implementation involves setting up the system you've planned. It is the last phase of the *change* effort (Figure 13.1). As we have noted in earlier chapters, although the diagram seems to show this happening after the system has been designed, in fact you must begin implementation activities quite early in the process.

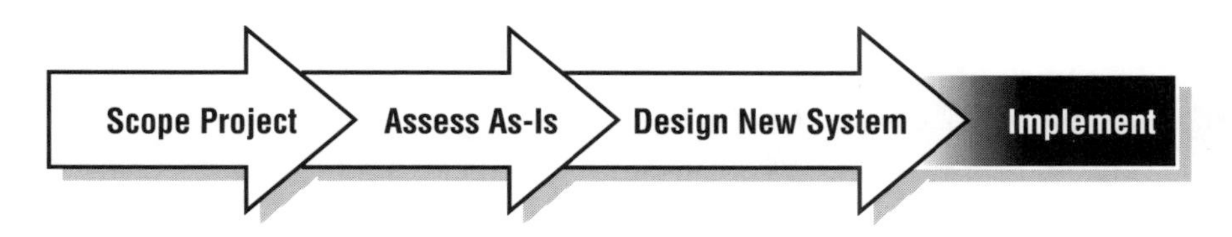

Figure 13.1 • Four Phases of the CEE Change Process

THE IMPLEMENTATION PHASE

To *implement* the new CE system means to install it so people in the company can use it to do new product development. There are six main activities in the implementation phase (Figure 13.2).

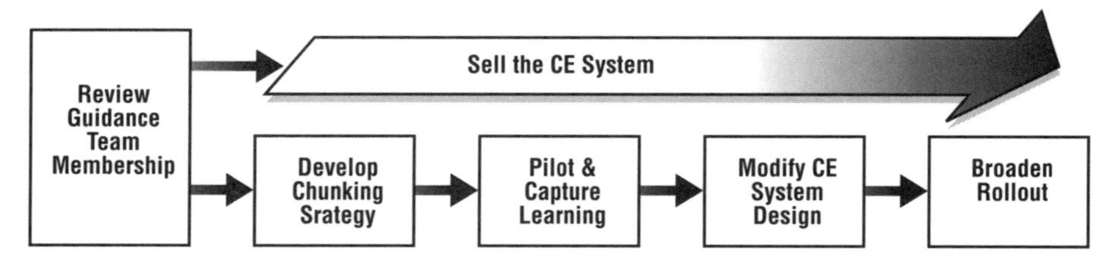

Figure 13.2 • Implementation Activities

1. ***Review Guidance Team Membership.*** The team that implements CE in your company may need to be somewhat different than the team that planned the new CE system. This is because you are now probably more focused on specific parts of the company and certain activities. You may wish to change the team's membership at this time.
2. ***Sell the CE System.*** This means to convince everyone involved that the new CE system should be implemented. This includes everyone from top management to the lowest level worker. You started this process in the very beginning (see Chapter 10), but there you were selling the CEE process, i.e., the process of planning the new CE system. If you did a good job of planning the new system, then selling everyone on the result should be much easier. But you still have to do it; and you have to do it continuously. That's why we have shown this activity as an arrow across the top of Figure 13.2, rather than within its process flow.
3. ***Develop Chunking Strategy.*** Since doing all of CE, in your whole company, all at once, is just too large a task for anyone or any group to take on, you have to divide it into parts, or "chunks." There are many ways to divide this up, but you have to choose. We will spend a lot of time discussing this issue, since your approach here is very important.
4. ***Pilot and Capture Learning.*** While it is certainly possible to just go in and have everyone start doing CE as regular practice, most companies choose to start with a pilot of some sort. This enables them to make lots of mistakes and learn from them at relatively low cost. You have to select a site for the pilot, usually a single design project. It is also important to make sure lessons learned from the pilot are captured so they can be used later on.
5. ***Modify CE System Design.*** Based on learning from the pilot, modify the design of the CE system so that other units which adopt the design can take advantage of the benefits of this learning.
6. ***Broaden Rollout.*** Deploy the modified CE system to other parts of the company. This may be done one unit at a time, or it may get

deployed to a substantial portion of the company — it depends on how confident you feel about what you have learned from the pilot and how much influence you have.

We will discuss each of these activities in turn, starting with the first step of reviewing the team membership and then moving on to selling, since it's ubiquitous.

Review Guidance Team Membership

The Guidance Team in this phase has to do many of the things the Guidance Team has already been doing, but perhaps targeted somewhat differently; that's why team membership needs to be reviewed at this point. In the earlier stages of planning the new CE system, you needed the kind of support necessary to engage in a planning activity — relatively small funding and staffing, protection from distracting assignments, help with collecting information, and the broad selling of ideas, rather than changing how people do their work. As we move to implementation, both the cost and the risk increase significantly, as does the number of people involved. With that in mind, you need to include the following roles on your Guidance Team.

Project Manager

The project manager focuses on the details of implementation — making sure all the needed selling happens, that the pilot is planned and executed properly, that deployment activities happen at the right time, etc. He or she has to be both a very high energy individual and a good manager. Depending on the size of the company and the CE effort, there may be a number of subordinate project (or sub-project) leaders who manage parts of the larger effort.

CE Technical Wizard

This is the enthusiast and expert on CE methods and tools. This person has a lot of outside contacts to help learn about new methods and tools, but he/she is also able to work with people inside the company to teach others about what is going on outside. In contrast to most other people in the company, the CE Technical Wizard probably reads books about CE and other technical topics.

CE Liaisons

The CE Liaisons serve to link the Guidance team with the other formal parts of the company. In most companies, the liaisons will be representatives of the various functions in the company. This of course implies that there are formal functions represented as departments in the company's structure. In any case, there need to be links to the various departments, whether they are functions or product lines. These individuals need to be fairly powerful within their departments — preferably the manager or someone who has been delegated significant decision making authority.

CE Executive Sponsor

In addition to people filling these roles, which are formally on the team, you also need to make sure that you've got an executive sponsor. The sponsor is a high level executive who ensures top level support and makes sure that there is funding and the other resources necessary for the project to succeed. The sponsor makes sure the project is in accord with company goals. You should have had a sponsor from the beginning, but now that you are moving into implementation, you may want to review the sponsor's role with that person. If he/she is willing and able to continue, fine. But, given that you now need much more in the way of resources and influence, you may find you need a different, more powerful sponsor. In fact, it may help to have more than one sponsor if you are recommending change in more than one part of the company.

Selling the CE System

At some level, the way you sell things in your company is a unique function of your company, who you're selling it to, and who you are. Inevitably, any advice we give you will be based on the experience of others, in different companies. Nonetheless, the advice we give should be generally useful, especially if you lack experience trying to sell complex projects inside your company. Even if you are quite experienced, we hope that the examples we provide will give you some new ideas and some added ammunition.

You have been through selling once before in the CEE Process — that was in the beginning, when you needed to find support for setting out on this journey. If you are now worrying about implementation, you probably found that support. However, the support needed now is much greater, and the risks are much greater. After all, the resources involved in analysis and planning for CE are much less than those required to actually implement and do CE. More importantly, the risk of analysis and planning is relatively low — it's a peripheral activity. The risk of actually changing is quite high — if the change doesn't work, new products may be late or may never appear!

What this means is that the selling of the CE implementation plan is vital. No change will occur unless the stakeholders involved are absolutely convinced it is in their best interests. Let's consider some of the stakeholders and their likely interests. From there we'll move on to considering what it might take to get their attention and to move them to your side.

Stakeholders and Their Interests

We identify three primary types of stakeholders involved in changing to CE. You could break these down further, especially by their discipline or function, but the fundamental interests seem to divide into these three types — top management, middle/lower management, and technical workers.

Top Management. In almost every case, a primary interest of top management is the bottom line — will this make more money for the company? For some top managers, this is the only interest — they just want to know if the company will make more money by doing this, as compared to doing something else. In a large conglomerate, you can expect that the corporate top management may *only* be interested in the money side of things. This is because each division of the conglomerate has different interests, so that when you get to the corporate level, the only common metric is money. Typically, the other major interest for top management is a vision of some sort. In some cases, this may be the latest corporate fad; but in many cases, the vision will be a bold, forward thinking view of how the company can succeed in the next decade. CE can be such a vision for top management.

To complicate matters, the attention span of many top managers is measured in minutes rather than months. A CEO with a big vision gets refocused on the bottom line very quickly if profits take a dive and the stock price tanks. In fact, you can assume that there will be a *cycle* to top management attention. No matter how much support you get for CE in the beginning, that support will change; top management attention will drift to other things. This means you not only have to continually work at getting support with top management, but you need to build up resources that you can use to carry over any dry spells in support that will come down the road.

Middle and Lower Management. In some ways, the interests of middle and lower management are more complex. They have to keep their bosses happy by accommodating whatever new things top management wants done, but they also have to keep things running smoothly on a day-to-day basis. They are also worried about keeping their jobs, and moving up if possible. Promotion is often based on being able to introduce change smoothly. These interests are often in conflict. Change tends to disrupt the smooth daily operation of a business; and those disruptions may place a manager's job at risk.

Technical Workers. The interests of technical workers (including engineers and designers) are similar to those of middle and lower managers in that they have to keep their bosses happy by accommodating change, while at the same time getting their jobs done. They too worry about keeping their jobs and getting promoted. However, because their scope of interest tends to be smaller, their focus is usually on getting the job done. Getting promoted often is simply the result of doing their technical jobs well.

What Does it Take to Move Them?

Given their somewhat different interests, it should not be surprising that different strategies are needed to move these different stakeholder groups. Let's consider what it takes to move each.

Top Management. The most convincing argument for top management is a dollars-and-cents analysis that will show how much more money will be made over a specific period of time. They need to see where the argument comes from in order for it to be convincing. They need to be convinced that the gain is reasonable, well supported, and that the timetable is possible. Unfortunately, for something innovative, like CE, this business case is usually very difficult to make, for the simple reason that the data to support your claims usually won't exist. This means you have to make estimates; and the more convincing the basis for your estimates, the better off you'll be. Consider the following example from a company we'll disguise (at their request).

> ***We Make Stuff Corporation***. *We visited the corporate director for CE at a large multinational company which we'll call the We Make Stuff Corporation (WMSC), since their products range so widely. He described some of the estimates he came up with to try and sell his program to the divisions (WMSC has highly independent divisions which would never "take orders" from someone at corporate). He gathered lots of credible data and calculated costs of such things as engineering changes, prototypes, and delays in production which would cost lost sales. He then went on to show how specific costs would be cut with CE; and to enhance his credibility, he also showed that some costs would increase (e.g., costs due to the learning curve for new techniques, and the costs of hiring contractors to replace staff who were off learning new things). In addition to the specific numbers he collected, he also asked division managers questions like, "How much would you save if you could get a prototype so much earlier?" The kinds of numbers he discussed seemed very convincing. They*

were convincing enough for a few of the division managers, because he was able to get pilots off the ground at three different divisions. Other divisions wanted to wait and see.

The alternative to making a convincing business case analysis is to appeal to the vision side (if it exists) and make a logical argument. One such logical argument goes as follows: it has worked elsewhere; we can make it work here. We suspect more big decisions are made based on these kinds of arguments than are made based on elaborate business cases. Here are two examples, one from Chrysler, and another from a company that we'll disguise.

Chrysler. *Consider Chrysler's move to platform teams. The company was in crisis, and top management knew something had to be done quickly. They had benchmarked Honda and other Japanese companies, but knew they couldn't do things exactly the same way. The platform team for Chrysler's LH program was developed to solve the problem, but its acceptance did not depend on some elaborate analysis of how much money the company would make as a result. The only analysis was that the company would fail if it didn't change drastically — and, based on their benchmarking of Honda and other companies, the platform team approach was the only viable response on the table. Fortunately for Chrysler, it worked.*

Technical Machines, Inc. *A few years ago, at a division of a large multinational high-technology company we will call Technical Machines, Inc. (TMI),*[1] *the then newly appointed engineering vice president was given the unenviable task of making drastic cuts in costs and headcount. While he was forced to make many personnel cuts, he also introduced a wide array of CE elements, including process changes, appraisal and reward changes, and organization design changes. These changes reduced product cost substantially and minimized the number of personnel cuts he had to make. In this example, the TMI engineering VP was driven substantially by bottom line considerations, but he was also driven by a vision of how CE could help his company.*

A critical point in making the sale to top management is that they may not really understand the technology or methods you are trying to sell them. There's a good chance they will view any technical discussion you make as just so much noise. At best, the technical discussion will help to reinforce your credibility, but the technology itself will probably not be convincing. Money *is* convincing; but in the absence of a lock-tight business case, an appeal to vision *may be* convincing.

[1] This describes real division in a real company, but it has undergone so many acquisitions and mergers over the past few years that we find it difficult to explain just who they are at this point.

Middle and Lower Management. The key to convincing middle and lower management to join the effort is to convince them that top management is serious about wanting it to happen. If they believe top management really wants to see the company develop new products using CE, and isn't going to change its mind next week, then they are much more likely to participate enthusiastically. The best approach to achieving this will be for top management to visibly and actively participate. For example, when Motorola began its push for six-sigma quality in 1985, top management (including the CEO) participated in quality training and joined in delivering training to their subordinates. Naturally, goals related to CE should also be included in each individual manager's goals and performance appraisals. Continuing emphasis is as important as the initial participation.

Technical Workers. As with the middle managers, technical workers need to be convinced the company is serious about doing CE in the long run. Thus, their managers need to participate in training, goals need to change, and technical support needs to be in place. However, and perhaps most importantly, the behavior of their managers needs to change. For example, if part of CE is allowing a cross-functional team to make decisions, then managers can't interfere in that decision-making process.

Stakeholders	Interests	Strategy
Top Management	cash flow, market share, beating Competitor A	financial analysis; appeal to vision will not work; need to gather plenty of data to make the business case
Middle & Lower Management	keeping things running smoothly; make top management happy	get top management behind it; show them how, after the learning curve, that things will run more smoothly with CE
Account Managers	selling more seat systems rather than components	show how CE leads to shorter time to market
Engineers	designing new features into new seats	show how CE enables them to design in newer features at lower cost and less time
Designers	getting to do more creative work; getting things done faster to make management happy	show how their skills will be upgraded in the new system

Exhibit 13.1 • Coseat Inc. Stakeholder Analysis Worksheet

In the Stakeholder Analysis Worksheet (Exhibit 13.1), Coseat Inc. has recorded their strategies for convincing each stakeholder group to participate in implementation.

Develop a Chunking Strategy

In general, it is difficult, if not impossible, to take on the task of introducing all of the changes we have discussed all at once into a large company. Probably the best reason is that it simply takes a lot of time to roll any large program out, with all the planning and training and learning curves involved. It is just too difficult to assemble the trainers and consultants needed to assist a whole company to do this. Furthermore, with the inevitable learning curve,[2] product development activities that are well under way could be brought to their knees for a long period of time if they are forced to change systems late in their process. The task is too large and the risk is too high. Even in a small company, the resources available are usually not sufficient to allow you to introduce all of this at once, and the risks are even greater. So, some kind of incremental change is necessary — the task has to be divided into chunks.

A *chunking strategy* is the way you divide the task up. There are two basic ways to divide things up — by the *CE Elements* involved, and by the *functions* or parts of your organization that are involved. Of course, you can also combine the two by doing some elements in some functional areas and not in others. We'll now discuss these three possibilities, along with some examples.

Chunking by CE Element

One way to chunk things is by CE Elements. For example, you might introduce the new CE technology, followed by changes in business process, followed by changes in jobs, followed by changes in project management, etc. This is a fairly common approach. But the tendency is to only do one thing at a time — i.e., implement a single element, see what happens, and then try something else. For example, many companies will introduce new technology as a means to solve a specific problem. Thus, in the 1980s, General Motors introduced what it called C4 Technology[3] to make product development more effective. But, by itself, the technology didn't help them very much. The next response was to look at changes in business process, but that has been a long and drawn out project. And that's the problem with introducing change one element at a time — the elements interact. You may get very little short-

[2] People who switch over to a new system are almost always much less productive for a period of time until they learn to take advantage of the new system.
[3] C4 stands for CAD, CAM, CAE, and CIM.

term benefit from introducing a single element. The absence of such a benefit may then endanger your ability to introduce later elements.

The solution to this problem is to introduce more than one element or sub-element, even if you aren't able to do everything all at once. For example, you might introduce changes in technology, business process, and job design; leaving organization design, culture, human resource, and project management changes for later. How do you decide which elements to "chunk" together? You can use three basic criteria.

- ***Choose elements or sub-elements that will be easy to implement.*** Early success leads to support for later efforts, so anything that will be easy to do (e.g., because of lots of internal support or because it's not much of a change from current practice) should be a candidate for early implementation. Ratings for ease of implementation can be taken from your To-Be Consolidation Matrix (Exhibit 12.26).
- ***Choose elements or sub-elements that will have high impact.*** Big success leads to bigger support later on, so anything that will "knock their socks off" when you are done should be a candidate for early implementation. Ratings for impact can be taken from your To-Be Consolidation Matrix.
- ***Choose elements or sub-elements which interact.*** Elements that interact (see Chapter 11 for a discussion of this) affect each others' success. For example, communication and cross-functional coordination are very tightly connected — you can't do much about cross-functional coordination without dealing with communication. This suggests trying to deal with *both* elements at the same time. In your To-Be CE Elements Consolidation Worksheet (Exhibit 12.25) you indicated which elements interact.

You can use the CE Elements Prioritization Matrix (Exhibit 13.2) to rate each CE Element on the three criteria. The Prioritization Matrix pulls a few items off of the Consolidation Matrix to make them easier to work with. In some cases, choices will not be obvious, partly because you may find conflict among the criteria. In many cases, elements with high impact may be among the most difficult to implement. You may also find an easy-to-implement element interacting with a much more difficult-to-implement element — yet, you'll need to use both in order to succeed. In any case, chunking by CE Elements is not sufficient — you also need to consider the organizational side of chunking before you are ready to proceed.

In Exhibit 13.2, you can see how Coseat Inc. used the Prioritization Matrix to record their ratings and make some *preliminary* decisions.

		Ease Of Implementation	Potential Impact	Interaction With Other Elements	Comments
Work Process	Activities	2	27	Sequencing & Coord., Comm., Supplier Comm.	Moderately Hard, But Interacts With Internal Org. Elements + High Impact
	Sequencing & Coordination	2	84	Cross-Functional Coord., Comm., Supplier Response, Supplier Comm.	Moderately Hard, But Interacts With Internal Org. Elements + High Impact
	Interfaces	2	21		
Internal Organization	Org Architecture	3	15	Project Mgmt., Cross-Function Coordination	
	Project Management	2	51	Cross-Functional Coordination	
	Cross-Functional Coordination	1	78	Comm.	Easy + High Impact
	Responsibilities	1	72	Comm., Motivation, Skills, Culture	Easy + High Impact
	Communication	1	90	Culture, Coordination	Easy + High Impact
	Culture	3	51		
Supplier Relations	Supplier Responsibilities	2	96	Supplier Roles	
	Supplier Communication	1	72		Easy + High Impact
	Supplier Roles	2	36		
People Systems	Job Designs	2	30		
	Skills	1	30		
	Motivation Systems	3	60		
Technology	CAD 1	1	72		Easy + High Impact
	CAD System 2	2	54		
	Interfaces	3	36		

Ease Of Implementation
1 = Easy
2 = Somewhat Difficult
3 = Very Difficult

Impact
Score = Importance Score From Consolidation Matrix

Interactions
List Only Strong Interactions

Exhibit 13.2 • Coseat Inc. CE Elements Prioritization Matrix

The tentative conclusion from Exhibit 13.2 is that Coseat Inc. should start with a set of internal organization elements, especially project management, cross-functional coordination, communication, and responsibilities, combined with work process changes, and possibly supplier communication and a new CAD system, depending on how much they want to take on at once. The easier people system and supplier relations elements could follow in a second phase of changes, followed by project management and some of the other elements later on. Note that no final decisions should be made just yet, until some of the other chunking has taken place.

Chunking by Functional Group

Another approach to chunking is by functional grouping in the company. For example, CE could be introduced in engineering, then in manufacturing, then in purchasing, etc. Indeed, if we look closely at

Chrysler, this is basically what they did. The platform teams at Chrysler are *within* engineering. While other groups, such as manufacturing and purchasing, are *represented* on the platform teams, both manufacturing and purchasing have totally separate organizations that are, in effect, matrixed with the platform teams.

Here is another example of chunking by function at the company we earlier called Technical Machines, Inc. (TMI).

> ***Technical Machines, Inc.*** *At TMI the Vice-President for Engineering exercised complete control over the engineering department. Based on his experience and reading, he decided to introduce CE practices and tools to his department. He did this by asking his 50+ staff members to resign and bid for five new staff jobs that would plan the new system. The other members of his staff had to find technical jobs within the company. In effect, he built a new engineering department from scratch, using existing personnel and facilities. As part of the new regime, he introduced changes in process, appraisal systems, reward systems, training, and many other CE elements. All of this went on within engineering, with considerable success. There was a substantial reduction in time and cost (e.g., 40–50 percent reduction in unit cost in one case) of new product development.*
>
> *No change was forthcoming in other departments, however, and in fact, the other VPs thought he was insane for taking such risks. For the most part, engineering has had to reach out to the other departments in order to foster cooperation. For example, engineering asks other departments to fill out a "report card" on project engineers on a monthly basis. This provides the other departments an opportunity to give feedback to the engineering VP. In practice, this usually means that the other department (e.g., manufacturing) must now complain to the engineering VP first, instead of just to the president of their division. This results in greatly improved communication between the departments as individuals resolve problems between themselves first, before involving higher management.*

At TMI, change has only really taken place within engineering since CE was only supported by their VP. There is creeping change in practice in other departments as they learn by osmosis from engineering, but it is slow.

Combining the CE Elements and Functional Approaches

In many cases, in order to limit your scope enough, you will have to chunk by *both* CE Elements and functions. Often this is because CE will have support within one or two functions, but not others. Since you have to start somewhere, you start where you have the most support. However, if your base of support is too narrow in either CE Elements or in the number of functions initially involved, you will have trouble demonstrating success. You can plan how you will combine CE Elements and functional implementation using the CE Deployment Matrix. Coseat Inc.'s CE Deployment Matrix is in Exhibit 13.3.

		Functions								
		Market Researchers	Design Engineers	Prototype Makers	Test Engineers	Tool Engineers	Tool Makers	Manuf. Engineers	Purchasing Agents	Suppliers
Work Process	Activities	X	X	X	X	X	X	X	X	
	Sequencing & Coordination	X	X	X	X	X	X	X	X	
	Interfaces	X	X	X	X	X	X	X	X	
Internal Organization	Org. Architecture									
	Project Management		X							
	Cross-Functional Coordination	X	X	X	X	X	X	X	X	
	Responsibilities	X	X	X	X	X	X	X	X	
	Communication	X	X	X	X	X	X	X	X	
	Culture									
Supplier Relations	Supplier Responsibilities									
	Supplier Communication		X						X	X
	Supplier Roles									
People Systems	Job Designs									
	Skills		X						X	
	Motivation Systems									
Technology	CAD 1		X							
	CAD System 2									
	Interfaces									

Exhibit 13.3 • Coseat Inc. CE Deployment Matrix

At Coseat Inc., the Guidance Team decided to focus on the coordination and communication aspects of internal organization and on the work process changes. It made no sense to anyone on the internal team to try and do these changes within a single function, or even in a combination of two or three functions. If they were going to the trouble of creating a cross-functional team, then they would try to include all the relevant parties. Naturally this would be dependent on obtaining support from those functions. They also decided that the work process changes had to proceed concurrently with the coordination changes — they were too connected.

Pilot and Capture Learning

Chunking tells you what you want to implement and where you want to do it. What it doesn't tell you is how quickly to start things off. You might conceivably try to transform the way all work is done from now on — in other words, change the way everyone in the affected functions does their work. Usually you don't want to do this because it takes too many resources, it is too risky, and you usually won't know enough to be able to do it well. The initial pilot, therefore, is usually a learning experience upon which you can build.

There are three kinds of issues to think about here. The first is the type of pilot you want to use, the second is how to plan the details of the pilot, and the third is how to capture results to use later.

Types of Pilots

A pilot can take many forms. We'll focus on three types which we think are most common and which demonstrate the advantages as well as the pitfalls of how most companies use pilots.

Demonstration Pilots. This type of pilot is done with low visibility using an artificial project or one that doesn't matter much. It will attempt to *demonstrate* the tools and methods of CE so other people in the company can see them in use, but it attempts to minimize risk by making sure that nothing vital is actually done using CE.

Disconnected Pilots. This type of pilot involves important project work, but is not integrated into the daily operations of the company. When the project ends, it leaves no trace; if it fails, the loss is only the loss to the specific project. If it succeeds, then the company has the option of deciding how to integrate the CE methods and tools which were used.

Integrated Pilots. This type of pilot not only involves important project work, it is also tightly integrated into the daily operations of the company. It is *high commitment* because the company has made a commitment to making CE work. If the pilot fails, the company almost certainly will have to try again. If it succeeds, the company is in an excellent position to deploy CE in the mainstream.

We will not take up any more space discussing demonstration pilots because we believe they are usually just a way to avoid actually doing anything about CE. While we strongly support the use of integrated pilots, and generally discourage disconnected pilots, we will discuss both because they are so widely used and because there are some times when a disconnected pilot is needed.

Disconnected Pilots

A disconnected pilot is done without connection to other parts of the company. For example, a special team might be set up to create a new product that doesn't fit within an existing product line. Or, a unit is set up to do "special projects" which serve to solve problems that can't be easily addressed by the normal business units. Consider the following fictitious example.

> *Office Furniture Inc. (OFI) makes wood and steel furniture for offices. They decide to introduce CE as a pilot by forming a cross-functional team within engineering. They assemble a design engineering manager, design engineer, designer, manufacturing engineer, and test engineer as a team with the charge to "design the R-4000 desk chair in half the time and for half the cost as the (similar) R-3999 chair." They are collocated in a suite of offices away from the other design staff. They are provided with the latest in CAD systems and trained in their use. Finally, they participate in a week-long training session designed to teach them about the fundamentals of CE and to provide a team-building experience. A design engineering manager is assigned to lead the team.*

If we had to guess, we'd say that this team would be fairly successful at their specific task. They might not meet the stretch goals set for them, but they'd probably design a fine chair in less time and at a lower cost than the predecessor. But there are some problems evident.

The main question we have is whether OFI can sustain this process over time and over many design teams. If we look at the CEE framework, it is clear that the company has done several things right. They have addressed organization design to some degree by forming a team; and they have addressed communication to some degree through collocation. On the other hand, they have not addressed any of the other CE elements, such a project management, people issues, culture, or technology. Indeed, they have not fully addressed communication (there are only engineers on this team!) or organization design (what is the link between this team and the rest of the company?). Finally, by setting this team off to the side somewhere and isolating them, the company has made sure that few others will know about and observe what is going on. This means that their ability to sell others on what they have done (even if they can document success) is severely weakened.

Two real-life examples of disconnected pilots are Ford's Alpha program and GM's Saturn Corporation.

> ***Saturn Corporation***. *General Motors created the Saturn Corporation in 1985 with the express purpose of developing a small car that could compete with the Japanese. They started from scratch with a totally new*

company and totally new processes, designed from the ground up. This meant that they could (and did) develop their own product development process (separate from GM's), select different technology from GM (which they did — they chose a different CAD system from the GM standards), select their own people (although many were transfers from other GM divisions), etc. Saturn developed a very successful line of cars, which far exceed any other comparable Big 3 car in reliability and consumer satisfaction. But, other than Saturn's sales, the impact of Saturn on the rest of GM has been almost negligible. There has been almost no application of Saturn's lessons to the rest of GM. This is primarily because Saturn is not only physically and organizationally isolated, but it also created tremendous resentment in other GM divisions, who felt starved for capital as a result of high spending associated with the Saturn start-up.

Interestingly, over the past few years, Saturn has become organizationally reintegrated into GM, with Saturn's former president being made head of GM's small car division and then of a management training organization. Whether this results in greater transfer remains to be seen.

Ford Alpha. *Alpha was designed as an advanced engineering group which used CE tools and methods to solve engineering problems presented by the operating divisions. As a CE pilot, Alpha did several things right. First, and most important, their projects succeeded and they collected data to prove it. So, if a project solved a big problem or saved several million dollars, the Alpha team documented the facts and made sure top management knew about it. Second, Alpha trained a large number of Ford personnel in CE tools and methods. This included the people assigned to work in Alpha as well as people from the operating divisions who worked on Alpha projects. Finally, Alpha rotated a large number of people back into the larger company after they had been thoroughly trained and indoctrinated in CE. We estimate that several thousand people were members of Alpha at one time or another. Ford disbanded Alpha in 1995 as part of a general reorganization (Ford 2000) that was expected to implement CE throughout Ford. Despite the reorganization, processes were fundamentally unchanged and CE was still not being done systematically throughout the company. As of this writing, it does appear that a major effort (the Ford Product Development System) is still underway to change the processes.*

We believe the problems at Ford are the result of not implementing enough of the CE elements in the CEE framework. Alpha primarily trained people in CE technologies and in a team approach to product development. They also emphasized changing the product development process to make it more concurrent and efficient. All of these are good things. But nothing Alpha did changed the larger Ford organization or that organization's processes. Thus, an Alpha veteran who transferred to an operating division returned to the same old processes and the same old organization and people systems. Hence, real change did not take place in Ford, and it appears that the thousands of Alpha veterans in the company were unable to foster enough change on their own without the institutional changes needed for sustained movement.

Again, Ford made a substantial financial and personnel commitment to Alpha, but made no commitment to change the *company* to CE. As a result, they are still struggling.[4]

INTEGRATED PILOTS

An integrated pilot applies CE methods and tools to a real project, but in a context in which there is substantial connection to the rest of the organization. Probably the best example of this is Chrysler's move to platform teams.

> ***Chrysler.*** *Chrysler introduced platform teams in 1988. As each platform got ready to introduce a new vehicle, the team concept was introduced, until the whole company had converted to the new system. Chrysler had the advantage that each product line only develops a single new vehicle platform at a time; thus Chrysler did not need to manage "old" development projects while the new system was being introduced. This process appears to have been very successful. In fact, the system was reinforced by construction of the new Chrysler Technical Center in 1992, which provided improved physical facilities in which the teams could operate.*

Note that Chrysler was placing a great deal at risk in this project. In fact, if the first project to use platform team approach, the LH platform team had failed, it is possible the company would have folded. More to the point, Chrysler took a business unit and said, "This is how you're going to do business from now on — make it work!" The level of commitment from a total company point of view was quite high.

Another example of an integrated pilot is the case of Technical Machines, Inc., which we described above. In that case, the engineering VP made a total commitment with a major change in many systems to support CE. Admittedly, the rest of the division was not behind the changes, but within the engineering department, the level of integration was quite high.

[4] For another example of a low commitment pilot, look at the Texas Instruments case study in the Appendix.

In our view, an integrated pilot is by far the best way to go. Anything else seems to result in a long, dragged out process in which most of the company just goes about its business the same old way.

At Coseat Inc., the Guidance Team agrees to start an integrated pilot. One of the people on the team, Fred Jenson, who fills a CE Liaison role, is also a project manager for a new design project which is about to get started. He volunteers his project to serve as the pilot. This is a big risk on his part because the new product is an important one with a major customer. Nonetheless, because of his involvement with the Guidance Team, he has enough confidence in CE that he believes there is no question that his project will benefit despite the risks associated with the learning curve.

PLAN THE DETAILS AND CONDUCT THE PILOT

There are obviously an enormous number of details that must be planned if the pilot you have selected is to be a success. To some extent, this needs to be planned much as you would any other project. In this section we will make several suggestions about some of the things that need to be planned. For many readers, much of what we suggest will seem obvious; for many others, much of it will not. Regardless of your background, we provide the following suggestions more in the spirit of a reminder, to make sure you don't miss important points in your planning.

The War Room. A war room is a separate conference space set aside for a project team's exclusive use. This enables the team to leave things around so they don't need to be cleaned up after every meeting. For example, most war rooms will have flowcharts and meeting minutes posted on the walls. This enables team members to refer to these public "documents" as needed, and for members to wander over to the wall and look at something that's posted. A war room saves time for a team which meets regularly since set-up time is reduced; but the most important function is to provide a space for informal interactions to take place. The benefits of such a space are hard to quantify, but most successful teams which are trying to run projects with relatively high uncertainty seem to use one. We'd call it good practice, but probably not essential.

Detailed Project Plan. A detailed project plan is essential and should include at least the following elements.

- ***Budget.*** The budget should provide for enough staff to carry out your plan, equipment and software to provide necessary tools, and travel budget if necessary.
- ***Staffing.*** The staff should have the right personnel to carry out

the plan, including the key people we described under the members of the Guidance Team.

- ***Responsibilities.*** Responsibilities should be well laid out, particularly, who will be responsible for critical activities.
- ***Schedule.*** The schedule will dictate when you expect to get things done. Shoot for having something measurable done within about six months.
- ***Activities.*** The activities will include what you expect to do and what the results should be.

Reporting Relations Between Pilot and Rest of the Company. Here you need to decide where in the organization the pilot will sit, who the project manager reports to, and how it will "fit" into the company in the long run. In the case of Chrysler's LH platform team, the pilot was part of a revamped organization structure, and the reporting relations were quite clear — the platform team leader reported to the VP of Engineering. At Ford Alpha's program, the program technically reported through the engineering structure of the company, but the head of the program was a long-time senior engineer who had the ear (and support) of top management. In the case of Technical Machines, Inc., the pilot *was* the engineering department. For the Texas Instruments Gen-X case (in the Appendix), the pilot was an unusual (for TI at the time) team based project organization which was outside the norms of the company. As with Ford's Alpha project, the Gen-X project might not have succeeded without the support of top management. Nonetheless, the lack of clear reporting relations with the "normal" part of the company may have affected its ability to deploy further.

At Coseat Inc., Fred's project team normally would report through design engineering. Since he's putting together a new kind of team (a cross-functional team), the team has to report not only through Product Design, but also to Manufacturing Operations, and Marketing and Sales. This is arranged at a meeting of the Guidance Team which has been empowered by the respective vice-presidents to make this decision. Normally he would not like to have three bosses, but in the case of this highly focused, temporary team, he feels it can work out well as long as the three vice-presidents continue their support for the basic goal.

Communication Plan to the Rest of the Company. Here you determine how the pilot will let the rest of the company know how it's being affected. This does not refer to the normal project work of the pilot — that plan was part of the design for the CE system. Think of this more as "public relations" — making sure others know what a success the pilot is and how it will affect them in the future. Ford Alpha did a wonderful job of this as we describe in the case study in the Appendix.

Initial Training for Kickoff. Naturally, training needs to be provided to everyone involved in the pilot. This includes the following.

- ***CE Introduction.*** This is an overview of the CE approach to product development, including the topics described in Part I of this book. This should be provided to everyone, including top management. Ideally, top management would receive it first, and then participate in the training of those who report to them.
- ***Team Building.*** This involves training the members of each team to work together as a team rather than as a bunch of individuals. This should be given to every team which is developed in the pilot.
- ***Problem Solving.*** This might include training in the various problem-solving techniques such as fishbone diagrams, quality function deployment (QFD), affinity diagrams, and others. All teams should receive an overview of the primary methods and then additional training on an as-needed basis.
- ***CE Methods.*** This is training in the specific product development methods used by a design team, such as Design for Manufacture and Assembly (DMFA). The CE system design should suggest which teams will need training in which specific methods.
- ***CE Tools and Technologies.*** This is training in the tools and technologies, such as CAD, used by the product development teams. The CE system design should suggest which teams will need training in which specific tools.

At Coseat Inc., all members of the pilot team get this training. Training on specific methods and tools is done on an as-needed and when-needed basis so the skills are not wasted and don't get stale.

Technical and Consulting Assistance. You can expect that the pilot will need a certain amount of help implementing some or all of the CE elements in the plan. You should have consultants (internal or external) available to provide assistance with all of the planned changes.

At Coseat Inc., Ed Cazotzky serves as an internal consultant to Fred's pilot team. They also have an internal consultant available from the HR department if additional skills appear to be necessary.

Capturing Learning

A critical aspect of the pilot project is to collect information about what goes right and wrong. In other words, you have to *evaluate* the CE

System design and the implementation process. To do this, you should capture information about the following.

- ***How the system design was implemented.*** This means you want to learn the story of what happened as the project team implemented the CE system. One thing you can count on — it won't happen according to plan (and if by some miracle it does, that story should be recorded carefully as well since no one will believe it), and you need to know just what did happen. This will give you a lot of useful information about what to do and what not to do when you try it again.
- ***If elements of the design were not implemented properly, what was actually done?*** As you record the story of how the system gets implemented, you'll find that some system elements were not implemented as planned, but that the project team did something else on the fly so they could get their work done. Find out just what they did. For example, if the project manager was supposed to evaluate staff performance and didn't, find out who, if anyone, evaluated staff performance. Whatever change they made may have been the key to their success (or failure).
- ***What went wrong in the project and why?*** Find out what problems the project team had in doing their work. For example, you might find that the team had problems making decisions about some aspect of manufacturing. Upon further investigation, you might find that manufacturing engineering was not included in that portion of the process — suggesting changes in later implementations.
- ***What seemed to work well in the project and why?*** Find out what seemed to work really well during the project. If you can find out why things seemed to work so well, you can try to replicate that next time.

At Coseat Inc., the Guidance Team contracted with a local university to assign several graduate students to collect information about how implementation proceeded. The students planned to visit Coseat every other week for a day to talk with people about what was happening and to attend meetings as appropriate. They and their professor also worked closely with Ed Cazotzky and Fred Jenson to make sure their observations were accurate and to provide quick feedback. The students prepared a series of interim reports to keep everyone up to date on what they were doing. Ed and Fred took the information provided by the students and worked it into a series of reports that they prepared for their management.

Modify CE System Design

Logically, the next step in this process is to take the lessons learned and use them to modify the CE system design. You would then use that modified design for the next pilot project or another part of the company which wants to implement CE. The actual process will almost certainly be messier than that. If the initial pilot looks successful, and you've been trumpeting that success around the company, then other projects will probably be trying to use CE, either as they start or in the middle. They almost certainly aren't going to wait until the initial pilot is done and you get your lessons learned integrated into a modified system design. That's OK. You should expect that to happen and prepare for it.

You can prepare for it by modifying the design *as you go* in the pilot. In other words, don't view the lessons-learned exercise as something that happens at the end of the pilot; keep track of the lessons as you go, and at least make notes about their implications for the design. That way, when another project comes to you and asks about doing CE, you'll be able to provide them with the latest and greatest.

Who Controls the CE System Design?

There are two views about "control" of the system design. One view is that the CE system design should be treated much like you would an engineering blueprint — engineering controls the "design" and takes input from other functions about how it should change. In this view, which we'll call an "engineering view" of implementation, there is a "master" design and everyone knows where it is and who controls it. In the other view, the CE system design belongs to the implementing units and there is no central control. The implementing units do what they want and other implementors can take any lessons they choose about it. We'll call this the "manufacturing view" of implementation, because it mirrors how manufacturing often views the product design they receive.

Not surprisingly, we mostly support the engineering view of this dispute. While we do believe each implementing unit should implement what most fits its needs, we do argue for central control over the design. The implementing units should have input to that design, but once it is agreed to, they should also use it. If they modify it out of necessity, and those changes turn out to be beneficial, then they should be reflected in the modified system design.

How Much to Modify?

The question here is whether to apply *all* lessons learned into a modified design, or just some. The answer usually is to only apply some. One reason is that many lessons are actually idiosyncratic to the unit in which the pilot takes place or to the situation of *being* a pilot. Of course, it's hard to know which lessons fall into this category and which would apply to all. Another reason to only apply some is that many lessons are not important enough to go to the bother of changing the design documents and training. They amount to tinkering around the margins with only minimal utility.

Obviously, important lessons should immediately be reflected in the design. Less important lessons should be carried in a design notebook that is available to later implementors and is prominently noted in training.

How to Modify?

We recommend that you first reflect modifications in your Consolidation Matrix and look at the impact on your objectives. Realize that your ability to rate the impact of the CE Elements on the objectives will have improved considerably based on your experience with the pilot. This helps you to make sure that the proposed modifications actually will help. You then need to plan the details, just like any other plan.

Broaden Rollout

The primary issue to be concerned with at this point is where to go next. In general, if your initial pilot was a disconnected pilot, your next step should be an integrated pilot. If your initial pilot was integrated, then your next step should be to one or more additional integrated pilots, perhaps raising the level of commitment as you go. If we look at Chrysler, they simply introduced platform teams to each succeeding product development project as it came along. You can use your Deployment Matrix to help you decide which other projects and which other functions should be included in your next effort.

Conclusion

Implementation has been less analytic and less systematic than the other steps in the CEE process. But that is natural, since all the other phases were concerned with *planning*, while this one has been concerned with *doing*. We know we have not told you everything you need to know, nor is it possible that we could. We have tried to present you with a series of *choices* you need to make during implementation, and to provide you with advice on how to make those choices. Ultimately, implementation is a dynamic process, like riding two horses at once, and the horses don't always agree on the same direction. Good luck!

Appendix

•

Case Studies

Chrysler Platform Teams

In 1988, it was clear to Chrysler that if they were to thrive as an automaker, they needed a replacement for the K-car in the mid- and full-size markets. So, Chrysler conducted a benchmarking study of the design organizations of their competitors. Honda was specifically targeted for benchmarking, and Mitsubishi, at the time partly owned by Chrysler, was also an important source of ideas. Based on this study, Chrysler top management decided the old functional organization could not deliver the quality or speed needed in today's competitive environment. They needed a team of people dedicated to one purpose — to design the best possible car with the customer as the focus. Ultimately, they decided to reorganize their entire product development activity around platform teams.

The first platform team was assembled to design the LH platform, which became the basis for all of Chrysler's new front-wheel drive mid- and full-size cars. Lee Iacocca and top management laid out objectives for the LH team, allocated $1.6 billion to finance it, made critical vehicle concept decisions, and monitored progress of the team. The platform team included all the people needed to design the car, build prototypes, and test it, and they all reported to the same manager, Glenn Gardner. At its peak, 841 people were on the team. Most members were located in the same facility, close to each other and the prototyping facilities. Their only job for the duration of the program was to develop the LH cars and engine. Included on the team were 65 manufacturing representatives who were on the team full time.

This radical experiment for Chrysler produced a car in a record 3.5 years — early, under budget, and meeting aggressive cost and performance goals. There were some quality problems in the first year of production and Chrysler is hard at work doing better for future models. Despite these early production problems, Chrysler seems to have hit the target in the styling and functional features designed into the car. The line of cars is selling extremely well.

Chrysler not only used this platform team approach for this one car line, but organized their total product development effort along these same lines. There are now four platform teams at Chrysler — large car, small car, minivan, Jeep/truck. Through continuous improvement of the development process, Chrysler is aiming for world class timing, quality, and vehicle performance at competitive costs. This case describes in some detail how the LH design process worked, and some of the benefits to Chrysler of the platform team organizational approach.

Organizational Structure

The platform team organization is illustrated in Figure 1. Each product line has a dedicated, collocated team with a focus on the total vehicle. Teams include all of the major functions necessary to develop the vehicle. Figure 1 shows that staff are drawn from functional groups (shown to the left) such as marketing, design, and engineering, but the platform team is not a matrix organization. In a matrix approach, individuals would have at least two bosses: a functional boss and a product

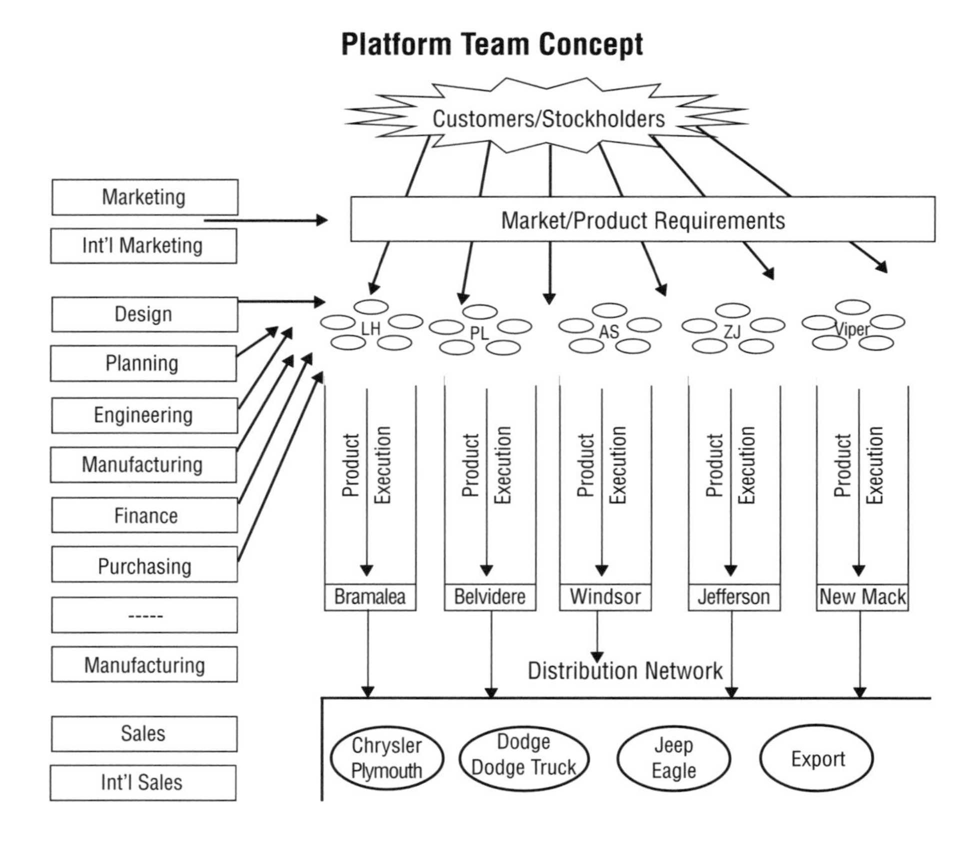

Figure 1 • Platform Team Concept at Chrysler

team boss. In the platform team approach, many individuals are assigned full time to the team and report only to the head of the team. Of course, there are some highly specialized support staff, such as international marketing who stayed in their home functions and worked part-time on the team.

Notice that the chart in Figure 1 does not look like a traditional organizational chart. It does not show reporting relationships from bosses to subordinates. Rather, it represents a process flow from customers and stockholders requirements at the top through product/process development to the assembly plants and finally to Chrysler's distribution network. This change in thinking from hierarchical control to a focus on the process of transforming the product represents a fundamental shift in Chrysler's management paradigm.

The platform teams are further subdivided into a set of interconnected subsystem groups. Each group also has a product focus, but on a major subsystem within the vehicle. Figure 2 illustrates where the electrical and electronics group fits in. These groups are represented in a circle around the core, again to emphasize that this is not a hierarchical organization, but a team organization. Each group was run as a small, autonomous business unit. It had to control cost and was responsible for the group budget. If it saved money in one area, it could spend a little more in another (e.g., putting in a premium suspension system).

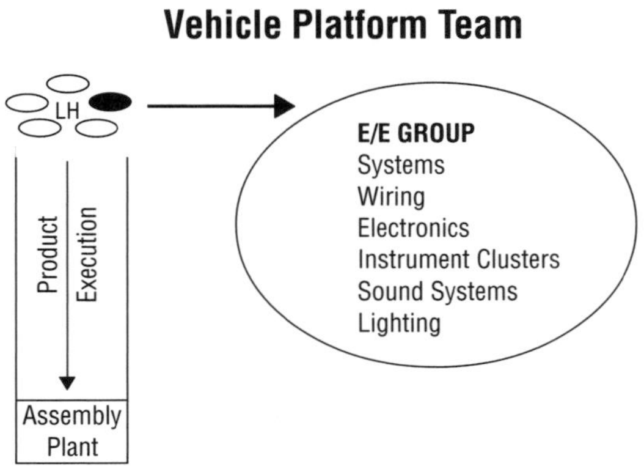

Figure 2 • Electrical/Electronics Group in LH Platform Team

The motivation for moving to a team concept was simple. As Chrysler moved toward a more lean engineering organization, there were not enough people to do things the old way. In the old functional organization, each discipline focused on its own thing in vertical chimneys. According to Vice President Thomas Gale, "we were building great components but not great cars" *(Ward's Auto World)*. The LH platform team structure is a horizontal structure that considers the entire process.

The team started small with Glenn Gardner as leader, his direct reports, and representatives of major support groups. The direct reports were managers from body, interior, electrical, chassis, powertrain, advanced packaging, timing, program administration, and product development. Support group representatives were manufacturing, procurement, finance, service, product planning, and sales.

It would certainly be wasteful to have a full complement of staff for the entire development cycle when not everyone is needed all the time. In fact, the team started with only a small complement of managers and ramped-up as they moved from general concept to prototype design to preparation for production. At its peak, the team had a maximum of 841 people, with 740 people considered ongoing members. During production ramp-up, a small number of people stayed on to work on engineering changes, while most people moved on to the next vehicle program.

In the past, products were designed that could not be efficiently manufactured. To remove the brick wall between design and manufacturing, 65 members from manufacturing were dedicated 100 percent to the LH line of cars early in the process.

In the past, there also was a brick wall between body stylists (considered to be artists) who developed the look of the car and then threw it over the wall to engineers who had to design the outer body so it was structurally sound and could be manufactured. Styling was not formally part of the platform team. But engineers from the team met weekly with the styling staff in the design studio while the styling was in process. By the time the full-scale clay model was approved, 80 percent of engineering feasibility was already there.

Glenn Gardner believes strongly that if team members are split among several teams and multiple projects, they will not be available when needed, will not have a strong focus on developing a world class car, and will not become dedicated team members. Thus, most members were located in the same facility and their only job for the duration of the program was to develop the LH platform.

Not only were people collocated, but the pilot production facility was located on-site (a five minute walk from the design offices). For example, when it came time to assemble the first prototype car, 110 product developers and 10 manufacturing people walked over to watch the car go together to be sure their part fit into the larger system.

COMMUNICATION MECHANISMS

A variety of communication mechanisms were used to ensure that top management was kept informed, and that the groups were informed about the total program as well as the decisions being made by other groups and outside suppliers. Among the communication mechanisms were the following.

- Weekly staff meetings for the total vehicle were held with an informational focus that was attended by the heads of all subsystem groups. The purpose was to keep each group up to date on the decisions of the others.
- Cost, weight, and schedule meetings for the total vehicle were held weekly. Each group presented on a five-week rotation. The engineers, not finance people, reported the numbers— no more secrecy about the price of components.
- Subsystem group meetings were held every day for communication, coordination, and decision-making/problem-solving purposes. Chrysler estimates that engineers on the team spent 90 percent of their time in contact with others, in formal and informal meetings (e.g., all stamping personnel and vendors met together every week).
- Technology reviews were held weekly. These were large meetings (up to 65 people) with the subsystem heads to review sensitive or controversial technical issues and to resolve these issues by consensus. Issues would be rostered by program management prior to the meeting and would not be left unresolved. The meeting would continue until each issue was resolved or there was a plan for how to resolve it. There were no minutes and no formalities. Usually participants would simply get handouts of any presentations made.
- Executive briefings were held with upper management (Vice President level and above) about every 4–6 weeks to review progress. The meetings were largely informational to report to upper management on the status of the program and on key decisions that had already been made.
- Technology clubs were formed for similar component people across platform teams. Since the functional organization was eliminated, there was some need for engineers from different teams in similar technical areas to communicate. They typically meet for one year to develop a white paper on where their technology specialty is going in the future. This is a voluntary effort. Some clubs meet regularly and are active, and others have not been active.

PROGRAM MILESTONES AND TIMING

The design process began with Concept Approval (CA) and then proceeded through a series of prototype stages as follows:

	Months to Production
Concept Approval	42
Prototype 1	24
Prototype 2	19
Production Prototype	12
Production Tryout	4
Pilot Production	1

A brief description of each stage follows.

- Concept Approval includes a financial plan, spatial and performance requirements, and the exterior concept (but not the interior concept). By this time, a full-scale clay model is built and approved.
- Prototype 1 was built fully functional.
- Prototype 2 was staffed entirely by 80 people from the future production plant at a pilot facility.
- Production Prototype was the first prototype done entirely from production tooling.
- Production Tryout was done in the Bramalea plant where the car would be built with all production tooling and processes in place.
- Pilot Production was delayed until 3 weeks before production from the usual 6 weeks because of all the prior up-front work. They did not need the extra time for debugging.

As can be seen, the design process centered around prototype development. Some believe that sophisticated engineering analysis can replace prototypes. However, analysis, at least in practice, is clearly not at that level of sophistication. All of the complex interactions between components simply cannot be anticipated. Prototypes also have the advantage of imposing a schedule discipline on what can otherwise be a very disorganized design process. When there is a serious date for delivering a prototype subsystem, everyone knows the prototype must be there on time or the omission will be painfully obvious. Schedules for completing prototypes became set in stone in the program. For example, management asked Glenn Gardner to commit in advance to a concrete milestone of having the first fully functional prototype 24 months in advance. In the past, the first prototype would be about 16 months before production and would not have the integrity of the LH prototype. Gardner accepted the challenge and the team delivered on time.

Not only were fully functional prototypes built unusually early in the program, but they were built using methods as close to mass production as possible. In the past, the prototypes were crafted using hand methods, and then when the final prototype was complete, manufacturing development started. In the LH development there was an overlapping of product and manufacturing system design as follows, for example.

- The first prototype body was made using the same scheme for positioning body parts in fixtures (i.e., part locating scheme) that was subsequently used in production.
- The paint shop verified that bodies could run at full speed at the first prototype stage.
- For the second prototype body, 19 months out, the same fixturing used to hold the body parts together for welding was later used in actual production. The second prototype was built by 80 in-house production people from the facility where the car would ultimately be built. In the past, at 16 months they were just making the first prototype and engineering was going off to an outside prototype shop to make it.

SUPPLIER MANAGEMENT

The LH team used supplier management methods that are often associated with Japanese automakers. Before LH, Chrysler used the over the wall method: engineers give the design and specifications to purchasing, then purchasing would outsource, usually to the lowest bidder. In this case, LH engineers identified qualified suppliers and gave a short list to purchasing. Suppliers were chosen on the basis of quality, service, and willingness to work with Chrysler in addition to price. Many were brought on board early in the development process and asked to design their own components.

In the past, pricing was set by first determining costs and then adding a profit. The LH group used a target pricing approach. A target price was set in stone for the overall vehicle early in the program. Each group had its own target price for their subsystem. If, as a team, they decided it was necessary to assume the risk of exceeding a target, it was up to them. But generally the difference had to be made up elsewhere.

The bidding process was more in the form of entertaining proposals from suppliers since cost was already defined. Because of the target price method, each component engineer had a checkbook and knew what they could pay a supplier. Like the component groups within the LH team, outside suppliers were given target costs for their components and told that if they added cost in one area, they would have to make up for it in another. Engineers, suppliers, and purchasing worked together to set prices. Everyone understood what was being spent. As a result, the car came in under budget.

In the past, Chrysler purchased small assembled components and individual parts, and did most major assembly in-house. For the LH, complete subsystems were turned over to suppliers, rather than small individual parts. This enabled Chrysler to reduce the number of vendors they had to deal with to 230 parts and material suppliers, and to 285 shipping points, compared with 456 suppliers, and 626 shipping points for the older Chrysler New Yorker. In addition, a larger share of the vehicle was turned over to outside suppliers than ever before — 70 percent of the car is built from components from outside suppliers. This number is typical of Japanese auto companies, but far greater than the historical U.S. average.

More of the design responsibility was farmed out to suppliers than ever before. The most design responsibility was given to suppliers with engineering capability, particularly in areas where Chrysler lacked expertise. Chrysler specified the technology, space, and performance requirements, and capable suppliers went off and developed the design. Thus, key suppliers were involved cradle to grave. This fostered dedication to the product. They have been moving toward key suppliers having a few people located on site at Chrysler. On a typical day, about 35 outside suppliers have engineering representatives at Chrysler.

In exchange for early selection, suppliers were expected to openly share cost information so Chrysler could set prices aggressively while allowing for a supplier profit. Suppliers were not used to opening their books and were not used to dealing directly with engineers on pricing issues, so there was some reluctance. It took time for Chrysler to build up trust with these suppliers. An important step to build trust was through sole sourcing. Almost all components, except for bearings, tires, and some other high-volume items, are single sourced. Thus, suppliers knew Chrysler trusted them and would be heavily dependent on them.

Project Management

Considering the complexity of managing the development of a complete automobile and engine, one might expect large notebooks of procedure manuals, detailed PERT charts, and other formal project management techniques. If anything, Glenn Gardner was at war against these formal systems. As he put it, "We tried very hard not to write things down — we avoided written procedures." Nothing was standardized or documented in great detail to avoid the paralysis that comes with too many rules.

Early in the program, they developed a list of product/process design targets and evaluated progress relative to those targets. There were no formal design reviews, but they did have many mini-design reviews. The general format was: here are our targets, and here's where we are relative to those targets.

LEADERSHIP

Glenn Gardner was selected to be the leader of the LH platform team because of his four years of experience with Diamond-Star Motors, a joint venture with Mitsubishi, and his reputation as a powerful leader who was respected by his superiors and subordinates alike. Mr. Gardner was chairman of Diamond-Star from 1985 to 1988.

Glenn Gardner is a strong believer in participative management. As quoted by *Ward's Auto World*, Gardner describes himself as the peacemaker. "That's an important part of team leadership. But I like the idea of having a bunch of mavericks. I don't want people to shut up." He went on to say he never had to pull rank during the LH development process.

All team members were empowered. Since there were fewer people to do the work, people were given more responsibility and influence. Lead engineers were treated as the president of their own business unit. They were empowered to do what they had to do to get the job done, including knock down walls, circumvent procedures, and create their own minimal procedures. For example, they streamlined the engineering change procedures that had been around Chrysler for years. According to Glenn Gardner's direct reports, they knew if they did battle with a corporate entity they would not get shot — their job was to rock the boat. Mr. Gardner encouraged risk-taking and stood behind people even when their idea failed. He understood that people need to be repeatedly shown that they are empowered. Gardner believes that someone who achieves 90 percent of a demanding task should still get praised.

Mr. Gardner is proud of the fact that he did not have to cut anybody from the team. As he explains: "Sure I got some deadwood; not everybody gave me their top people. Once in the team environment some of the deadwood didn't turn out to be deadwood after all; they really performed well." *(Ward's Auto World)*

For their part, top management at the V.P. level and above were much more consistent than in the past. They kept decisions constant, e.g., they did not change the assembly plants, vehicle styling, etc. The main job of top management was to provide broad goals. For example, they wanted the first working prototype at 95 weeks before production (30 weeks earlier than ever before). They wanted a BMW ride. Vehicle cost and weight targets were carved in rock in November 1989 — they would go over at their own peril. On the other hand, when it came to the details of how this would be achieved, top management gave up decision-making power and did not make late-stage changes in the vehicle. Changes made were driven by customer feedback, not by management.

Design Tools Used

When pressed, Mr. Gardner and his direct reports describe how they used the contemporary toolkit of design tools and methods. And they used modern tools quite extensively. But they always emphasize that these are just tools — people design the car. The design tools and methods used included the following.

- QFD to bring in the voice of the customer.
- A Centralized CAD database so that everyone had the same reference point. Suppliers brought on early were expected to have the same CAD system as Chrysler (CATIA) and use the centralized database on-line. This allowed for interference checking.
- Geometric Dimensioning and Tolerancing, which helped them design parts for minimal spring back.
- Variation simulation to simulate stack-up tolerances so they could determine the fit of the body panels before they were produced.
- Finite element analysis to determine structural capabilities which was all done by Chrysler engineers on the team. The cab-forward design made structural analysis particularly critical. There was much more glass in the windshield, at a precarious angle, and structural forces on the glass were far greater than in a traditional design.

Accomplishments

The LH team brought the vehicle in on time and under budget, even though targets were very aggressive compared to past Chrysler designs. The development cycle went from 4.5 years to 3.5 years (from concept approval to production). They also met or exceeded all of their goals for cost and quality — goals which were also very aggressive.

The entire development cycle was accelerated. The first prototype was at 95 weeks before production (30 weeks earlier than ever before) and was of better quality than the normal 65-week build.

By front-loading the development process, they were able to fine tune the product and process before full volume. Historically, manufacturing got involved at 22 weeks and this resulted in expensive changes, even after launch. Because of up-front involvement, they were fixing problems between 30 and 50 weeks before production. They solved over 1000 problems before production, and were only finessing the car design later in the program. There was also a major reduction in the number of changes made at each of the prototype build stages; whereas in the past, it was level or even increased with prototype stages.

There were hundreds of examples of production savings because of teamwork between design and manufacturing. For example, production people recommended minor design changes that had no real impact on the look of the vehicle, but allowed them to adapt major elements of the body-framing system used to build the Eagle Premier and Dodge Monaco formerly built at the Bramalea plant. Steel body dies and fixtures were designed so no shimming was needed. That is, there was no need to compensate with shims for poor quality body parts that did not fit together.

Lessons Learned

The lessons below were extracted from a talk given by Glenn Gardner who was looking back at the LH development process. We are not asserting that these are the one true way to design products. But they are the reflections of one leader of a successful design effort on the success factors for that case.

- Team members should be under one roof, with one job, with one common set of goals — otherwise it won't work. Totally dedicated employees with a focus on the LH platform was the key to success. All team members should be under the same top management.
- Push decision-making down. Decisions should be made by consensus. When John puts his hand up, none should take it down but John.
- Strong leadership is needed that supports a culture of teamwork and cooperation.
- Define a target cost and give budget responsibility for the program to each subsystem group.
- Top management should provide a set of challenging goals, which they stick to, and then get out of the way. Mr. Gardner did not take requests for decisions to them, he took status reports.

- Mr. Gardner would like to eliminate from the English language the words coordinators, liaisons, follow-up, and the like. He believes that if you do not have the resources on the team, and have direct communication across functions, liaison roles will not work.
- Prototype production should be done in-house using facilities approximating actual production. In the LH case, the pilot production facility was a 5 minute walk for design engineers.
- Teamwork is needed between product engineers and manufacturing. They can point to hundreds of improvements as a result of design and manufacturing guys working together.
- Suppliers for major systems should be selected early, given design responsibility, guaranteed the business to produce the product, and given a target price and broad performance requirements.

Eaton Engine Components Division as a Full-Service Supplier

The Engine Components Division of Eaton Corporation has long been a parts supplier to Ford Motor Company's Engine Division. In fact, Ford has been the largest customer of the division; and in some years, it has been the largest customer of Eaton Corporation as a whole. Eaton played a traditional role in design in the past. Ford's engineer responsible for the engine component would come to them with specifications after the concept design for the engine was complete, and ask for a design for specific individual components, usually valves and lifters. Eaton would develop a design which would be modified by the Ford engineer, and the drawing would be sent to purchasing who would ask Eaton and possibly others for a price quote. After some haggling, a contract would be worked out. Thus, it was a radical change for Eaton when, in 1986, Ford asked Eaton to play the role of a full-service supplier in the early stages of developing their new V8 engine to be built in Romeo, Michigan.

What is the role of the full-service supplier? There is certainly no black and white definition. The concept seems to have developed from observations of Japanese companies, relationships with their large first-tier suppliers. In that case, certain key suppliers are given responsibility for a complete subsystem (e.g., seating system, air conditioning system, starting system, etc.). They are brought on board in the early stages of development program when the vehicle concept is still being worked out, and have some influence over the specifications for their subsystem. They may even have a resident engineer working full-time in the development offices of their customer. They then have responsibility for seeing that the subsystem is developed, appropriate engineering analysis is done, prototypes are made and tested, and prototypes are delivered to their customers for assembly into prototype vehicles. They work intensely on cost reduction in the product and process development stage, and cost-savings are shared with their customer. From the time the full-service supplier is asked to work on the design, they are generally confident they will ultimately get a production contract, though in Japan they rarely get the actual contract until shortly before the start of production. In short, the supplier acts as an arm of the customer responsible for the design, development, production, and timely delivery of an entire subsystem. They also have sub-suppliers for parts of the subsystem and are responsible for coordinating the activities of those suppliers.

For the Romeo engine program, Eaton was given responsibility for the design and development of the valve train assembly which included the valves, rocker arms, hydraulic lash adjusters, valve guides, and seat inserts. They coordinated some additional suppliers, e.g., spring keys and retainers suppliers. They were not responsible for complete assembly, but they were responsible for supplying a kit on a plastic pallet with all parts appropriately placed on the pallet so they could be robotically assembled into the engine at the Ford Romeo engine plant. In this case, we describe the process of working with Ford as a full-service supplier. But first we provide some background on the Ford Romeo engine.

The Ford Romeo Engine

Eaton's early involvement in design for the Ford Romeo engine program was part of a broader change by the Ford Engine Division toward an unprecedented level of cross-functional teamwork and supplier involvement. Prior to 1980, whenever Ford wished to develop a new engine for one of its vehicle programs, they would routinely purchase it from the Ford Engine Division. In the 1980s, in its effort to get the best engines possible at the lowest cost, Car Product Development at Ford began to consider outside sources as engine suppliers. For the 1991 Lincoln Town Car, one strong contender was an engine designed by Nissan that was available in 1987 in prototype form. To ensure Ford Engine Division was chosen as the engine source, they had to commit themselves to meeting targets in the areas of quality, cost, and machine efficiency equal to or exceeding the "best in class" measures in North America for V8 engines. These targets had not yet been attained using a traditional plant structure. For example, they signed up to attain a cpk (process capability) of 2. This means the processes would use $^1/_2$ the variation permitted by the product engineers in their tolerances. This is expected to lead to a defect rate of 1 defect per billion parts produced which means virtually no defects. (Later, as we will see, Eaton signed up to this same aggressive goal.)

Engine Division concluded that to achieve these ambitious targets, a new approach was needed to design and manufacture the engine. First, concurrent engineering would be used. Historically, the design of the product occurred prior to the design of the manufacturing process. On the Romeo engine, the product and process were designed concurrently. Second, a modular approach to engine design and manufacturing would be used. To complicate the design task further, it was decided the design and manufacturing system should be flexible enough so multiple sizes and engine configurations should be manufacturable on the same line, with many common parts, and with very short set-up times. At first this included the ability to make V6 and V8 engines on the same manufacturing equipment, but this was later modified to make various sizes and configurations of V8 engines only on the same line (e.g., number of valves per cylinder, crankshaft size, etc.). Third, it was decided a team approach would be used throughout the design and manufacturing process. This included involving suppliers as members of cross-functional teams.

The very early stage of engine design, when the overall specifications of the engine were being set (e.g., power, size, and weight targets), was handled by Engine Division product engineering. However, very soon in the design process, by historic standards, process engineers from divisional offices were brought on board to help design the engine so it could be manufacturable. Detailed engine design began in early 1987, and components process engineers from the divisional offices and representatives from suppliers were assigned to cross-functional teams to work with product engineers. In fact, the process engineers were physically relocated to the product design offices. At an early point in the engine design, detailed design of the manufacturing plant began. It was at this point that Romeo was identified as the location for the new engine, and a launch team was formed who would eventually assume leadership roles in the new plant.

It is important to note that Ford used a concurrent engineering process for this engine program to a much greater extent than they had in the past. Figure 1 shows Ford's traditional engineering process. Specifications were developed by Ford with no supplier or manufacturing input, and some suppliers were brought in at the first prototype design stage mainly to design individual components (as opposed to subsystems). Suppliers involved in this early stage did not know whether or not they would ultimately get the manufacturing contract, but hoped that by designing the product to suit their manufacturing systems they would gain a cost advantage. Ford design engineers strictly controlled the design process, and even Ford's own manufacturing groups did not give input until after the second prototype when most important design decisions were already made. Finally, after the product was completed, Ford went to potential suppliers for competitive bids. The suppliers involved in the first prototype design had to bid along with other suppliers who had not incurred those costs.

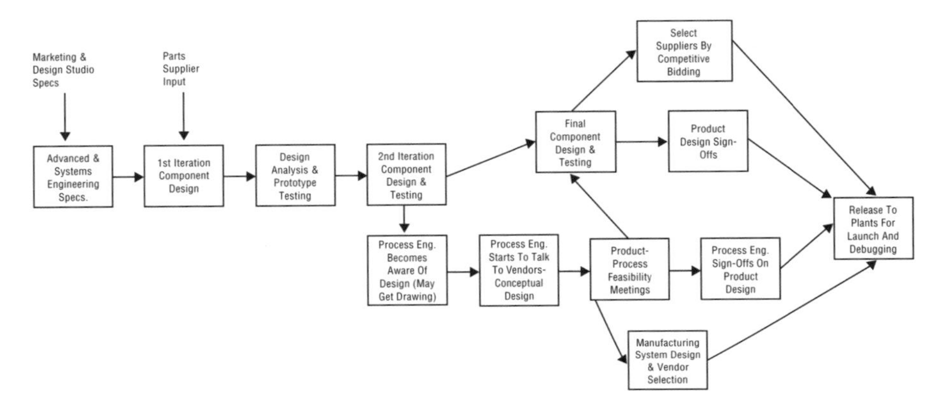

Figure 1 • Traditional Design Process at Ford Engine Division

In Figure 2 we see the concurrent engineering process used for the Romeo engine. In this case, collocated cross-functional teams for each major subsystem were created including Ford design and manufacturing engineers. Suppliers selected to participate in the design were given verbal agreements they would get the initial manufacturing contracts, and these suppliers were included on the appropriate cross-functional teams prior to the prototype stage. As we will see in the case of Eaton, suppliers actually had considerable influence over specifications.

The plant manager was chosen to lead the launch of the new engine plant in January 1987. In 1987, George Pfeil began to assemble a launch team beginning with his operating committee and certain key process engineers. At the same time, Ford decided to transfer the bulk of its tractor manufacturing outside the United States. Ford's primary U.S. tractor manufacturing site in Romeo, Michigan, was going to be idled. In July 1987, an innovative operating agreement was signed with the UAW designating Romeo as the site for building the new engine. The operating agreement included historically unprecedented (for Ford) flexibility in work rules.

Very early on, George Pfeil became convinced that in order for the plant to achieve its objectives, it needed to be organized utilizing a team concept. The first step in planning the process was to develop a launch team. In addition to the traditional complement of managers and

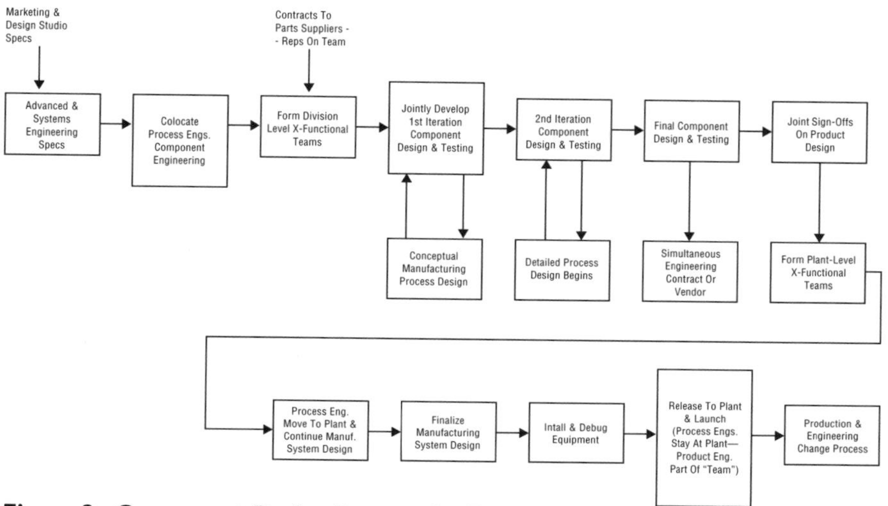

Figure 2 • Concurrent Design Process for Romeo Engine

engineers, Pfeil decided to include hourly production people and a training coordinator. By the summer of 1987, the full-time launch team included approximately thirty people, including managers of production, quality, finance, personnel, the training coordinator, key process engineers, and a small number of hourly skilled trades and production employees. The process engineers were mainly from the group that had worked with product engineering to concurrently design the product and process so there was continuity from the early design stages. An important decision of the launch team was that the plant needed to operate with "wall-to-wall" shopfloor teams. The teams included hourly operators and skill trades, as well as engineers. In fact, the great majority of engineers were assigned to teams with offices in team facilities on the shopfloor.

In traditional engine plant launches, two or three process engineers designed and purchased a line (e.g., a block machining line) without much input from production employees. At Romeo, a decision was made to involve machine operators, assemblers, and maintenance employees in the selection and build phases of the equipment being purchased for their lines, as well as the processes being considered for assembling engines. This decision allowed the three areas — engineering, manufacturing, and maintenance — to have input not only into equipment and process selection and design, but training as well. From the viewpoint of process engineers, this meant they would stay in the plant and become part of the teams that operated the equipment they had helped design. While this gave them responsibility for living with the results of their design work, it also meant leaving their comfortable divisional offices and embarking upon a non-traditional, and potentially more risky, career path.

Equipment began to be installed at Romeo in the Fall of 1988. The first time a complete engine was actually built in the plant, called "product validation," was in January 1990. At Ford, thirty engines are built and tested to validate the product. The quality level of engines built at Romeo at this stage was recognized as being much higher than previous engine launches. At launch, there were 80 percent fewer defects compared to the prior four engine launches at Ford. In fact, the quality of the Romeo engine at launch was the best in the company even though the other engines had the benefits of learning and working out bugs. One year after launch, the data indicated the engines produced at this plant were among the highest quality Ford had ever produced. At that time, the 4.6 liter Romeo engine was considered the third best in the world after the Lexus and Acura Integra. Since that time, quality has continued to improve, and in 1992 the Romeo engine passed the engine in the Lexus 400 as the highest quality engine produced (in terms of number of defects in the field). Since then, the two engines have generally been running neck and neck on quality, with both of them going through a number of quarters without a single defect in the field.

Quality Building Blocks at Eaton

Eaton's ability to respond to the new ways of working and aggressive quality and cost targets of the Ford Romeo engine program was greatly enhanced by almost a decade of quality training and programs at Eaton. In the early 1980s when Ford embarked on a quality journey under the guidance of Edward Deming, Eaton joined Ford on that journey. Back in 1982, about 90 percent of Eaton management went to Deming seminars. By the time Ford asked Eaton to achieve a 2 cpk level for the Ford Romeo engine, Eaton already had a mature statistical process control and cpk program. Nonetheless, at the time, Eaton management was skeptical that the 2 cpk level could be achieved.

In the early 1980s, Eaton picked up almost all of Ford's valve business when Ford decided to shut down its in-house valve plant. In 1983, Ford asked Eaton to be one of the first suppliers to go through Ford's rigorous Q1 supplier certification program. Certification was like a mini-version of the Malcolm Baldridge review and involved documenting evidence of quality systems, use of statistical quality methods, proof that processes were capable, and evidence of customer satisfaction with quality and delivery. By 1984 Eaton was Q1 certified. They made considerable investments in quality, including investing in a goal of providing 40 hours of training per employee per year.

Even with the Q1 accomplishment, Eaton was not satisfied with their quality levels. In fact, the Q1 process brought to the forefront weaknesses in quality systems that still needed to be addressed. The Romeo engine program acted as a prism, focusing attention on making a quantum leap forward in quality systems from design through manufacturing.

Working with Ford on the Romeo Engine

Traditionally, Eaton made large-volume valves and lifters for Ford as their largest customer. In fact, they made over 90 percent of Ford's valves and lifters worldwide. Eaton made large investments in R&D for a long time and used their facilities for basic research, product development, testing, and prototyping to stay at the leading edge of their field technically, and to retain the high level of business they had with Ford. So it was no surprise when Ford came to them during the fourth quarter of 1986 and invited them to work on the development of the new V8 engine. What was a surprise was the scope of the program and the level of responsibility Eaton was asked to take on.

The valve train for the Romeo engine was evolutionary from a technological standpoint. The rocker arm was more complex than usual and there were some specific problems in the details which needed to be worked out, but these were still incremental innovations. What was revolutionary were the extended requirements. Ford wanted Eaton to be responsible for the design, development, and manufacture of the total valve train system, even to the extent of working with suppliers of system components Eaton does not produce. Eaton was asked to deliver a kit to Ford in a form that could be robotically unloaded onto a conveyor at Ford's plant, and be completely picked and placed by automation. The kit provided by Eaton is in the form of a plastic tray that, after it is emptied, is returned to Eaton for more parts. In fact, the design of the returnable plastic trays turned out to be one of the more challenging design tasks for Eaton.

In addition to asking for the total package from Eaton, Ford provided a target price for the package. Ford said they didn't care how Eaton allocated the price among the individual components of the system as long as it added to the total target price. This meant Ford could not then pick out individual components and competitively bid those components. (Though later, for this and other engine programs, Ford did ask for individual component prices and Eaton had to provide a price breakdown.)

To describe their level of responsibility, Eaton makes a useful distinction between being a "full-systems supplier" and a "full-service supplier." A full-systems supplier is responsible for integrating the entire subsystem and becoming a supplier of that subsystem. A full-service supplier is also responsible for all aspects of service, including design and development of the system, putting a full-time resident engineer at Ford to provide follow-up service, being responsible for engineering changes, following up on manufacturing and packaging concerns, continuous improvement of the product and delivery, etc. Eaton describes the Romeo engine program as somewhere between the two. For example, Ford has full-time valve design engineers in-house who still had to formally approve all of Eaton's work. On the other hand, the degree of partnership between Ford and Eaton on this program was without precedent in their relationship.

An important example is ownership of the blueprints. In the past, Eaton's blueprint said it belongs to Eaton and it prohibited use by the customer without Eaton's permission. And Ford had blueprints on Ford paper that could not be used by suppliers. In this project, the blueprints were jointly owned by Ford and Eaton. Legally Ford could take the blueprints produced by Eaton to other suppliers, and Eaton had the right to sell the parts to other customers.

In describing the process of working with Ford, it is useful to distinguish between the early product development phase and the manufacturing process development phase. Each phase was in fact handled by different personnel at Eaton.

Early Product Development Phase

In the third quarter of 1986, Ford executive engineers said they wanted to build all engine sizes and configurations on one line and needed common parts to accomplish this. They later modified this requirement to making only V8 engines, including two-valve and four-valve engines using the same rocker arms, lash adjusters, and valve guides. In the process of discussion, Ford also shifted the type of valve system to be used from a direct acting system to one with rocker arms leading to quieter and smoother performance. So there were some technical challenges from the start.

Early in the program Ford and Eaton realized that an unusually high level of commitment to this program was needed from Eaton. Thus, Eaton decided a full-time program manager had to be appointed just to handle all of the technical and administrative issues. In the early conceptual design stage, Eaton's chief engineer was given the program management job. Though he retained the chief engineer job, working on the Romeo engine became his primary responsibility for over one year. His job was to coordinate the engineering resources at Eaton and work with Ford engineering and purchasing. Formal monthly meetings were held between Ford and Eaton as well as more frequent informal meetings.

Eaton had made investments in DFM/DFA training, and because of the degree of responsibility granted to them by Ford, they were able to incorporate this into the design process. The chief engineer and his direct reports, based on their in-depth knowledge of Eaton's products and processes, developed a number of important innovations in this design stage. These innovations led to reduced cost — all of which were passed on in full to Ford.

As one example, from long experience Eaton learned the material composition of the valves was a key factor in wear and subsequent field problems. Eaton had already been working on the material composition issue as an R&D project before the Romeo engine came along. So they suggested an innovative composition which reduced the need for special lubricants and significantly increased the life of the valve train. Interestingly, this innovation significantly increased the cost of the valve guides, but so reduced warranty costs and customer complaints that Ford was pleased with the change.

As a second example, Eaton was able to convince Ford to eliminate chrome plating from the leak down plunger inside the hydraulic lash adjuster. Up to this point, all of Ford's plungers were chrome plated. In the past there was some evidence this helped reduce field failures, with fuels and lubricants that are not in use anymore. Eaton proved through validation tests the chrome was not needed,which saved money for each part and also avoided the environmental concerns associated with chrome plating. Based on Eaton's tests, Ford also dropped chrome plating from all other existing engine programs.

A third example was that Eaton convinced Ford to drop the process used for hydrogen relief of valves. When valves were hard chrome plated, hydrogen could get trapped which could potentially cause failures. The solution in the past had been to slow bake the valves so the hydrogen would seep out. With the Romeo engine, Eaton convinced Ford that was no longer necessary because of the special process they had developed for chrome plating the valves which also saved cost.

Yet another example of cost reduction was that Eaton convinced Ford that instead of investment cast and machining, the rocker arms could be formed by stamping. The resultant savings were approximately $7 million/year.

All of these examples illustrate the advantages of utilizing the design and manufacturing expertise of suppliers, which is only possible if both customer and supplier view their transaction as part of a broader long-term relationship.

Manufacturing Process Development Phase

When the basic design concepts were worked out, after about one year, Eaton appointed a new program manager with expertise in manufacturing and industrial engineering, and primary responsibility was shifted from product design to manufacturing. This appointment was based not only on technical abilities, but also on project management and interpersonal skills. This appointment was successful and is now considered the benchmark for all program managers.

In late 1987, Ford gave Eaton a letter of commitment for Romeo as the supplier of choice for the valve train. At that point Eaton began committing capital for the plants involved. (Various sites made components and the kit was assembled in Marshall, Michigan.) Eaton made these investments without a firm contract from Ford, but because of their well-established business relationship with Ford and deep involvement in this project, they felt confident of getting the business.

Monthly meetings continued with Ford through the launch and into production. These meetings included whatever cross-functional representatives were needed for the particular issue at hand. They were working meetings — lists of action items, responsibilities, and due dates were developed at every meeting. All meeting notes were distributed to manager levels at both organizations.

The program manager was responsible for coordinating the Eaton manufacturing process to meet Ford's exacting requirements. This included working with major equipment suppliers, developing a returnable package, developing a just-in-time delivery system, using self-directed teams in the plants, and overseeing the overall project at Eaton. The development of the returnable packaging system was itself a challenging design project. Eaton worked with Ford on a major value engineering effort to develop the kitting, including a thorough analysis of every element of the design, taking out cost wherever possible.

Another issue handled by the new project manager was certification of Eaton suppliers. Ford did not want any receiving inspection by Ford or Eaton. They wanted a certification process for ensuring all suppliers were capable of shipping good parts on time. Thus, Ford worked with Eaton to develop a process Eaton could use to certify all of their suppliers. The same strict demands for quality and on-time delivery Ford placed on Eaton were placed on Eaton's suppliers.

Communication with Ford remained intense throughout the design and launch phase and into production. As Ford's modular engineering group matured, Eaton continued close ties with this group and moved right into Ford's subsequent V6 modular engine development. Eaton placed a resident engineer full-time at Ford to work with the engine plant and the modular design group. Eaton committed to weekly visits to the Romeo engine plant after launch to check on the quality of the parts being shipped and continuously improve the quality and service provided.

Even though the Romeo engine (2- and 4-valve configurations) has been successfully in production and the V6 program is now well under way, Eaton has maintained the program manager role. A new program manager took over — a product engineer. This role is full-time, coordinating all program activities focusing on new programs and launch sites. It is intended to extend such program/project management to other major customer programs.

For the Romeo engine there was a high level of integration between the design of the modular engine and the design of Eaton's valve train. Decisions were highly interdependent and Eaton was developing some new technology. However, this level of integration has not characterized all new engine programs with Ford. In one case, a new four cylinder engine was engineered in Germany and there was no similar interaction with Eaton, in part because of the distance. In the case of the new V6 engine, the architecture is so similar to the V8 engine that a high level of coordination was not necessary. There was less need for monthly meetings as structured as with Romeo.

At the time of our interviews in 1994, Eaton was in the process of implementing a structured design development and launch process with 7–8 stages and gates, called Simultaneous Product and Process Development (SPPD). This is to be used for the 29 new design programs in the division. The challenge for Eaton was getting from concept to customer for revolutionary new products. They are very good at evolutionary modifications. The focus is on reducing lead time, knowing when to kill products, reading the market accurately, and in general adding discipline to the design-through-launch process.

Success Indicators for Eaton

Eaton regards Romeo as a seminal experience in their development as a full-service world-class supplier. Romeo focused Eaton and helped integrate various programmatic pieces and became a spring board to new levels of quality and teamwork. Some of the indicators of success of acting as a full-service supplier by Eaton include the following.

- Achieved a cpk of 2 for critical characteristics — 1 part/billion defective. They have also expanded this beyond critical characteristics to all measured characteristics.
- Eaton was able to design the valve train so it could be manufactured at reduced cost with benefits to Ford. In addition to the design innovations mentioned earlier, Eaton was able to structure the design around Eaton's core competencies. For example, testing techniques for engines and components utilizing state of the art electronics which Eaton had developed at one of their corporate research centers was simplified. The testing techniques were then brought to the engine component's division as an engineering tool to expedite the design and development process.
- In general, Ford expected from its suppliers a 5 percent cost reduction/year for all products (including design cost reduction ideas). On the Romeo engine, Eaton maintained to Ford they had driven out all the fat up front in the design stage.
- As a result of these design innovations and Eaton's quality record, their products and services are in demand by other customers (e.g., Chrysler, Caterpillar, Renault, GM, etc.). In some cases, these companies are asking Eaton to act as a full-service supplier. For example, Eaton is approaching 100 percent of Chrysler's business acting as a full-service supplier. Eaton has an advantage over most of its main competitors which lack full systems capabilities. Eaton can design and/or coordinate full systems, while most of its competitors can only supply individual components. Its chief competitor is in the process of forming alliances with other suppliers who design and make lifters, guides, and rocker arms to complement their product lines. Eaton has already proven this capability.
- The quality processes enhanced from the Romeo experience are being incorporated into all of Eaton's programs.

Summary

Eaton developed a valve train for Ford's V8 engine built in Romeo, Michigan, that — if not revolutionary technologically — became a benchmark for quality and reliability, and a totally new way of working with Ford. The result was a highly successful engine and quality levels Eaton never thought possible. Eaton took responsibility for the complete subsystem (rather than individual components), coordinated suppliers of some components in the subsystem, was involved in the conceptual design stage, influenced specifications, worked toward a target price for the subsystem, and coordinated very closely with Ford engineers designing the engine. Strong program manager roles were created by Eaton, and these individuals were dedicated full-time to the project.

Eaton's organizational approach is interesting in that they did not use a dedicated, collocated design team that stayed with the project from start to finish, but made the best use of their core competencies as the project evolved. The project management role shifted from a chief engineer responsible for design in the early stages to a manufacturing engineer in the manufacturing process design and launch stage. What seems to have made this approach work for Eaton is the high focus on their valve train components, and the fact that engineering personnel of all areas are located in one technical center in Marshall, Michigan. The chief engineer and his staff obviously had a good understanding of Eaton's manufacturing processes and were able to make significant improvements in design for manufacturability. And they had easy access to the prototype shop, test labs, and individuals with manufacturing expertise in the technical center. The approach used was cross-functional but a solely dedicated cross-functional team was not used. The project managers acted as the main liaisons to Ford and brought along representatives of other functions only as they were needed for various action items. Again we believe this was possible because different functions were close at hand, and there was a considerable shared understanding of Eaton's products and manufacturing processes across functions.

The benefits to Eaton of acting as a full-service supplier went considerably beyond their successes with the one engine program for Ford. Certainly having control over the subsystem design and manufacturing enabled significant improvements in design and manufacturing from cost and quality standpoints. But the experience also greatly expanded their competencies, and there is evidence that the full-service role is in demand by other customers.

Ford Alpha Simultaneous Engineering

Ford Alpha Simultaneous Engineering (Alpha, for short) is an organization within Ford Motor Company which provides a project-based technology development function for the rest of the company.[1]

Alpha's mission is to champion and accelerate improvements, in partnership with operations, to support the Company mission of being a worldwide leader in automotive and automotive-related products and financial services. Alpha Simultaneous Engineering is a multi-disciplined organization that develops innovative product and process concept improvements for all elements of Ford's business.

One way Alpha accomplishes this is by transitioning new processes and technologies from R&D providers, through proof of principle into a pilot. Once the process or technology pilot is completed and verified, it is up to the operational divisions to implement. Alpha reports through Technical Affairs to the Chief Technical Officer and is thus organizationally quite separate from the car divisions. Alpha has its own budget.

Alpha has more than 300 staff representing most, if not all, of the functions within Ford. These include product engineering, manufacturing, electrical/electronic engineering, sales and marketing, quality assurance, purchasing, strategy, communications, employee relations, and the United Auto Workers (UAW) union. Staff are assigned to Alpha for a 2–5 year period, after which they move on to another assignment within Ford. An assignment to Alpha is viewed as a plus for one's career.

Alpha was established in 1985, initially in response to General Motor's establishment of the Saturn program in 1984. The original idea of Saturn (and Alpha) was to create an organization that could compete with the Japanese in the development and production of small vehicles. For Alpha, this original mission (and the name) evolved to emphasize bringing simultaneous engineering principles and practices into the com-

[1] After this case was written (in 1994), Ford Alpha was disbanded as a separate organization as part of the Ford 2000 reorganization.

pany. By 1989, Alpha was effectively demonstrating these principles and practices through a series of projects conducted for the operating divisions. Through these projects and close collaboration with the operating divisions, Alpha has shown the effectiveness of the practices and has trained hundreds of members of the Ford technical community in their use. The rotation of personnel through Alpha has also served as a mechanism for transferring knowledge directly into the operating divisions. Thus (and we will see this more clearly later in this case study), Alpha has served as an outstanding example of a transition mechanism, designed (implicitly if not by intent) to move simultaneous engineering into the day to day operations of the company.

We can contrast Alpha's success with Saturn's relative failure as a transition mechanism. Although a huge success as a car company, Saturn has so far had only limited impact on General Motors operations. The clearest known impact has been the Oldsmobile Division's adoption of Saturn's marketing and dealership approach. In contrast, Saturn's approach to concurrent engineering and its approach to supplier relations have found much less acceptance in the rest of GM.

The Alpha Process

Alpha goes through a four step process to identify and execute projects.

1. ***Identify Needs/Opportunities.*** In this step, Alpha matches external "opportunities" with customer needs. External opportunities are new processes and technologies that are developed either in or outside of Ford. Customers are the operating divisions (including North America Automotive Operations, Allied Component Group, International Automotive Operations, and Jaguar). Alpha searches for opportunities from vendors, universities, industrial consortia, and Ford Research. The match between these external opportunities and identified customer needs results in a strategic direction for Alpha. This technology direction is represented in an unusual (and useful) way using a chart such as Figure 1. This figure shows only a small part of a much larger chart. The complete "rainbow chart" is a full circle populated by dozens of projects and technologies. At a glance, these charts depict the relationship between programs and technical developments, and illustrate likely paths for technical developments over time.
2. ***Selection/Prioritization.*** In this step, potential projects are proposed and customer acceptance and support (including provision of resources) is sought. Included in this step is formal budget approval for the project. A formal process for showing who is re-

sponsible for what actions, called "Shared Objectives and Interlocking Actions" (SOIA), is used to ensure continued support. SOIA will be discussed in greater detail later in the section titled "Tools and Methods."

3. ***Management/Execution.*** The project is actually executed in this step. Typically this includes a pilot plan in which the new process or technology is developed and actually implemented on a pilot basis. For example, new product ideas may be tested on a sample of vehicles that are sold to customers and then monitored.
4. ***Implementation/Migration.*** In this step, the pilot results are evaluated, costs and benefits analyzed, and migration planned. Alpha may or may not support the customer in the migration plan, depending on customer needs.

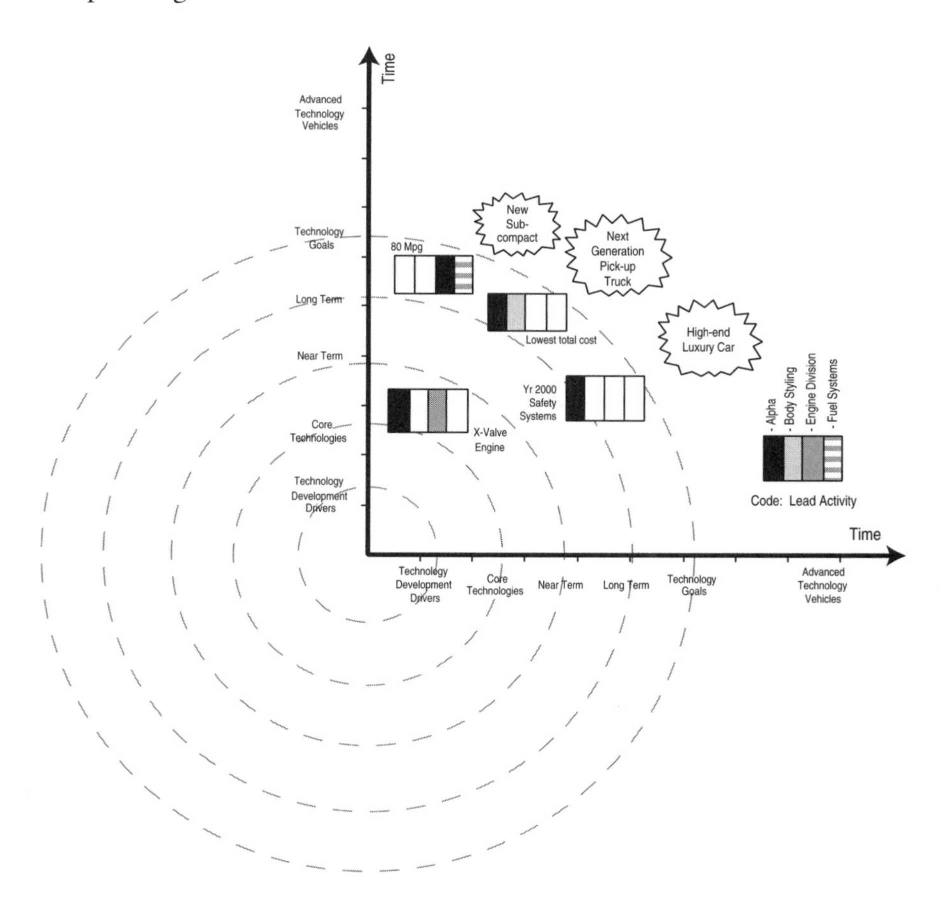

Figure 1 • Technology Planning Chart

By itself, this process is not very different from a "typical" process undertaken by any R&D group in any fairly sophisticated company. Not surprisingly, it is the details which make Alpha's use of this process more effective than that used by most companies. Alpha appears to go to unusual lengths to make sure that projects have the support of its internal customers. Thus, the potential for an "orphan" project, which is completed successfully, but is of no interest to operating divisions, is reduced. Alpha also goes to unusual lengths to insert the results of the project into a "live" pilot in which the new process or technology is tested under field conditions. This ensures that when the customer proceeds with full-scale implementation, there is little potential for surprises to occur. Finally, as a cross-functional group itself, Alpha ensures that the concerns of multiple disciplines are considered during the course of the project. This increases the probability of success. In contrast, many corporate R&D groups are composed of deep functional specialists who rarely interact.

Full-Service Supplier Process

Since 1980 Ford has evolved its relationships with suppliers from an adversarial, "throw it over the wall" relationship to one in which suppliers are closely allied with Ford and heavily involved in product development. In the early 1990s Ford identified subsystems as either core or leveraged. Core subsystems are those which Ford could and should do better than an outside organization. Examples might include body panels and powertrains. Leveraged subsystems are those which might better be done by an outside company. Internal resources were to be concentrated on core subsystems, while outside suppliers would be asked to do more in regard to the leveraged subsystems. This evolved into the concept of the full-service supplier which has responsibility for R&D, advanced engineering, new model engineering, and production. The proportion of parts which are leveraged has steadily increased over time — as of 1994 it stood at 50 percent of a vehicle by value and 80 percent by part count. It has also resulted in a drastic reduction in the number of direct suppliers to Ford — from 7000 worldwide in 1980 to 1700 worldwide in 1994. Ford expects this number to drop to only 1000 by 1998. Clearly the technology strategy which is demonstrated in the Technology Planning Chart (Figure 1) must include consideration of the need to leverage suppliers.

An example of the full-service supplier process in operation can be found in the Eaton Full-Service Supplier Case Study.

FORD'S WORLD CLASS TIMING PROCESS

World Class Timing (WCT) is Ford's structured product development process which is used for all vehicle programs. It includes eight stages and milestones as follows.

1. ***Program Definition.*** Milestones include definition of strategic intent, assessment of alternatives, development of module targets, and approval by various committees.
2. ***Program Implementation and Approval.*** Milestones include commitment of major product development resources, approval of hardpoints for theme development, approval of strategy, and long lead funding by various committees.
3. ***Interior and Exterior Theme Decision.*** Milestones include re-definition of target ranges, implementation of structural prototype program, approval of theme, and long-range funding by various committees.
4. ***Program Confirmation: Objectives and Total Funds.*** Milestones include approval of program objectives, release of tooling and facilities funding to operations from design, and approval of total funding by committees.
5. ***Prototype Readiness: Structural Prototype Assessment and Confirmation Prototype Readiness.*** Milestones include completion of structural prototype testing, completion of CAE analysis, assessment of designs for confirmation prototype readiness, and product design at Job # 1 level.
6. ***Engineering/Manufacturing Design Sign-off.*** Milestones include engineering and manufacturing agreement that the confirmation prototype is to design intent, customer intent, and is manufacturable.
7. ***Launch Readiness.*** Milestones include demonstration of design, process, and quality intent in production parts; and confirmation that the material control system is ready to support mass production.
8. ***Job # 1 (First Production Vehicle Off the Assembly Line).*** Milestones include launch completion, in-process and final quality indicators meet program objectives.

Alpha projects often (but not always) occur prior to the launch of this cycle. Sometimes Alpha projects will proceed in parallel with a specific vehicle program in support of that program (Figure 2). Presumably, this only occurs in cases when the new process or technology is already reasonably well understood and can be confidently timed to achieve completion in time for the vehicle program to proceed according to plan.

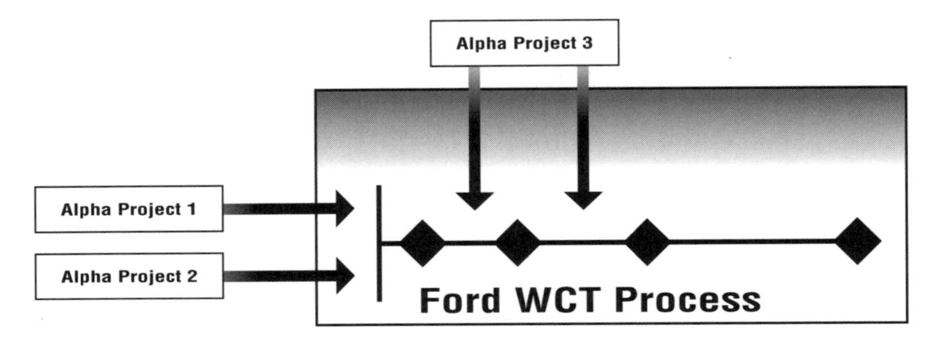

Figure 2 • Alpha Projects in Ford's WLT Process

Tools and Methods Used in Alpha

Alpha stands out for its use of a wide variety of concurrent engineering tools and methods. In addition to more common methods such as Design for Manufacture, Design for Serviceability, Failure Modes and Effects Analysis, they use *Shared Objectives and Interlocking Actions*, *Quality Function Deployment*, *Robust Engineering*, and *Process Improvement*.

Shared Objectives and Interlocking Actions

SOIA is a process by which different functions in the company show their agreement to both a set of shared objectives and the actions needed to accomplish those objectives. One objective of the SOIA process is completion of a SOIA form in which representatives of the various functions formally sign-off on their agreement to perform certain actions and to the whole process. An example SOIA form can be found in Figure 3. Note the following.

1. The shared objective is explicitly stated (upper-left-hand corner).
2. The sponsors and stakeholders for the objective are explicitly called out (upper right hand corner).
3. The major actions are described (leftmost column).
4. The involved functions are listed (column headings).
5. The level of involvement for each function is described. (See key at the bottom — note this is very similar to the RASI model described in the Functional Involvement Matrix used in organizational assessment. Also note that the cells in the matrix are used to specify the level of involvement for each function in each action.)
6. The individuals who made these agreements are specified and they sign the form. This not only confirms the commitment for the individuals involved, but it makes it clear to any successors (people do move around!) that the commitment was indeed made.

SHARED OBJECTIVE
To create a standardized systematic optimization procedure for vehicle sound package development, design and specifications.

SHARED OBJECTIVES AND INTERLOCKING ACTIONS

SPONSORS/STAKEHOLDERS
F.E. Johnson J.A. Jones
P.K. Samuels G. McKinnon
V.L. Smith

MAJOR ACTIONS	ALPHA	VDO	AVSE	PURCHASING	NVHAT	APTE	BOBY	EFDH	TRUCK	SUPPLIERS	EAO
DEVELOP METHODOLOGY Adopt appropriate existing Fort Test Procedures, Specify Regions, & Establish Standardized Protocol	Lead (1,3,4) Initiate Test Protocol	Involved (4) Review Protocol	Involved (4) Review Protocol		Involved (1,2) Review Protocol	Assist (All) Review Protocol	Assist (1,2)	Assist Review Protocol	Involved (1,2,3) Review Protocol		Assist (1,2) Review Protocol
DESIGN SOUND ANALYSIS KIT Develop Instrumentation Package To Facilitate Vehicle Acoustical Data Acquisition	Lead (1,2) Initiate Design	Involved (1) Recommend Improvements	Involved (3) Review Design		Assist (3) Review Design	Assist (All) Review Design	Assist	Assist Review Design	Involved (3) Review Protocol		Assist
ESTABLISH MATERIALS LIBRARY - TESTS & DATA BANK Create Material Acoustical Performance Library For Use In CAE Design Of Vehicle Sound Packages	Lead (1,2) Initiate Tests And Data Bank	Involved (4) Review Data	Involved (All) Review Data & Set Up System	nvolved (All) Assist In Obtaining Sample	Involved (1,2) Review Data	Involved (All) Review Data	Involved (2,3,4) Review Data Set Up System	Involved Review Data & Set Up System	Involved (All) Review Data	Involved (All) Provide Samples	Assist (All) Review Data
WRITE SOFTWARE VEHICLE 1 Develop Automated Data Reduction & Vehicle Sound Package Simulation Software	Lead (1,2) Initiate Tests Develop Software	Involved (3) Review Software & Recommend Improvements	Involved (3) Review Software		Involved (3,4) Review Software	Assist (All) Review Software	Assist	Assist Review Software	Assist (1,3) Review Software	Assist (2) Facilitate Assistance	Assist
CONDUCT PROVE-OUT STUDY	Lead (1,4) Direct & Conduct Project	Involved (1) Observe Software & Review Data	Assist (1,3) Review Data	Assist	Involved (1,2) Review Data	Assist (All) Review Data	Assist (1,2,4) Review Data	Assist Review Data	Assist		Assist (All) Review Data
CONDUCT PILOT STUDY	Lead (1) Direct & Conduct Project	Involved (2) Observe Process & Review Data	Assist (1,3) Review Data	Assist	Assist	Assist (All) Review Data	Assist (1,2,4) Review Data	Assist Review Data	Assist		Assist (All) Review Data
IMPLEMENT METHODOLOGY DEW-98	Lead (1,4) Direct & Conduct Project	Assist Review Data	Involved (All) Observe Process & Review Data	Assist	Assist	Assist (All) Review Data	Assist (1,2,4) Review Data	Assist Review Data	Assist		Assist (All) Review Data
IMPLEMENT METHODOLOGY LIGHT TRUCK	Lead (1,4) Direct & Conduct Project	Assist Review Data	Assist (1,3) Review Data	Assist	Assist	Assist (All) Review Data	Assist	Assist Review Data	Involved (All) Observe Process & Review Data		Assist (All) Review Data
SOIA AGREEMENT	1 Ceee 2 Deee 3 Ieeee	1 Rmmmm 2 Xxxxx 3 Ooooo 4 Emmmm	1 Ayyy 2 Jaay 3 Eff 4 Elll	1 Geee	1 Uuuuu 2 Geeee 3 Jaay 4 Emmmm	1 Emmm 2 Xxxxx	1 Ceee 2 Rmmmr 3 Aay	1 Ieee 2 Geeee	1 Emmm 2 Uuuuuu 3 Deeee	1 Elllll 2 Efff 3 Emmm	1 Deee 2 Xxxxxx 3 Ooooo

A.B. Ceee ______ D.E. Eff ______ G.H. Ieee ______ J.K. Ellllll ______ M.N. Ooooooo ______
P.Q. Rrrrrrrrr ______ S.T. Uuuuuuu ______ V.W. Xxxxxx ______ Y.Z. Aay ______ B.C. Deeee ______
E.F. Geee ______ H.I. Jaay ______ K.L. Emmmm ______

Lead: Total responsibility to lead the team memebers in the complexion of the stated task.

Involved: High level of involvement with commitment to the completion of the specific task.

Assist: Lower level of involvement with commitment to complete tasks when required, possession of knowledge, or a particular skill.

Not Involved: Not directly involved, but committed to help and advise if required (blank).

Figure 3 • Shared Objectives and Interlocking Activities Form

The SOIA process is nearly identical to the Functional Involvement analysis and matrix. The major differences are that it is focused on a specific objective as opposed to more general organization design issues, and that it places major emphasis on obtaining a commitment from individuals.

Quality Function Deployment

Ford was one of the earliest users of QFD in the U.S. It is used widely (if somewhat inconsistently) throughout the company.

Ford' s version of QFD includes four phases.

1. ***Product Planning.*** Relating customer wants to substitute quality characteristics.
2. ***Parts Deployment.*** Relating substitute quality characteristics to part characteristics.
3. ***Process Planning.*** Relating part characteristics to key process operations.

4. ***Production Planning***. Relating key process operations to production requirements.

Ford has recognized some of the difficulties involved in using QFD in an American environment — in particular, the tendency to get so caught up in the process of filling out the matrices that the design team loses track of its purpose. They have attempted to develop simplified forms of QFD but, at the time of this writing, it is not clear how successful they have been with using them.

Robust Engineering

Alpha in particular makes extensive use of robust engineering methods, particularly analysis of Taguchi's signal/noise ratio in a design of experiments context.

Process Improvement

Ford engineers make extensive use of a seven step methodology for process improvement:

1. identify the opportunity
2. define the scope
3. analyze the current process
4. envision the future process
5. verify and pilot the proposed changes
6. implement the changes
7. improve continually.

Training in this methodology is provided widely, and the method is indeed used for most, if not all, process improvement projects.

Examples of Alpha Projects

Automatic Transmission Thermal Management Project

The purpose of this project was to "develop methodology to analyze and optimize the Thermal Management of automatic transmissions and to determine alternative(s) to the existing approach of using an in-tank cooler along with auxiliary air/oil coolers for improved reliability/reduced weight and lower cost."

- ***Team.*** The project had stakeholders from light truck cooling, climate control division, and transmission, as well as several parts of Alpha. The work team included representatives from climate control engineering, climate control purchasing, powertrain programming, light truck development, transmission design, UAW, and several parts of Alpha (including quality assurance, business and product strategy, and climate control).

- ***Responsibilities.*** An SOIA was completed using the quoted purpose above as its objective.
- ***Tools and Methods.*** The team used benchmarking, QFD, simulation (CAE) models to plan and evaluate alternatives.
- ***Results.*** The team analyzed the cost of reduced transmission temperatures under various climate and load conditions, and considered the potential warranty and customer satisfaction impacts. The team had to consider the total system, rather than just the transmission, since simply providing additional cooling capacity to the existing cooling system might just dump the heat elsewhere in the engine compartment to the detriment of other components. The team considered 39 alternatives and had to develop a CAE tool to simulate them. Additional cooling requirements were developed for various vehicles with substantial savings resulting.

Central Lighting System Migration Planning

In an attempt to reduce cost and increase reliability of lighting systems, Alpha started a project to develop a system with a single central light source for all lights, using fiber optics. After the system was developed, it was first piloted in a single 1994 model in limited production and only for the headlights. For 1995, the pilot will be expanded to include a larger number of lights on a non-pilot basis on the same model, and on a pilot basis using a slightly different lighting method on a different, limited production model. By 1997, the plan is for the system to be in full production for the first model, and in 1998 in full production for yet another model. By 1999, the system will be in full production for three Ford models.

Conclusions

The Alpha organization at Ford is a successful example of a migration strategy for CE methods and tools. Alpha not only provides a live demonstration of the benefits of CE, it serves as a training ground for practitioners who are later dispersed throughout the company as part of a regular rotation process. Alpha serves as a living demonstration, not only of the methods and tools, but of the day to day practice of CE. Thus, the many thousands of Ford engineers and other workers who came into contact with Alpha projects had the opportunity to experience working on a cross- functional team, and to see the benefits (and difficulties) of making such a team work. This, in addition to the regular rotation of Alpha staff back into the mainstream of Ford, serves as a training mechanism for Ford personnel.

If we may again make the comparison to GM's Saturn strategy as a migration path to CE, we can see significant differences. Although Saturn was promoted as a mechanism to transform GM, it was deliberately kept as separate as possible from the rest of GM — with the intent that it should not be "contaminated" by GM's culture and bureaucracy. There was little contact between Saturn and the rest of GM. In fact, once someone from GM transferred to Saturn, it was extremely difficult to return. In contrast, Alpha was kept integrated with the rest of Ford, through its project structure (Alpha did projects for the rest of Ford), its staffing of projects (all projects included both Alpha and operating division personnel), and its rotation practices. As this is being written, GM appears to be moving Saturn closer to the core of the corporation by merging it into a new small car division. Whether this will succeed in bringing Saturn's lessons into GM, or whether it will lead to the demise of Saturn, remains to be seen at this time.

Motorola Government and Systems Technology Group (GSTG)

The Motorola Government and Systems Technology Group (GSTG) develops and produces electronic components and systems, primarily (but not exclusively) for government customers. Examples of their products include Global Positioning Systems, electronic security systems, radar systems, and receivers for satellites. A new commercial product will be the IRIDIUM, a satellite-based, wireless personal communications network. The Group employs about 5,000 people and is divided into four divisions: Communications, Diversified Technologies, Tactical Electronics, and Satellite Communications. Its sales for 1993 totaled $858 million.

Motorola as a company was in serious decline in the 1970s due to increased competition from overseas competitors. For example, in 1974 it dropped out of its once core television business, and in 1980 it dropped out of the business which started the company — car radios. In the early 1980s competition from Japan forced Motorola out of the Dynamic Random Access Memory (DRAM) chip market and severely reduced its hold on the U.S. market for pagers and mobile phones. In 1981 a conscious attempt was made to renew the company, specifically through an increased emphasis on measurement and goal setting. At that time, a seemingly impossible goal was set to improve product quality tenfold within five years. This goal was reached, and the sense in the company was that it hadn't been too difficult to do so. Another goal was set for a further increase in two years. This led to the award of the first Malcolm Baldridge Quality Award in 1988.

Motorola has continued to work on improving quality to what it calls the six sigma level — which means a defect rate of no more than 3.4 per million, i.e., approaching zero defects. The bottom line of all this effort has been a complete turnaround to a widely recognized, highly successful company. Motorola's revenues have increased to $17 billion in 1993, from only $5 billion ten years earlier. Debt is low and profit margins are relatively high. It is the world's leading supplier of cellular phones, pagers, and two-way radios. As a recent article in *Forbes* (Sept. 13, 1993, pp. 139–145) notes:

> *Motorola has become much admired as a role model for American business. Its excellence lies in good part in a deeply bred ability to continually move out along the curve of innovation, to invent new, related applications of technology as fast as older ones become everyday, commodity-type products. This is precisely what IBM failed to do. Having skillfully dominated the mainframe computer business, it faltered when its business entered a new phase, permitting upstarts to dominate the PCs that were fast replacing mainframes for many tasks.*

In the past few years, Motorola has taken its approach to improving quality and attempted to apply it to its product development process. Motorola is now attempting to reduce its product development cycle by ten times, using what appears to be fundamentally the same types of activities they used to improve product quality. Motorola focuses on seven factors to achieve Concurrent Product Development.

1. the human aspect, rooted in technical competence and the human-technology interface;
2. people and vision;
3. equal partnership in design responsibility;
4. collocation;
5. cross-disciplinary teamwork with clear team objectives (as one manager said: "team culture is an unnatural act in our society, so we all have to do it");
6. understanding different functions;
7. timely, well documented communication.

Motorola staff are careful to note that CPD is a process, not an end in itself. Within GSTG, Motorola is working on CPD through a variety of methods. These include the following:

- principles and goals embodied in a highly structured product development assessment process
- empowered teams focused on both operations and improvement
- methods and tools to aid the technical work of product development
- training for employees on both technical aspects of product development and teaming
- structured design processes

- organization and project management structures which provide for leadership throughout product development
- measurement and reward systems which reinforce achievement of product development goals
- involvement of suppliers in product development
- following a consistent product development process.

PRODUCT DEVELOPMENT ASSESSMENT

Motorola's Product Development Assessment (PDA) was developed at the corporate level in response to the drive to reduce product development cycle time. It is based on a year-long benchmarking process in which 18 firms worldwide were analyzed. The assessment covers ten topics:

1. cycle time reduction
2. new product development process
3. new product drivers
4. engineering metrics
5. robust design and training
6. engineering tools
7. electronic networks and interfaces
8. empowerment and customer visits for engineers
9. research labs and interface to product groups (creativity and innovation are priority)
10. technology sharing and reuse/recognition (e.g., patents, awards).

Each area is covered by a series of questions, each of which has an explanation and a set of scoring anchors. A sample question under new product drivers is: what percentage of new products were developed in the last 5 years? Each question is weighted, with the weighting determined individually for each division. This keeps the assessment relevant to the specific needs of the division. Each division in the company is rated by a 3- to 4-person team from outside the division, but inside the company. Each person on the rating team collects data by interviewing approximately 10–12 people inside the division. The result of the assessment is a set of goals for improvement which are developed by the division leadership in consultation with the assessment team. At the end of a year, the division is rated on its achievement of these goals.

Motorola has long had a Baldridge-like quality assessment system called Quality System Review (QSR). The QSR is conducted every two years for each division. The QSR has been revised twice in the past ten years. Each time, the bar is raised, in the sense that anchors are changed to make it more difficult to get high ratings. This is in response to changing external benchmarks. Motorola expects the PDA will follow the same model over time.

Teaming

Motorola has created two different types of teams: high-performance teams, and total customer satisfaction (TCS) teams. High-performance teams are a form of empowered project team, although it does not appear to fall into the category of self-directed since teams do have managers. The high-performance team is one of the fundamental work units in the organization, hence most "line work" is done in the context of such teams. Training for teams is provided from a curriculum of 80–120 hours which includes job analysis, job design, team design, organization structure, problem solving, meeting management, selection, and appraisal. Team leaders receive additional training in topics such as listening skills, and planning.

TCS teams are ad hoc problem-solving teams which form around a specific problem, much like what a quality circle was supposed to be like. TCS teams typically form within the context of a functional group or project team. The responsible manager provides resources to the team (both time and funding) as well as acting in a coaching function. Motorola has a goal of 80 percent participation on TCS teams and has achieved this. Motorola has a competition each year for TCS teams. This competition provides considerable recognition for both the members of the competing teams and their divisions.

Training

There is a strong emphasis on training and professional development. Each individual receives a minimum of 40 hours training per year. Training is also provided for teams and includes job analysis, job design, team design, problem solving, meeting management, selection, and appraisal. GSTG provides in-house training in several areas. New engineering employees participate in the Concurrent Engineering Training Program (CETP) which is focused on the ten areas from the PDA. This course started as a solid-state design course, but has been expanded to include the culture of Motorola (including key beliefs such as respect for people, integrity, goal to be best in class), training on tools (including SPC, DFM, DFA, DOE, Robust Design, Design for Total Unit Production Cost — DTUPC). The CETP is for the GSTG Group only. Motorola has developed a CE course of its own in conjunction with Georgia Tech which is used by the rest of Motorola.

Organization and Project Management

The GSTG uses an organization structure which it describes as a matrix. Individuals report within a functional structure. Functional managers "own" staff and are responsible for both performance and bringing in business. Within each functional group, there is a set of projects, and project managers and individuals work on these projects. In some sense, individuals feel matrixed in this structure since they report to both functional managers and project managers. In terms we use in this book, however, we would probably call this a functional organization (since there is only a functional side to the permanent reporting structure) with a heavyweight project management structure, since the project managers do control their staff.

An additional "structure" in both the group and the company as a whole is what Motorola calls "councils." These are high level groups within each organizational entity which champion various initiatives such as 6 sigma or the 10x cycle time reduction. These councils develop strategies, ensure their execution, and share success stories. As an example, in the early 1980s, the Corporate Quality Council developed and recommended actions relevant to quality initiatives. The CEO was their sponsor. When they first started on quality improvement, they trained everyone in the company on SPC. Later they found new employees had entered without knowing this, and old employees had forgotten or not kept up — so, they found they needed procedures to maintain skills and to train new employees, and recommended new training courses to fill this need.

Methods and Tools

Motorola regularly uses a wide variety of CE methods and tools in their design work. These include SPC, DFM, DFA, DOE, Robust Design, and Design for Total Unit Production Cost — DTUPC. While Motorola makes use of many externally developed tools and methods, some are developed in-house. For example, a DTUPC software tool was recently developed in-house by a TCS team using commercial spreadsheet software. The development costs for the system were quite low, and it has already generated widespread interest and use.

Measurement and Reward Systems

Motorola is one of the most measurement oriented companies we have ever seen. As one of the people we met said, "We as a company believe in keeping score." At each management level, a strong attempt is made to measure and keep track of anything which might be a useful indicator of either success or of problems. For example, at the Division level, the Engineering Manager of the Communications Division in GSTG tracks a wide variety of aggregate measures, including such things as mean total cycle time, lines of code, and solder passes per hour. Interesting, he recognizes that measures such as this are not only potentially misleading, they may lead to counterproductive behavior. For example, more lines of code might lead a manager to believe that more was better, when in fact, more compact programs might be superior. Individuals might well deliberately write long, inefficient programs in order to look more productive. The engineering manager we spoke with was very sensitive to this issue and made it quite clear that he used this only as an indicator, and only when it was appropriate.

A substantial portion of the labor budget at Motorola is used to reward individuals for unusual performance. These awards range from about $25 to $1000 and are mostly awarded by the teams of which the individuals are members. There are also a variety of team based awards ranging from small monetary amounts to cookouts and plaques.

Supplier Involvement

Motorola has recently developed an Early Supplier Involvement (ESI) program. This is seen as an enabler for Concurrent Product Development (CPD), and indeed in one division the CPD Champion reports to supply management. The ESI program has four elements.

1. Early involvement in the conceptual stage of new product development — this includes a commitment to the supplier who has been involved early in concept development.
2. Single sourcing for difficult or complex parts.
3. Identification of critical parameters in suppliers' processes, and maintenance of a database of supplier capabilities.
4. General price agreements — the result of a cost analysis in which Motorola and the supplier agree up front on a target price for the component or part. These suppliers generally open their books to Motorola.

Motorola has reduced the number of its suppliers from 6000 in the early 1980s to only 600 today.

Consistent Product Development Process

Motorola has documented design processes for individual disciplines (e.g., mechanical engineering, manufacturing engineering, software engineering). The documentation for these processes includes objectives, entrance criteria, exit criteria, validation criteria, and activities. The activities show concurrency and coupling mechanisms between activities and between disciplines. The documented design process is tailored for each project. The documented process has six phases:

1. requirements
2. concept
3. initial design
4. final detailed design
5. fabrication and quality
6. product readiness.

As tools to assist in completion of this process, Motorola has a series of CE checklists, a Design Review Checklist, and a Product Readiness Review Checklist (e.g., percent complete, goal dates, responsible individual). Each checklist has the task with level of completion. Formal design reviews, including both functional managers and project managers, are conducted at each exit, while informal reviews occur with greater frequency.

Conclusions

This case has been a bit different from the other cases contained in this book in that it has not attempted to tell a story about a particular project or program which benefited from CE. Rather it has described CE processes at Motorola GSTG more analytically. The overall sense one gets from this organization is of a very disciplined, structured company that uses its discipline to permit its people tremendous freedom — freedom within limits. That discipline includes the measurement of almost everything, and a strong emphasis on teams and empowerment.

Newport News Shipbuilding

Newport News Shipbuilding (NNS), in Newport News, Virginia, has been building ships since 1886. This case study looks at the advances in ship design that NNS has put in place during the course of designing the U.S. Navy's Seawolf, which is the first in a new class of nuclear attack submarines. By tradition, the Navy refers to a class of ships by the name or number of the first one of the class. The Los Angeles class attack submarines are the predecessors to the Seawolf. The Los Angeles was launched in the late 1960s, so the Seawolf work, which began in 1982, has been able to take advantage of more than 20 years of newer technologies, both for the ship and for the design and manufacturing processes. Many aspects of this massive program (nuclear submarines are incredibly complex) are quite different from what NNS had previously done in ship design. Customer requirements, which are substantially beyond those met by the Los Angeles class submarines, coupled with a significant segment of the design technologies and design staff new to submarine design, have led to substantial improvements in the design process.

Customer Requirements

The customer — the U.S. Navy — imposed significant requirements on NNS. The new requirements fell into two areas: performance requirements and contractual requirements. Compared to any previous submarine, the Seawolf performance requirements include substantial reduction in noise, increased ability to withstand shock, and increased operational depth. Together, these requirements ensured that little equipment from earlier submarines would be of any use. Effectively, and for the first time, virtually no parts or systems from previous designs could be reused. The Seawolf has been designed virtually from scratch.

Contractual requirements on NNS have also led to changes in the way they work. The Navy divides ship design and construction into the following four separate contracts.

1. *Preliminary design* generates the specifications.
2. *Contract design* develops major structures, including cross-sections, plate thicknesses, major machinery placement, and preliminary system diagrams establishing operational requirements.
3. *Detailed design* develops all the details and documents required to construct a working ship to the original specifications.
4. *Build* — this is the contract to construct the ship.

The Navy has required that NNS work with its primary direct competitor, the Electric Boat Division (EB) of General Dynamics, for the Seawolf design and construction. The intent behind this unusual step was to keep both companies alive in a time of shrinking defense budgets. This need for substantial cooperation between two companies, which are normally quite competitive, has posed significant challenges. NNS, EB, and the Navy worked out a division of design effort based on the two primary regions of a nuclear submarine. The propulsion section (the aft half) is EB's responsibility. The living and operations section (the forward half) is NNS's responsibility. These two halves really are substantially separate, so the interface between them is about as simple as could be found in such a complex design problem.

Even though they had to share the design work, each company had the goal of getting the contract to build the first ship. Once EB won that contract, NNS then had the goal of building the second ship. Given the changes in the world's politics and economy since the Seawolf was conceived, there will probably be only two or three of the class produced altogether rather than the 20 or so originally envisioned. EB is now virtually certain to build the class, so NNS has revised its goals to being recognized as extremely able in the design and construction of submarines (they are currently building the last few of the Los Angeles class subs), as well as the Navy's principal resource for ship engineering, construction, overhaul, refueling, and repair. They also hope to become the commercial U.S. company of choice in the global market for shipbuilding, ship repair, and marine electronics.

Contributing Technologies

With the division of design responsibility between two companies, building a Seawolf class submarine requires each company to obtain a great deal of design data from the other. Accordingly, the issue of data exchange has been and continues to be significant. The Seawolf is a very complex ship, dramatically expanding the boundaries of submarine design. From the start, the necessity of using computer-aided design tools was obvious. Two specific technologies have provided especially important contributions.

The foremost contributing technology is NNS's proprietary solid modeling system, known as VIVID. VIVID is a true solid modeler based on the constructive solid geometry (CSG) approach, and is capable of handling very large models. It was envisioned and first developed in the late 1970s. It was thus in the right place at the right time for the Seawolf program.

The large models are the key to VIVID's value. It supports the construction of a single comprehensive solid model of NNS's entire half of the Seawolf (truly a large model, with over 700,000 parts). This large-model capability is an important contributor to the changes in how NNS is designing ships (see Design and Manufacturing Process Changes below). Note that the information sent by NNS is not limited to representations of individual parts. They also use the VIVID system to generate and send Sectional and Construction Drawings, which are the equivalent of assembly manuals, the series of drawings that show the incremental fabrication and assembly of the various parts of the ship.

Piping, electrical cabling and wireways, and ventilation systems are only modeled in VIVID. No other electronic models of them are needed, as VIVID can feed the necessary information directly to any numerically controlled machine tools involved in their fabrication and assembly. Not everything is done in VIVID, however. Structural parts and machinery components are modeled as wireframes in CADAM, a commercial CAD system. They are then converted to CSG models for use in VIVID. CADAM is used to maintain the master versions of the structural parts because some of the manufacturing processes used for those parts read directly from the CADAM files. The purchased machinery components are modeled in CADAM because they only need simple wireframe models — only the basic shape is required. The details required for manufacturing are minimal for the purchased machinery.

CADAM is used for one more purpose — the generation of manufacturing drawings. Manufacturing drawings are two-dimensional drawings with substantial notations that start as views from VIVID. These are then taken into CADAM for cleanup, formatting, and plotting. In this role, CADAM basically serves as an output formatter and device driver. This makes sense because CADAM already has the capability to generate 2-D manufacturing drawings and there is no reason to go to the expense of adding that capability to VIVID when the CADAM approach works well.

The Navy's requirements for the Seawolf program included the electronic exchange of three-dimensional CAD data plus drawings and text. The problem there was that NNS uses VIVID and CADAM (now owned by Dassault Systems) while EB uses ComputerVision's commercial CAD system and a proprietary piping system, PIPER. CADAM and the ComputerVision system cannot directly exchange data, nor can the companies, proprietary systems. To solve that problem, NNS, EB, and the Navy agreed to use the second major contributing technology, IGES (Initial Graphics Exchange Specification), a U.S. national standard that supports the exchange of geometric product data between different CAD systems.

At the start of the Seawolf program, IGES was relatively new and had not been applied to the kind of major data exchange problem posed by Seawolf. IGES provided a standardized basis upon which NNS and EB could build their data exchange, but it needed a lot of work to become truly useful. Both companies and the Navy put a great deal of effort into improving IGES itself, including development of the first IGES Application Protocol (AP) for 3-D piping (MIL-D-28000, Class V), plus working with their CAD vendors to improve the quality of the exchange. They were successful in their efforts. IGES fulfills their CAD data exchange needs for the Seawolf program. An article entitled "Development of an Initial Graphics Exchange Specification Capability" by D.J. Wooley and M.L. Manix in *The Journal of Ship Production* (vol.3, no. 4, Nov. 1987, pp. 264–273) describes this in detail.

The only change from the initial data exchange concerns is that most of the design data flow is only one way, from NNS to EB. NNS has not gotten a construction contract, but is still responsible for one-half of the design. Not all data are sent electronically. Much is on paper, especially the manufacturing drawings. However, the IGES work has been very important to the success of the program, especially in the area of three-dimensional piping systems, structural piece parts, and the selected record drawings for the ship. If a whole new program were to start now, STEP would be a bit farther along than IGES was then and would probably serve the needs very well.

By comparison, a similar project to build a new destroyer for the Navy took a different route. The same kind of shared design between two shipyards was set up, but data exchange was handled differently. They chose to go the route of a direct translator between their two different CAD systems. That worked for a while, but eventually they had to give up the process because the translator got out of date (as new versions of the CAD systems were adopted) and the contract did not have the money to support the development of new translators. In other words, by relying on the neutral standard, NNS and EB have been able to exchange data throughout the life of the program, even though their software has continued to evolve.

NNS Organization

The basic organization of NNS has long been along product lines. They currently have product teams for the Seawolf, the Los Angeles class submarines, the Nimitz class aircraft carriers, and for commercial ships. Each product team is, for all practical purposes, totally independent of the others.

The primary organization of the design staff within each product team is by functional groups. These groups are based on the four major elements of ship design: hull (structure), electrical, piping, and mechanical (including heating, ventilating, and air conditioning). While there are subgroups within these four, they are the traditional "homes" people use to describe themselves. Needless to say, the designs from all four groups have to come together in the actual ship.

Traditionally, ships were designed by the four systems groups separately, with minimal interaction. That is, each group (electrical, piping, mechanical, and hull) would design its entire system for the whole ship almost independently of the others. Coordination was done through occasional meetings, composite drawings, and a document called the Space Control Book. The Space Control Book included verbal descriptions of what each group was doing and where they were doing it. The composite drawings (developed and maintained by compositors) attempted to show everything in a portion of the ship. This was only practical to a point (maybe two-thirds of the way through the design work). After that, the drawings became unreadable, even for the compositors, which led to a great deal of "cut to fit" instructions. Throughout this process, design decisions for each system were inevitably made based on what was best for that particular system, with no thought of the effects on other systems.

Manufacturing planners design the manufacturing process and serve as an interface role between design and manufacturing. Typically the planners had little information on which to base the construction of the ship until they received the drawings from design. Then they had to figure out how to make it. Since the drawings were done at separate system levels, this was not an easy job. Most of the actual space allocation took place during construction or through the construction of a full-scale wood model of the entire ship.

The Seawolf Organization

The Seawolf organizational structure has been a significant departure from the traditional system. With the capabilities of the VIVID system and the commitment of the head of the Seawolf program to make full use of those capabilities, the need for someone (i.e., compositors) to try to develop combined drawings is eliminated by the comprehensive model in VIVID. A new role was created to take advantage of VIVID. "Space Kings" were put in place, each with space allocation responsibility for one or more well-defined regions of the submarine. The Space Kings are drawn from the design staffs of all four major functional areas, based primarily on which system appears to present the most difficult design challenge for a given space.

Space Kings generally do not have line authority over the design staff working in their space, rather they have control and authority over the design database in VIVID. Thus, they control authority over the work being done, rather than the people doing it. In the early days, Space Kings were all supervisors, although not necessarily of the people contributing to their spaces. This was done to enhance the standing of the position. As the design evolved, the demands on the Space Kings became too great for part-time work, and non-supervisors took over most of the positions.

To address conflicts at the interfaces of separate spaces and to give the Space Kings some teeth, Space Czars also came about. These are high-level functional managers with the identified role of solving problems that could not be worked out at the Space King level. As managers, they do have direct leverage over the people involved. Beyond the Space Czars, the overall program manager is the final arbiter. The current program manager could only think of one or two cases where a problem had actually gotten to the top level. Most decisions are made at relatively low levels. Through the common model, necessary information is available to all. One role of managers is to challenge the staff to do the best possible job. This includes asking questions of the staff, especially in areas of special importance or concern.

As the design process evolved from a focus on components to a focus on ship systems, the space issues became more and more dominant. On top of this, the electrical design work got behind, especially in getting it into the composite model. To improve the focus and allow greater flexibility in personnel use, the Space Kings and their supporting staff (compositors) were gathered into a new organizational unit, "Arrangements," in early 1993. Top management believes this has strengthened the overall space allocation process, because those with space responsibility have a common identity that is separate from the functional units. Interestingly, this change is seen as a logical part of the process. That is, in early stages, there was a real advantage to having the Space Kings in the regular functional groups, where the direct contact with the design staff is important. Only in the later stages was a separate group seen to be appropriate.

The VIVID system is not just used by the Space Kings. The manufacturing planners also use the system extensively. Manufacturing planners come from all the manufacturing disciplines. As with the Arrangements group, they act as a cross-functional team. Planners use the visualization capabilities of VIVID to determine the detailed assembly sequence for the ship. This assembly plan then drives the content of each drawing to be produced. Since NNS is not building a Seawolf-class ship, the planners primarily serve in an interface role. If NNS had gotten a contract to build a ship, the planners would have been more of a bridge between manufacturing and design. They do provide some feedback to the design team, occasionally pointing out potential problems. Their primary focus, however, is "how do we build this thing?" and what parts should go on which drawings to make that construction most efficient. Because of their liaison role with EB, the planners shifted from an independent department to a home in administration. The product model also supports concurrent development of logistic support products and studies. Lifting pad locations and equipment maintenance access are verified. Equipment removal paths are also developed and demonstrated by logisticians using the product model.

Whether this particular organizational approach will be used in future programs is not clear, although the basic concept of Space Kings will probably be maintained. The idea of building in an organizational change in the middle of a program may appear a bit unusual, but for a decade-long program organizational change should not be a surprise. Part of the identity of the current Arrangements group is bound up with their being the real VIVID users. In future programs, the software will be used by more and more of the company, and this identity characteristic is likely to become less significant.

Underlying all this is the explicitly stated principle that all decisions are to be made based on what is best for the ship. This concept was not easy for many people to understand early on. At the current stage of design, however, this message has generally been thoroughly internalized. Many examples can be found of people from one functional group volunteering to make sacrifices in favor of another functional group. Needless to say, this principle has required a high level of communication between the different groups. The Space Kings have provided the primary focus and encouragement for those communications.

TEAM APPROACHES

Outside of the day-to-day design activities, NNS has two programs that have contributed to the cross-organizational communications: Task Teams and Quality Improvement Teams (QIT). QIT is a formal structure and a seven-step process for identifying problems, especially company-wide problems. All managers have gone through QIT training. Task Teams are informal, short duration teams of design and manufacturing staff pulled from various groups. They are constituted to solve a specific, identified problem, and dissolved when that problem has been solved. Because they are focused in purpose and cross-organizational in nature, they have contributed greatly toward opening lines of communication across functions. About four or five of these teams form in a year.

There is also an active suggestion program called "Opportunities for Improvement." People who come up with successful ideas are publicly recognized. About 30 percent of the staff have participated, and a substantial portion have made multiple contributions. Other activities encourage information exchange and communication, such as newsletters and training sessions.

In general, NNS is attempting to move from an autocratic, military model organization to one which is much more team based and empowered. The team efforts described above are seen as only a first step in a much larger and longer-range process. All staff have received quality training, and a substantial portion have had training in communication issues. Perhaps the best example of empowerment is the Space Kings. They have been provided with authority and power in the form of control over the information (the VIVID model) and knowledge in the form of access to the entire VIVID database. They are thus empowered in a way that is totally new to NNS.

People

When the Seawolf team was initiated in the early 1980s, supervisory and management staff were pulled from other parts of the company. The majority of the design staff was new hires, many of them young people recently out of college. While this may seem risky, it worked because the whole submarine was to be newly designed. It also allowed better use of the CADAM and VIVID software systems, because the incoming design staff was not yet set in its ways. The new approach was not a threatening change for them. Some of the supervisors and higher level managers were not so comfortable with the new approach, but with top management support, they have generally adapted quite well.

Design and Manufacturing Process

In general, the overall design process for the ship's components and systems has not changed a great deal from previous ships. CAD has made some changes in day to day design work, but not a great deal. What has significantly changed is that, with the comprehensive solid model, manufacturing people have gotten involved at an earlier stage. That is, the systems have supported a substantially greater level of concurrency in the design of the ship and the design of its manufacturing processes.

Each functional design group has its own specific design process, but overall coordination comes through the VIVID system and a single master schedule. The functional design groups develop their own schedules based on the master schedule. The master schedule includes such activities as design reviews by the Navy and when specific information will be needed for the particular sectional construction drawing. Much of the design work is scheduled based on the overall construction plan. That is, if the ship is assembled stern first, then the aft sections must be designed first. In a sense, much of the detailed design is "just in time" for manufacturing.

The VIVID system allows real planning of the ship construction and assembly processes in a way never before possible. Because of the ability to visualize the results, check for interferences, and quickly change the model, the ship can be "assembled" in the computer first. Before the VIVID electronic mockup, this was done by actually building a full scale wood mockup of the ship. The visualization capabilities of VIVID are further supported by the accuracy of the model. Rather than specify many elements (piping, supporting brackets, etc.) within a couple of inches and require cutting and bending to fit at assembly time, parts can be accurately dimensioned to provide proper fits and locations. This has eliminated the need for modification during construction.

Furthermore, the VIVID system has allowed the generation of a series of detailed assembly instruction drawings. With them, putting the ship together is much less likely to run into major errors and delays. Costly mistakes such as installing components in the wrong order (thereby making some of the installations impossible) are dramatically reduced. Servicing will also be improved, as access requirements are built into the positioning of components as well. As a result of these changes, the Seawolf is going together far more easily and efficiently than any previous ship of its complexity. According to EB, it is going together more easily than one would expect for the fifth or sixth ship in a series.

Lessons Learned

- Putting early effort into component design helps reduce effort later. This is based on making the best use of the VIVID system to put the components together. The sooner that can be done, the better.

- It is necessary to keep design work advancing roughly equally in all functional areas. If one group gets significantly behind the others, it holds up the whole design process because the comprehensive model cannot be built. Three levels of detail were identified for this purpose: general arrangement, first level of detail, second level of detail.
- Having a single comprehensive model is an extremely powerful tool. That tool and the management support for its use have been key to the success of the program.

Points of Pride

- NNS is very proud of their work on the Seawolf program. They believe it may well be the finest submarine ever built. They expect future submarines to be simpler due to the changes in world politics.
- The coordinated, common design model is seen as a first for such a complicated product.
- The Seawolf design team has become a team of people that support each other, with a common goal focusing on what is best for the ship.
- At least a quarter of the staff and all the management have had training in quality improvement. Managers have been trained to listen to their people.

The Texas Instruments GEN-X Development Project

In 1992 the Defense Systems and Electronics Group (DSEG) of Texas Instruments won the Malcolm Baldridge Award. The Baldridge committee, in their feedback report on the design and introduction of quality products and services, found no areas for improvement and noted the following strengths.

- DSEG has developed and deployed a strongly integrated product development process that is built on organizing and using concurrent engineering teams that include both customers and suppliers.
- All product designs are validated through formal testing, multiple design reviews, and audits that have direct customer involvement.

In February 1994 we visited DSEG to learn about their approach to concurrent engineering. As it turned out, the program that had most successfully used concurrent engineering teams was the development of the GEN-X decoy. This program was considered exemplary within DSEG, and in 1992 and 1993 it won the Presidents Gold Award, the highest honor within DSEG. We also learned about the Integrated Product Development (IPD) process, which was named best in industry by the Navy's 1991 Best Manufacturing Practices survey. IPD was just being launched in actual projects at the time of our visit. We report here on the GEN-X project and provide a brief overview of the IPD process.

GEN-X is a generic expendable decoy developed by the Naval Air Systems Command to provide endgame protection for tactical aircraft. It is designed to protect against the latest surface-to-air and air-to-air radar guided systems. These small devices are dropped from planes and deflected by the air stream; as they drop to the ground, they attract the radar systems of enemy missiles, thus leading the enemy missiles away from their targets. In September 1987 DSEG was contracted to design and test a prototype decoy. While they developed a successful prototype, it was too heavy and too expensive for the Navy. At the price quoted, there were other less expensive alternatives for accomplishing the same thing. So TI bid on and won a second contract to develop and build a new version which was far less expensive than the original. TI's proposal promised the achievement of very aggressive targets for price and delivery time. For example, they agreed to reduce the cost by a factor of 4. Amazingly, they met these aggressive targets! The GEN-X group accomplished these seemingly unreachable goals through a team approach and concurrent engineering. For DSEG, this was a pioneering effort in achieving such demanding goals and using a team approach to this degree.

The GEN-X Project TEAM

GEN-X consists of a small metal casing with fins which houses micro-integrated circuits and a battery. GEN-X has a full-fledged jammer inside. It has a unique battery, small enough to fit inside, which provides three separate voltages and up to 6 amps of current. Missiles with radio frequency seekers are fooled into thinking GEN-X is the aircraft and follow the decoy. While the basic concepts behind GEN-X were not new, the main task was to design it so it was small, light, and could be made at much lower cost than the original prototype on which it was based. To accomplish this, the GEN-X team, with the help of outside suppliers, had to develop all new components — nothing was used off the shelf for GEN-X.

The project to redesign GEN-X so it was smaller and cheaper officially began in May 1992. By that time, there had already been about nine months of work in anticipation of getting the contract. For this first nine months of anticipatory work, a cross-functional team was set up but it was mainly consultative to the project leader who used a traditional top-down project management approach. Because of the aggressive targets, the project manager realized they needed to do something different to have any hope of succeeding. He began to look around at successful cases of product development and they all seemed to have in common the use of cross-functional teams. Thus he decided to use an empowered team-based organization after hearing a presentation by a consultant. The team was formed and trained in May 1992 when the project was officially launched.

DSEG has historically had strong functional departments. Project managers were used to pull together teams, but the project manager had no formal supervisory responsibility for the project staff; so, by and large, different functional groups went back to their corners to work on their piece of the system. The project manager primarily acted as a coordinator and scheduler. Thus, when GEN-X began, it was clear to the project manager that he could not form a dedicated, collocated cross-functional team. Team members would be part-time, not full-time on his program.

On the other hand, this project was seen as strategically important to DSEG, and the project manager had a high level of top management support. So he was able to pull strings and get a substantial portion of the time of a core group of people located in the same building. In terms of the coordination methods described in this book, this approach was closest to a lightweight project management structure, though with strong executive sponsorship from top management. The program manager did not supervise team members outside the project and had only limited influence over their performance evaluation. But the project manager had top management support and a budget against which team members could charge their time. As we also saw in the Whirlpool case, team dynamics worked to develop a highly committed group of people who were highly dedicated to the project even without formal external rewards. Team members worked overtime and weekends, often despite lack of encouragement from their functional bosses.

The first two months of the project felt to the group like an aimless search for direction. No one in the group, including the project manager, had experience with a team approach, though there was clearly a sense of enthusiasm and a belief that they were doing something new and innovative and important. The team spent 12 hours/day, 6 days/week planning and figuring out where they were going, including developing metrics and a project plan.

Another important task in the first two months was to figure out how the team should be organized. There were too many people involved to act as one big group — about 110 people total. Thus, they decided that the total program would be divided into three smaller teams corresponding to the components of the device — the module team (four microwave modules), the antenna team, and an airframe group responsible for overall systems integration (including the battery, wings, flexible circuits, interconnections to the electronics, scheduling, testing, delivery). Each of the three smaller teams had a program manager who reported to the general program manager for the contract. The antenna and airframe teams followed similar directions. Each started with a single team and, over time, broke into subteams. The module team also tried subteams but had problems coordinating them, and eventually created a larger team for coordination.

As the program progressed, it was a continual struggle to integrate the activities of the teams. One of the problems created by this division into smaller teams was communication across the teams. At first a great deal of time was spent in large group meetings. But as production launch approached, it became increasingly difficult to find time for face-to-face meetings. Particularly in this later stage, members found e-mail to be the most useful and efficient communication medium.

Any effective concurrent engineering program must find ways to bring the voice of the customer into the process. This program had an exceptionally high level of customer involvement compared to conventional practice at DSEG, and in the defense contracting industry generally. Every major defense program has a representative of DPRO — defense procurement representative organization. In the past, this representative played a very adversarial role. DPRO was viewed much as taxpayers view an IRS audit. You avoid the IRS except when you are called on. You only show the IRS auditor what he/she asks for, and even then you reveal only cleaned up documents that portray your case in a favorable light. In the defense contracting business, DPRO generally did not audit a program until late in the program after all the important design work had been done, and then the DPRO's goal was to expose problems. Engineers prepared for these audits by trying to clean up documents and hide any problems.

The GEN-X project team decided to use DPRO in a more proactive way. In fact, they made the DPRO representative a kind of ex-officio team member. The DPRO representative was used to represent the voice of the customer, and sat in on early meetings almost every week. At first, team members were very nervous about discussing problems and cost issues. As time progressed, the DPRO representative became one of their biggest supporters. In retrospect, the program manager wishes they had used a formal method like Quality Function Deployment to build the voice of the customer into the process. However, using the DPRO representative in a collaborative, proactive way clearly went a long way toward accomplishing what QFD is intended to accomplish more formally — customer wants and needs were reflected in each stage of the design process.

Design for Manufacturability for GEN-X

The functionality of GEN-X had already been proven with the first prototype. It worked fine. It was simply too expensive. Thus, the principal goal for the revised design was low cost. The team realized that the key to accomplishing this was to carefully think through manufacturability issues in the product development phase.

Most of DSEG's products are low volume. By DSEG standards, GEN-X is high volume production at 25 units/day. So DSEG did not have a tradition of worrying a lot about manufacturability. Once again, the GEN-X group needed to be innovative and develop their own methods as they went along.

The first step was to include manufacturing engineers and hourly workers on the team, which was a break with tradition for the DSEG group. But the GEN-X team was not satisfied with simply having manufacturing make suggestions in team meetings. They wanted a systematic approach to concurrent design of the product and manufacturing process. At that time, process reengineering was becoming a hot industry buzzword. So about nine months after the May launch one manufacturing engineer was charged with business process reengineering (BPR). There was a group at the division level working on a generic approach to process reengineering, but the GEN-X program could not wait. So they developed their own business process reengineering at the project level for GEN-X.

The manufacturing engineer first formed a cross-functional team of 17 people (e.g., industrial engineering, information systems, 3 assemblers, etc.) and read available literature on how to do BPR. They developed a very detailed description of the manufacturing process in the product design stage. In addition, the team developed a design-to-cost model since the project was so driven by the 4× cost reduction goal. This model, set up in a simple spreadsheet program, estimated the costs of various steps in the manufacturing process. In this way, alternative product designs could be evaluated for their cost implications. To build the factory throughput costs into the model, a factory simulation package was used which estimates defects as part of the simulation output.

In the process of developing the model, the team had to think through many design alternatives and assign manufacturing costs to them. This process crystallized, early on, many different design considerations that traditionally would not be thought about until much later in the design process. This early planning also led to many improvements in the manufacturing process that may never have been considered in the heat of production launch.

An example of a DFM innovation that came out of this process was the reduction of soldered joints from 49 to 6. Originally product engineers had designed the device with the 49 soldered joints. They apparently did not realize that soldering is a prime source of defects and did not understand the costs involved. But the cost equation clearly showed substantial savings by reducing the number of soldering joints.

Other innovations came simply from having production workers as part of the team. An example is the design of a clamp to hold the internal modules together. In the prototype stage, an assembler tried to assemble a unit and discovered that with the existing design, the modules would move around during assembly causing damage. This led to a redesign of the clamping mechanism. Traditionally prototyping has happened in the engineering lab, and engineers may not have understood the implications of their design for the assemblers.

There were also manufacturing process innovations during the design phase before any equipment was ordered. An example was the design of the vibration testing fixtures. Originally a mechanical engineer designed a single fixture so all 25 units could be tested in one shot. This assumed they would build all 25 units and hold them in work-in-process inventory and they would then be tested. This would lead to a great deal of work-in-process inventory which is undesirable from the perspective of process throughput. The BPR team came up with smaller fixtures that could separately test four groups of four units. This would give them the flexibility of running lots of 4, 8, 12, or 16 depending on demand, and help in moving toward a pull system. In addition, the new fixtures were lighter and more ergonomically sound.

Early involvement of suppliers was also important to the success of GEN-X. This was particularly true for the battery supplier as the battery was a unique design. They made the engineer from this supplier part of the GEN-X team and he participated in the design from the beginning of the program. The lightweight battery turned out to be of great help in reducing the overall weight of the device.

CAD Systems at DSEG

The GEN-X group had access to excellent CAD systems for the design. DSEG has a design center that provides support for design automation. The main tool used is ProEngineer™ which allows for integration of drafting, design, analysis, and stereo lithography from the same database. They can also send databases directly to the customer or translate them using IGES. DSEG has had a lot of experience with IGES and is generally able to make it work using appropriate flavoring algorithms.

The purpose of the design center is to act as an interface between functional groups (e.g., electrical, manufacturing, suppliers, customers) to make sure data flows correctly between places and to support use of design automation tools. Suppliers do not have ProEngineer, but since they are generally only concerned about exchanging the outer shape with TI, IGES often is fine. Otherwise DSEG may take their blueprint and enter it into ProEngineer.

Results of the GEN-X Project

At the time of our visit, GEN-X was not yet in full production. But some interim results were available, comparing GEN-X to the DSEG average.

- Development cycle time was reduced by 30 percent. The total design was done in 2.5 years, and 90 percent of the design was done in the first 2 of the 2.5 years.
- Cycle time was reduced by a factor of 5.
- Cost was reduced by a factor of 4.
- The number of design changes was reduced by 30 percent.
- The design was innovative. In this short time, over 90 percent of the design was changed to bring down the cost. And no off-the-shelf parts were used.
- Part count was reduced in half for the airframe.
- The prototypes and early trial devices produced were all delivered to the customer on-time.
- The original goal was to get to 6 sigma quality. Based on a simulation, at the time of our interview they forecasted the manufacturing process would achieve 4.7 sigma, and continuous improvement efforts were still underway.
- Many of the teaming concepts developed through GEN-X are incorporated into the IPD process.

The next challenge for GEN-X at the time of our visit was to successfully move to production. Since the development phase was almost complete, there were no funds to keep the teams together. As team members went back to their functional organizations, they were left with the question of how to support assembly. As one team member put it: "How can we get a string on an engineer to pull him/her back from a new project to work on assembly problems?"

Spreading Concurrent Engineering Through the IPD Process

In parallel with the GEN-X project, DSEG has developed a structured development process for concurrent engineering called Integrated Product Development (IPD). The approach was developed in-house by TI, and in fact is now being marketed as a product to other companies and government laboratories. IPD was recognized in 1991 by the Navy as best in industry. As of October 1992, it was mandated that all new design projects at DSEG must use IPD. Even ongoing projects must go through the relevant remaining phases of IPD; though at the time of our visit, it was just being rolled out to actual projects.

There are thick training manuals and many short courses offered by TI on IPD. Ultimately every engineer in DSEG must go through a

3-hour overview. There are also 2-day tailoring courses to teach teams how to tailor the process to their specific project. As of February 1994, 2500 people had attended a 3-hour overview in 135 offerings, 760 had attended a 2-day tailoring course in 45 offerings, and 450 had attended a concurrent engineering workshop in 25 offerings. An IPD Process organization of full-time staff has been assigned to conduct the training, provide support to teams, and sell their services to other companies.

IPD is a structured development process with five high-level phases, three milestones, and eight program review points (reviews are as mandated by DoD). An overview of IPD is shown in Figure 1. The process is intended to be tailored for each project and to meet customer needs. Each of the five phases is broken down into traditional flowcharts, much as is described in COSAT in the information/workflow model. In all there are four levels of flowcharts (top level, intermediate level, detailed level, and task descriptors).

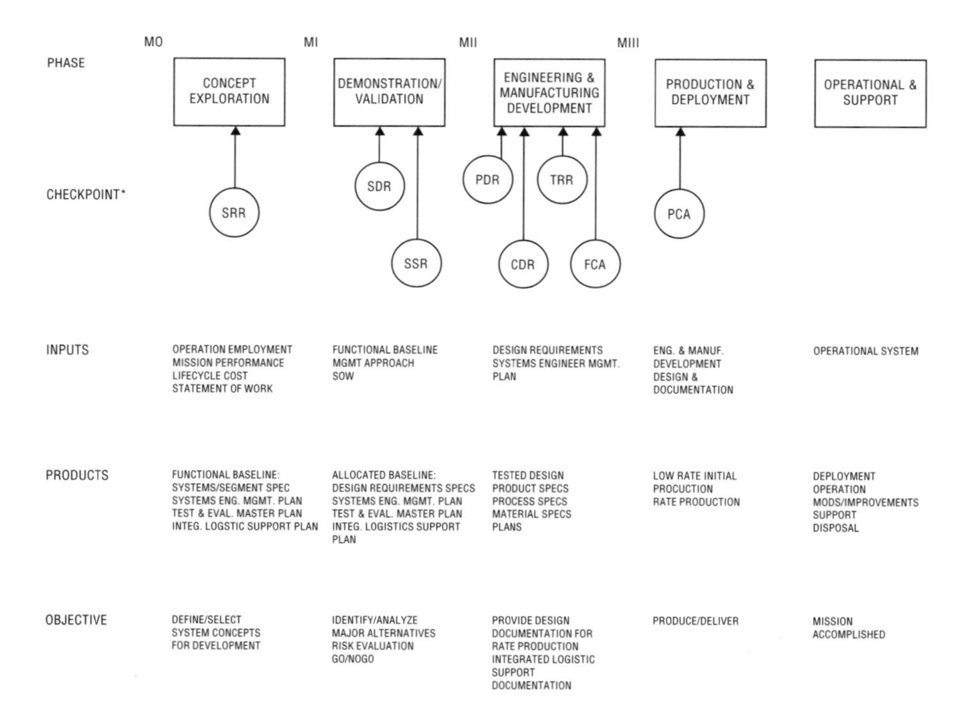

Figure 1 • Overview of IPD Process

IPD includes approaches to teaming and a recommended organizational structure for the IPD process. For example, an IPD champion should be identified who is a team member but also acts as a guardian of the process. This person is a member of the IPD Process organization whose role is to provide deployment leadership for specific programs in the adoption of the IPD process. This person becomes the adopting program's single point of focus for all aspects of IPD Process training, tailoring, implementation, and tracking. There is no cost to the program for this person's support. Presumably this is a short-term role as programs are just starting out using the IPD Process.

In addition to the structure given to the process over time, IPD places a great emphasis on standardized procedures and design guides for many aspects of design. Figure 2 provides an overview of the large number of standardized procedures and guides at DSEG.

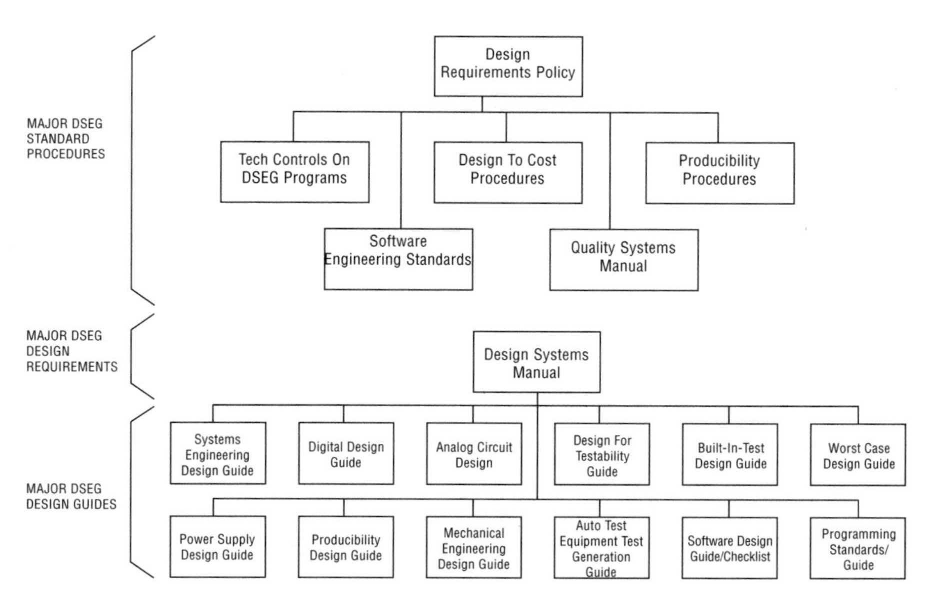

Figure 2 • Standard Procedures in IPD

IPD is a more front-end-loaded process than has been true in the past. This is expected to lead to a higher bid price for the design phase but a lower cost for the manufacture of the product. In defense contracting, this is significant as there are generally separate contracts for the design and manufacturing phases. To make this work, the customer must change their funding profile and be willing to give more money up front. The government has in principle been very supportive of this approach.

IPD was developed after GEN-X began so it was not used for this project. Training started in October 1992. But GEN-X had a significant influence on the development of IPD and is used as a case example in training. In fact, one of the quality managers responsible for developing and disseminating IPD was a major player in the GEN-X project.

One of the challenges of GEN-X was getting the time of team members since most members had primary homes in functional groups (e.g., process engineering, mechanical engineering, quality engineering, electrical engineering). Most team members were only on loan from the functional groups, and often functional group goals clashed with GEN-X goals. Performance appraisals are done completely by the functional group. In fact, there were no external rewards for participating on the GEN-X team even though it made great demands on time during and after regular hours and was considered a great success by the company.

Another issue in treating GEN-X as a pilot was the feeling by team members that they were outsiders back in their functional organizations. They described GEN-X as a bubble in a big hierarchical organization. They said they were often referred to as outsiders or rebels. One team member described the frustrations she experienced in moving from GEN-X team meetings, where she felt like part of the team and could comfortably express sometimes strong opinions to functional group meetings, where she had to act very formal and repress many of her ideas.

The GEN-X team members felt that the functional organization would have to weaken to effectively implement IPD. They said some functional managers are uncomfortable with teaming because it threatens their control. In fact, many of the tasks of the functional groups are created by government regulations and do not really add value to the product. As DSEG makes the shift toward operating more like a commercial operation, the more bureaucratic tasks of the functional groups may not be needed. There were a number of important developmental questions being asked about the future. How can you take people in functional organizations with outmoded skills and move them into team-based organizations? How do you flexibly assign people to teams and then reassign them when they are not needed by those teams? How do you make program managers aware of who is available and what their skills are? How do you maintain functional expertise while assigning individuals to cross-functional project groups? Clearly, moving from a pilot to changing an entire organization is a complex process. But DSEG has taken some critical steps on the journey.

SUMMARY

GEN-X became an important benchmark for DSEG. They started with aggressive goals that no one thought were attainable. A team was assembled and the level of teaming surpassed anything done before in product development for this company. Team members worked long hours, worked under great stress, and dedicated themselves to the team above and beyond their normal work hours. The teaming process and challenging goals combined to motivate the team members despite the fact that their formal performance review and external rewards were controlled by their home functional organization. The results were achievement of great performance goals. These are goals that surpassed minimum requirements and characterized a truly great performance. The concept of great performance goals is also being applied to other projects.

DSEG is trying to replicate the commitment and teamwork of this group throughout the company. They have developed a world class formal process for concurrent engineering in IPD, have top management commitment, have mandated the process, and invested in training for all personnel. The challenge is to move a functional bureaucracy to a flexible, networked, team-based organization.

Toyota's Product Development Process

Durward Sobek, University of Michigan

As arguably the best automaker in the world, Toyota is certainly the most benchmarked. The Toyota Production System has received the most attention, but Toyota also has a very efficient and effective development process. It is difficult to compare lead times, but by most accounts Toyota is the fastest company at designing vehicles using the fewest number of engineering hours. Toyota vehicles top most lists on quality. How does Toyota accomplish this? Do they use all of the modern concurrent engineering tools we hear about, such as dedicated collocated teams, QFD, Taguchi experiments, and sophisticated engineering analysis? Surprisingly they violate many contemporary precepts about best practice in design.

1. They use a matrix organization with very strong functional groups, not platform teams like Chrysler.
2. Individual engineers do not reside on teams, but are primarily assigned to functional groups reporting to a functional manager and are loaned out to development teams run by chief engineers.
3. Design and manufacturing engineers are not collocated but each belong to separate divisions reporting up to separate vice presidents.
4. They rarely use QFD, Taguchi experiments, or very fancy engineering analysis. The main voice of the customer is the chief engineer, and many of the decisions he makes are intuitive, based on prototype tests.
5. Written communication is the preferred first line of attack, followed by meetings when written communication proves inadequate.
6. Toyota does not use a uniform and highly structured development process, but prefers a simple timeline with strict deadlines tailored to each project by the chief engineer.

7. Toyota explores many alternatives in ways that others may regard as wasteful (e.g., a large number of different prototypes), but manages to have a fast and very efficient overall development process.

Since Toyota is so widely benchmarked, this case goes into some detail about their organization and development process. We particularly emphasize the effective use of a matrix structure, the often misunderstood role of the chief engineer, and examples of Toyota's tendency to consider *sets* of alternative designs throughout the process.

Organization

The Toyota Technical Center, most of which is located at Toyota's headquarters in Toyota City, Japan, has four Vehicle Development Centers (VDC):

VDC 1 — rear wheel drive passenger cars
VDC 2 — front wheel drive passenger cars
VDC 3 — commercial and recreational vehicles
VDC 4 — components and systems development.

VDC 4 is primarily concerned with the advanced development of body systems, chassis, engines, transmissions, electrical systems, and materials. VDC 4 also does some vehicle development activity, particularly in engines, transmissions, and electrical systems.

Support activities comprise the rest of the technical center (i.e., a "support center"), including human resources, prototyping, proving grounds, advanced styling studios, basic research, etc. Essential functions such as manufacturing (or production) engineering, purchasing and marketing/sales are physically and organizationally separate from the R & D Group. These "support centers" are a common pool of support for the four development centers.

	Styling	Body Engineering	Chassis Engineering	Powertrain Engineering	Vehicle Evaluation	Planning
Camry CE						
Celica CE						
Corolla CE						
etc.						

Figure 1 • Toyota's Vehicle Development Center Organization.

Each vehicle development center is a matrix organization (Figure 1). A chief engineer leads a vehicle development project, coordinating and integrating the work of functional divisions inside each center. The functional divisions, which do the vast majority of the design and engineering work, include styling, body engineering, chassis engineering, powertrain engineering (actually concerned more with engines than with transmissions), and test. Each center also has a planning division that does long-range strategic planning and facilitates coordination among the different vehicle projects. Center 4 and other research groups transfer advanced technology to the vehicle development centers for application to particular vehicle programs.

Chief Engineer System

The heart of Toyota's new product development process is the chief engineer (CE).[1] Clark and Fujimoto (1991) refer to the CE as a heavyweight product manager, but we prefer Toyota's term because at Toyota the CE, more than being a project manager, is also the system designer. Chief engineer at Toyota also differs from the conventional U.S. definition of the term.

Clark and Fujimoto claim that outstanding heavyweight product managers (PM) combine the roles of internal integrator and concept champion into one, thereby simultaneously achieving cross-functional integration and external integration by infusing customer expectations into design details. To do this, heavyweight PMs have the following characteristics.

- Wide-ranging coordination responsibilities across functions.
- Coordination responsibility of the project from concept to market.
- Responsibility for concept creation and concept championing.
- Responsibility for specification, cost target, layout, and major component choices.
- Responsibility for making sure that technical details properly reflect concept.
- Frequent and direct communication with designers and engineers.
- Direct contact with customers.
- Multidisciplined abilities (including an ability to speak/understand jargons of different disciplines) in order to communicate effectively with various functional groups.
- Managing conflict, and initiating conflict when necessary to maintain concept integrity.

[1] Note that in this case study only, the acronym CE refers to the Chief Engineer rather than to the concept of concurrent engineering.

- Market imagination and ability to anticipate customer expectations.
- Circulating among the people working on his project and not doing a lot of paperwork and formal meetings ("management by walking around").
- Engineers by training; possess broad, if not deep, knowledge of vehicle engineering and process engineering.

Toyota's CEs embody all of these characteristics, but are different from Clark and Fujimoto's characterization in an important, though perhaps subtle, way. The CE is not only responsible for concept creation, he personally writes the vehicle concept and presents it to upper management for approval. He also designs the system, deciding the major dimensions of the vehicle which determine its basic character, which components to use, and how they will be arranged. More than just having project responsibility from start to finish, Toyota's CE creates the project plan and decides all major events and their timing. In other words, he is not just responsible for seeing that these things get done and done well (i.e., a project manager), he is responsible for *doing* them (i.e., a chief engineer).

In addition, while Toyota's CEs have direct contact with customers, the implications are more pervasive than just getting the voice of the customer into the vehicle concept. The CE represents the customer in the vehicle program, particularly for product engineering. Engineers typically do not have direct contact with either customers or marketing (although they do get warranty data and do study the competition extensively). The CE and styling meet with marketing and sales quite frequently to discuss marketing data, market trends, and customer preferences and expectations. So the CE views the market through marketing's lens and through styling's lens in addition to direct observation.

In many respects, the CE has a remarkable degree of latitude and responsibility for vehicle design, but in two important respects he is quite constrained. Toyota has an incremental development strategy. Most vehicles are engineered based on an existing platform, including a given powertrain. While the chief engineer can specify some modifications to an existing powertrain, such as multi-valves for a high end performance version of the vehicle, mainly the powertrain is predetermined. Thus, the chief engineer is focused largely on interior and exterior styling, engineering, and the integration of components in the vehicle. More recently, when Toyota organized into the current system of vehicle centers, each headed by a high-level manager, some additional constraints were placed on the chief engineers. There is a major focus on parts commonality across vehicles, and chief engineers must attempt to use common components wherever possible. Despite this, the systems integration responsibilities of the chief engineer seem almost overwhelming given the large number of parts in an automobile.

The CE is an engineer by trade, and at Toyota he will have spent at least 10 years as a working level engineer without supervision responsibilities. Then he typically starts to receive broader exposure, working as a manager or assistant manager,[2] and eventually as an assistant CE. Clark and Fujimoto seem to imply that deep knowledge is great if you can get it, but broad knowledge is key; whereas at Toyota, the CE first attains deep knowledge in one area, as a prerequisite to being able to understand the broader issues of vehicle design.

The CE does not do his job alone, however. A dedicated staff of 5 to 15 "assistant CEs" help him perform many of his duties. They are all engineers — no marketing or finance or purchasing representatives. Each staff member typically has responsibility for overseeing the development of a major subsystem like body exterior, along with some vehicle-wide responsibility like tracking vehicle weight or overseeing the prototype build process. Their main job is system integration, and having two sets of responsibilities helps foster integration.

The CE solicits information, advice, and help from functional groups. For example, in deciding the major dimensions of the vehicle, the CE typically consults styling on how different dimensions affect the look of the vehicle or the "feel" of the interior. He also talks to many groups concerning areas of improvement over the current vehicle, such as warranty and manufacturability issues, and about new technology that might be appropriate for his vehicle. This information gathering is not only important for informing intelligent decisions, the process of communicating with so many people also helps build consensus for his concept — an element often cited as germane to the success of Japanese business practices.

Process Description

Toyota has a standard development process in that every program has the same milestones, or major events, in roughly the same order. However, since each CE decides the plan for his vehicle, he can customize the plan for the particular needs of his program. Thus every program is different. For example, most programs have two vehicle prototype stages, but some have more and, unbelievably, some have

[2] At Toyota, typical of Japanese companies, engineers are promoted to assistant manager (kakaricho) after about 10 years with the company. Assistant managers typically supervise about 5 engineers. After about 5 or so years as assistant manager, the engineer becomes manager (kacho) who supervises about 5 assistant managers (or 25–30 people total).

fewer. Also, the time to go from styling approval to start of production varies from program to program.

The process description which follows first lays out Toyota's overall development process in terms of defining milestones. Then the process is broken into five phases which are discussed in turn. The phases are not Toyota's, but rather our convention. The transitions between phases are rather fuzzy with considerable overlap, but they provide a logical breakdown in which to discuss the simultaneous activities of the various functions. The section concludes with a discussion of how suppliers are involved in the development process.

Figure 2 provides a graphical display of the vehicle development milestones. Toyota's process begins with the chief engineer (CE) *concept*, which espouses the CE's vision for the vehicle. The concept contains keywords or phrases that capture the character and image of the vehicle, information about the target market and customer, basic vehicle dimensions (such as width of the car and wheelbase), engine and transmission combinations, and general targets such as "costs less." The CE presents his concept to upper management for informal approval (official program approval comes later). There may also be an earlier, preliminary presentation to upper management, before some of the details have been worked out.

Following the CE concept is *styling proposal*. Styling makes a number of 1:5 scale clay models of different styling designs. They are shown to a review board consisting of representatives from upper management, the Technical Center, marketing/sales, and manufacturing. Two or three designs are chosen for full-scale clay models.

As the full-scale clay models are being developed, the CE submits a development proposal for official program approval from upper management. In addition to the updated CE concept, it includes what the model variations will be, which engines and transmissions will be offered for each model variation, numbers for cost, weight, and other targets, where the vehicle will be produced, volume forecasts, and start of production and program milestone dates. Upon receiving approval, the CE issues an official *development order*. Shortly after the development approval is the program's cost planning authorization, or upper management approval of the program budget. Technically this is a separate event, but it can be thought of as part of the development approval process.

Approximately concurrent with the development order and styling proposal (sometimes earlier), most Toyota programs have an advanced prototype, called "pre-prototype." This prototype build generally uses production vehicles modified to reflect major new changes in the CE concept.

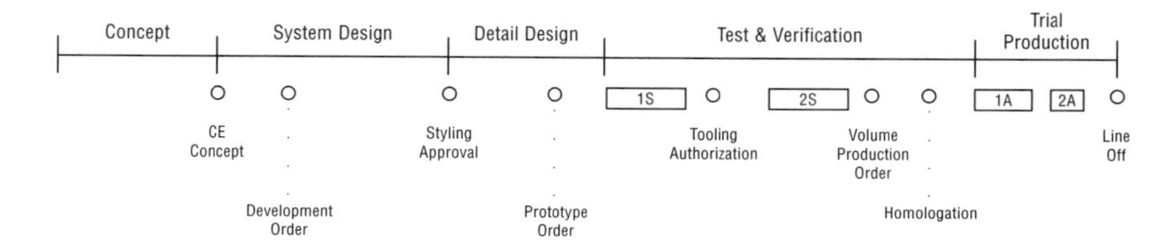

Figure 2 • Toyota's Overall Process

Next, full-scale clay models culminate in one of the biggest events of the program — *styling approval.* At styling approval, the CE, product engineering, manufacturing engineering, marketing, and upper management agree on the final style. The skin is "frozen," which means the styling design cannot change significantly from this point forward without another formal styling approval. Traditionally this event kicks off the detailed engineering work. A few months after styling approval, the CE issues a detailed specification of all the options and variations that will be offered on the vehicle.

Next come two *prototype* build and test phases. Tooling authorization occurs after the first prototype, which officially releases funds for tooling development. Volume production order occurs after the second prototype, officially notifying the company (plants, etc.) that the vehicle is ready for volume production.

Several months elapse from the end of the second vehicle prototype phase during which the vehicle undergoes homologation (certification testing for government regulations), which may require a third prototype build, and manufacturing engineering certifies the tools and dies. Then the program enters two phases of *trial production* in the plant, followed by start of production, or "*Line-Off*" at Toyota.

To facilitate the remaining detailed discussion, Toyota's process has been divided into five broad phases. The *concept* phase starts from the end of the previous program and ends with the CE concept. The *system design* phase covers the period from concept to styling approval. From styling approval through the making of the first vehicle prototype is *detail design*. *Test and verification* carries us up to the start of *trial production*. Figure 3 summarizes how the phases map to Toyota's milestones. The following is a discussion of each phase.

Concept Phase

The culmination of the concept phase is the CE concept. The CE drives the concept but seeks input from many sources. One critical area is market and customer information. The CE visits dealers and talks to

Phase	Toyota's Milestones	
	Start	**End**
1. Concept	end of prior program	CE concept
2. System Design	CE concept	Styling approval
3. Detail Design	Styling approval	1st prototype
4. Test & Verification	1st prototype	1st production trial
5. Trial Production	1st production trial	Line-off

Figure 3 • Development Phases

customers and users about the current product in the different markets, to see customer reactions first-hand. The CE also works closely with styling when formulating his concept. In some programs, styling may actually lead the concept creation process. The RAV4 (Toyota's recently introduced small sport-utility vehicle), for example, was initially a concept vehicle created in an advanced design studio. However, for most projects, the CE takes his initial ideas to the styling group, and together they discuss many proposals. Besides styling and marketing issues, the CE talks to different engineering and advanced development groups about technology. He's primarily concerned about three things: what components and subsystems are best suited for his concept, ideas for improving the current design (especially regarding warranty and plant issues), and new technology that will make his vehicle better. For example, the CE may give chassis engineering his ideas or vision for the vehicle and perhaps some rough dimensions. Chassis then consults with the test group to find out about design weaknesses, with marketing to learn about field issues, and with manufacturing to find out about plant issues. Chassis also consults with advanced suspensions R&D group, and with brake suppliers about new technologies. Chassis formulates several proposals about the types of chassis systems that might be appropriate for the CE's vehicle. They then discuss the proposals with the CE to help him develop the suspension section of the CE concept.

The resulting concept includes descriptions of the target market and customer, key words and phrases that describe the vehicle's image, basic dimensions, basic power train technology, and general targets. The CE presents his concept to all the functional groups for their review. Upon receiving informal acceptance from these groups, he discuses his concept with upper management to get its informal stamp of approval.

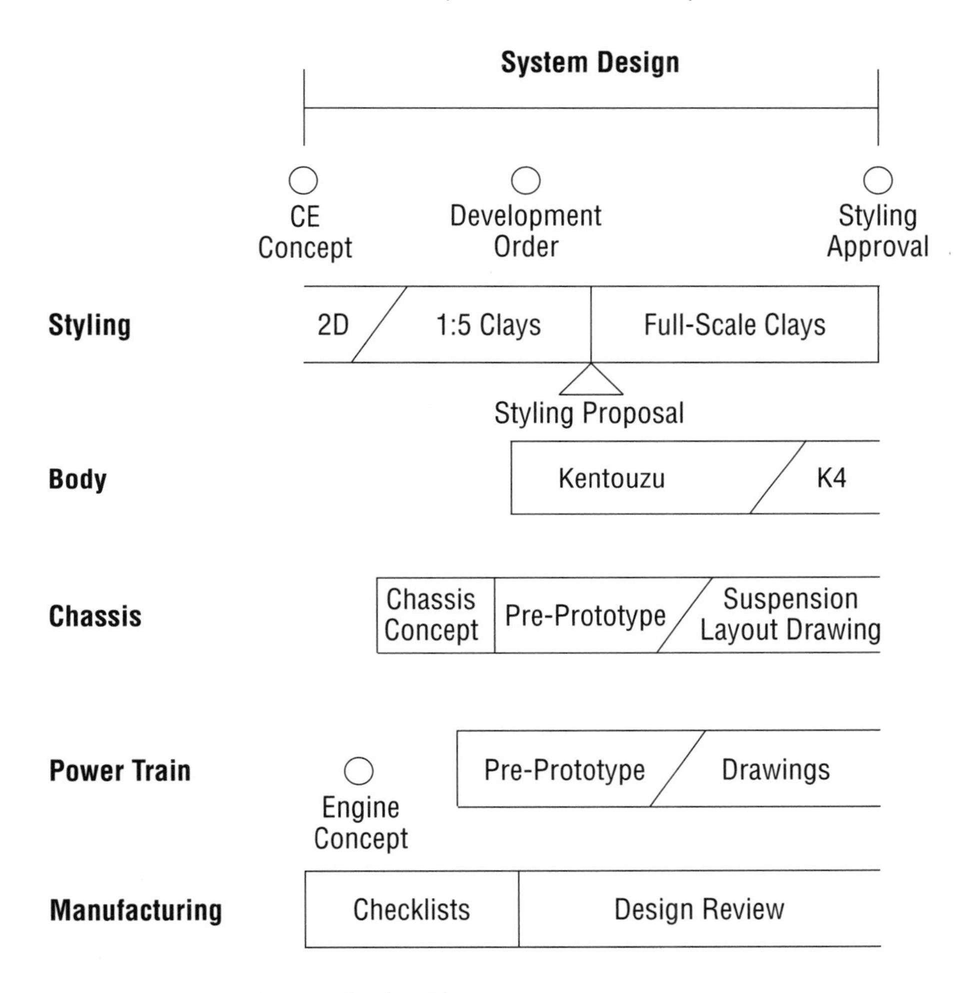

Figure 4 • Toyota's System Design Phase

SYSTEM DESIGN PHASE

From the CE concept, styling begins intensive design work on the vehicle (see Figure 4). The styling team continues to develop additional ideas in two dimensions (sketches, renderings, tape drawings) and increasingly in 3-D computer models. From the multitude of ideas, the styling team chooses their best ideas, typically numbering 5–10 (in extreme cases, up to 20), and shows them to the CE and marketing for discussion and feedback. One or more of Toyota's advanced design studios usually submits a design(s) in the idea competition at this stage. The ideas are then transformed into 1:5 scale 3-D clay models, all of which conform to the same basic dimensions outlined in the CE concept, for further evaluation. Many groups, including product engineering and manufacturing engineering, review the clays which may result in modifications.

At some point, the 1:5 clays are submitted in a formal styling proposal to a review committee consisting of representatives from product engineering, manufacturing engineering, marketing, and upper management. In some cases, the styling proposal may be made with renderings and drawings instead of clay models. The purpose of the proposal is to get feedback and buy-in from the corporation and to narrow the field to 2–3 ideas. The decision is made jointly by the CE and the review committee. If an advanced styling studio's idea is chosen, the design development responsibility now shifts to the design division within the VDC, usually with one of the lead stylists reassigned to the VDC design division.

From styling proposal, the selected 1:5 clays are scanned to create a mathematical surface.[3] Stylists can adjust the math data as needed, then use it to drive a numerically controlled (NC) milling machine to mill full-scale models out of modeling clay. The clays' surfaces are smoothed by hand, then are reviewed and modified as necessary. A final model approved by management becomes the "master clay."

Engine Compartment Layout. Soon after the CE concept, simultaneous with styling's 1:5 scale clays, the CE forms an engine compartment work group. A member of the CE's staff leads the cross-functional group comprised of engineers from body, chassis, power train, electrical, one or two key suppliers who supply a good portion of under-the-hood parts, manufacturing, and planning division. This group is tasked with deciding how all the parts that need to go in the engine compartment will be organized. Typically the work group completes an engine layout drawing by the end of the 1:5 clay stage. The layout, issued to all concerned parties, allocates space to each component or subsystem and roughly indicates interfaces. Of course, the layout must be able to accommodate all the styling designs under consideration at that time.

Body Engineering. While the clay models are developed, body engineering (hereafter called body) studies the styling alternatives and creates *kentouzu*. *Kentou* in Japanese means "study" or "investigate," and *zu* means "picture" or "drawing." The most intense *kentou* activity gets started once the full-scale clays are created, though some activity may start earlier on special issues or on designs that show a lot of promise. Body receives CAD data from styling, analyzes it, and begins

[3] At the time of writing, Toyota's rendering system was not integrated with its CAD system, making it necessary to create the 1:5 scale clays by hand, and modify by hand as appropriate. Styling then creates a mathematical representation of the design by laser scanning the clay surface. However, some stylists can create their ideas directly in CAD, making it possible to NC mill the 1:5 clay since the data exist *a priori*, but stylists with these skills are a decided minority. Currently Toyota is working on integrating its computer systems. When this happens, Toyota will not make hand-made clay models.

planning the body design. Planning entails significant design work: dimensioned cross-sections of body structure panels, preliminary structural analysis, clearances, joint definition, relation of section to glass, part layout, wire harness routing, head room, approach angle, crash issues, fasteners, and fastener locations, to name a few. The design work is all done in CAD by the design engineer. Body has on occasion created more than one design plan for a styling alternative. And since they study 2–3 styling alternatives during the *kentou* process, body essentially creates a number of body designs for any given vehicle.

Body engineers use engineering checklists to aid their analysis and design work. The checklists contain data and experience accumulated over years of body engineering work. Each part on the body has a detailed checklist that provides guidance on how it should be designed so that it performs well and can be manufactured at low cost; the checklist also provides the number and types of fasteners, bounds on curvature radii that indicate what is producible, what geometries, dimensions, and metal thicknesses give the required stiffness, standard locating and fastening points, and how to avoid problems that have occurred in the past. The checklists are not just lists of design rules, but part-specific recommendations based on abstractions of past experience.

Kentouzu not only contain preliminary body designs and plans, they also identify any problems they discover with the styling designs, analysis of those problems, a number of proposed solutions, and body's recommendation. Since any one kentouzu covers a relatively small subproblem (e.g., hood interface with front grill and head lamps), body generates many of them during the full-scale model stage. Once completed, *kentouzu* go to the CE, styling, manufacturing, and any other party than may be affected for feedback on body's plans, analysis, and proposed solutions. The feedback may generate more *kentouzu*, and the process continues, gradually building consensus on what the final body design will look like.

Once the final style is chosen (usually a few weeks or a month before final styling approval), body begins to create the *kozokeikaku* (K4[4]), or body structures design plan — the culmination of the *kentou* process. The K4 includes dimensioned cross-sections and specifies clearances, but is not the detailed part design. For example, the K4 may indicate that the hood-grill clearance is 6 mm and that the hood-headlight clearance is 4 mm, but it would not indicate how the hood design makes the transition from 6 to 4 mm clearance — that would be left to the detail

[4] The term "K4" was coined by American engineers in Toyota's U.S. design operations because the Japanese word was too complicated. *Kozokeikaku* has four K's in it, hence the term. It's gaining widespread use in Toyota's domestic operations as well.

design stage. The K4 is then circulated to all concerned groups for approval. Once approved, it becomes the basis for detailed design work.

Chassis Engineering. Chassis engineering (hereafter called chassis) creates a chassis concept based on the CE concept and communication with the CE. Then chassis creates an advanced prototype of the suspension. In most cases, this prototype is an existing system modified to reflect the new chassis concept. (In the case of a new suspension or new underbody, the advanced prototype occurs much earlier, before the CE concept.) Upon test and evaluation, chassis begins work on the suspension layout. Suspension layout contains critical points and dimensions, tire size, spring locations, suspension travel (or bounce clearance).

Power Train Engineering. Power train engineering at Toyota is really concerned with engines. Vehicle Development Center 4 (VDC 4) develops the base engine design, then power train engineering customizes the design to particular vehicles. Design work in the VDC can be quite substantial, possibly including the type of cam shaft, number of valves, displacement, and cylinder bore size. It also includes all the peripheral components, such as fuel system, cooling system, etc. Transmission design is handled in the drive train division in VDC 4, or by suppliers in some cases.

Advanced Prototype. Most vehicle programs at Toyota have an advanced prototype, as already mentioned. This prototype is extensive, and may involve building a fleet as large as 50 vehicles, which are usually modified versions of current production vehicles. These vehicles are tested in many areas: from noise, vibration, and harshness (NVH) and ride and handling, to crash and buildability. The two most important purposes are to develop the power train matching and system integration. Power train matching is tuning the engine and transmission to make sure shift levels and shift points, gear ratios and RPM levels, etc., fit one another perfectly. System integration concerns ensuring all the parts of the vehicle fit together into a cohesive whole. It also serves as a final check on any major new product designs or technology. The VDC's vehicle evaluation division does the testing, and issues reports of the test results to the CE and relevant groups.

Manufacturing Engineering. Manufacturing engineering's role in this stage of the design process is twofold. At the CE's request for participation in the development process, usually at about the CE concept, the manufacturing engineers generate pre-product checklists. These checklists describe current manufacturing process capability in terms of product design, indicating how to avoid problems in production. Similar to body engineering's checklists, they are stored in a central location and updated regularly. The manufacturing engineers access the appropriate file and update the checklist based on their most current experience and on problems with the current vehicle. They then meet with the appropriate product engineers early in the process, discuss the important points, and hand over the checklists. Many of the product engineering groups use manufacturing's checklists to update their own internal design standards.

The second role is that of design reviewer. As alternative product designs are proposed, manufacturing analyzes the proposals to ensure that the designs are economically producible. When they find a problem, manufacturing suggests several alternative solutions, meeting with the product engineers as necessary to discuss them. As already mentioned, manufacturing engineers sit on the styling review board.

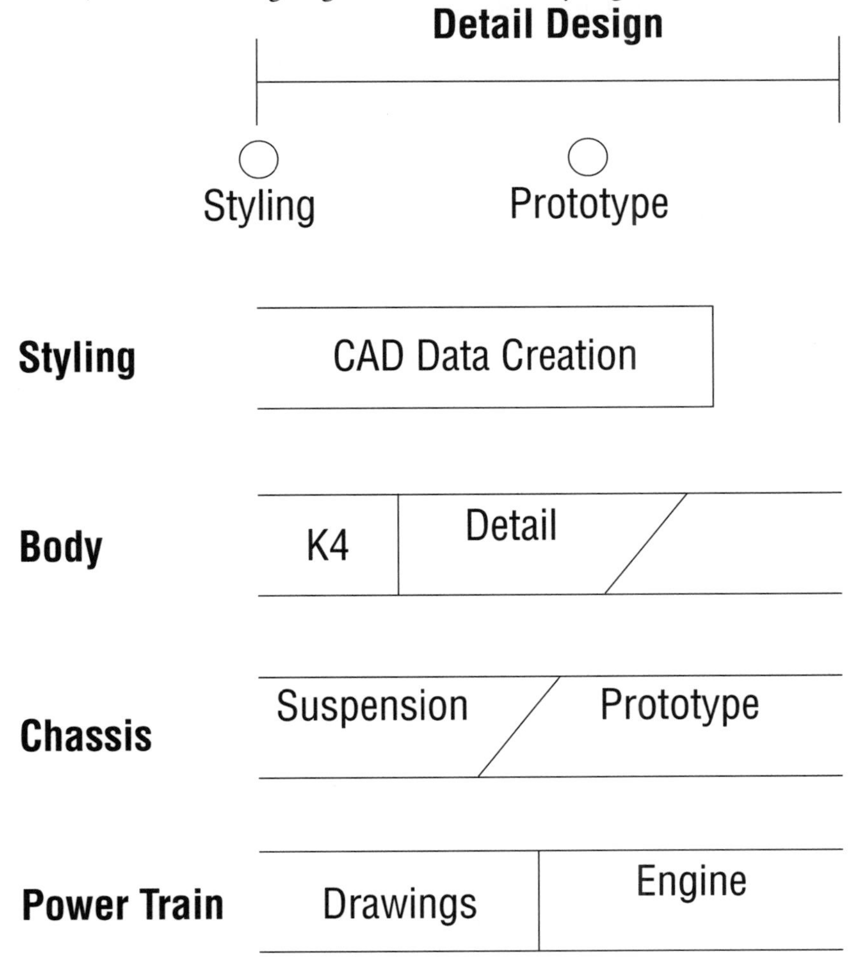

Figure 5 • Toyota Detail Design Phase

Detail Design Phase

By the time Toyota enters the detail design phase, substantial engineering and design work has gone into system design: styling has an approved surface, vehicle layout has been established, body has its K4 (body structures design plan) complete or nearly complete, chassis has its systems layout drawings (nearly) complete, power train has a preliminary design. The development team now works out the details needed to bring the vehicle into being (Figure 5).

In body, the K4 is released about 1–2 months after styling approval. After receiving approval for the K4, or corrections to it, they set to work filling in the details left out of the K4 based on the data received from styling. They perform analytic checks on the detailed design such as finite element analysis and stress and strength analysis. The designs are communicated to manufacturing and other groups for review. Generally speaking, the designs follow the K4 fairly closely. Large deviations from the K4 spell big trouble because this may require system level changes and threaten to throw the program off schedule. The detailed designs are refined based on feedback from analysis and other groups, then released to manufacturing engineering and prototype. Body's prototype drawing release is staged, with parts being released as they are completed. In general, the K4 is not updated based on the detail design work.

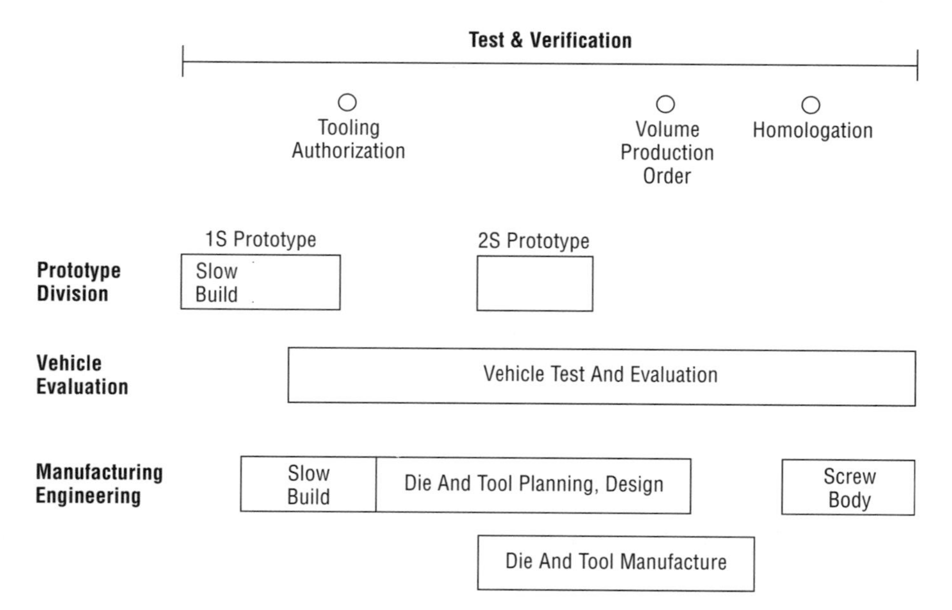

Figure 6 • Toyota's Test and Verification Phase

Test and Verification Phase

The test and verification phase begins with the first vehicle prototype (see Figure 6). In preparation for the prototype build, Toyota's prototype division (separate from the VDC's) creates soft tools and dies from the prototype release drawings. The first prototype stage (1S) begins with a "slow build." Prototype division's engineers and technicians take a great deal of time and care in building the first few vehicles from soft tools. They carefully study any problems with the build to find root cause, suggesting design changes and implementing process changes when necessary. Manufacturing engineering observes the slow build to note any issues that crop up and to help with problem solving if necessary. Slow build vehicles are prototype build study vehicles, and usually do not enter the test process.

Once most of the design and build problems have been ironed out, prototype can very quickly build the rest of the 1S vehicles and send them off for testing. After observing prototype's slow build, and simultaneous with the 1S fast build, manufacturing engineering does its own slow build to examine manufacturing and assembly issues with the vehicle design. They feed this information back to product engineering for design changes as needed.

1S is not just a design verification prototype — it's an integration prototype. As the program enters 1S, many questions remain open that other auto companies would have decided by this point in the process. For example, which power steering module to use may be narrowed to 2 or 3, but the choice between the inexpensive, medium performance model and the costly, high performance quiet model may wait for 1S test results. The test group tries both models in a vehicle(s), then product engineering decides which is best for the vehicle based on total vehicle test results. Similarly, Toyota may try several air conditioning units, brake master power boosters, and other functional parts. Even more extreme are items such as exhaust systems and suspension springs which require subjective evaluation (e.g., sound timbre of exhaust system, handling 'feel' of suspension, etc.). Toyota routinely tests 10 or more different prototypes for one vehicle model before deciding which is best suited for this particular vehicle.

For Toyota to do this, its designs must be modularized (i.e., have common interfaces which allow parts to be interchangeable). This would not be feasible if, for example, the entire engine compartment has to be rearranged to change the power steering pump. Toyota maintains consistent interfaces between parts from model to model.

Prototype division uses robot welders to assemble the body rather than manual welders (the industry norm). Robot welders yield a higher quality weld, similar to production welds, creating a body that better simulates a production body. A better body means that the prototype will act more like a production vehicle during crash testing. Toyota does not need as big a safety margin in its body designs, and thus can make them lighter and cheaper, and still confidently exceed crash requirements. However, for robot welders to be feasible, the vehicle weld lines must be the same from vehicle to vehicle. Again, this emphasizes the importance of standardized interfaces and incremental model to model change in Toyota's process.

Once built, the prototype vehicles go to the test group for testing. Styling views the vehicles to make sure the styling design is properly reflected in the vehicle. Based on the results of prototype's build, manufacturing's slow build, and testing, final design decisions are made. Product engineering makes design changes which are then communicated to suppliers and manufacturing. Provided the vehicle performs sufficiently well, vehicle cost is reviewed with upper management who gives authorization for tooling design and development to begin.

The second vehicle prototype stage (2S) follows tooling authorization. Its primary purpose is to verify the design changes from 1S. The number of vehicles in 2S is decided based on the vehicle's performance in 1S — not predetermined — and may be as few as half a dozen compared to the 30 to 50 vehicles normally built in 1S. 2S ends with the volume production order issued to plants and suppliers.

After 2S, the vehicle goes through homologation, or verification of conformance to government regulations. This can be quite extensive, depending on how many markets in which the vehicle will be sold. It can also result in significant rework. A third prototype stage (3S) may be required.

The last step in the test and verification phase is manufacturing process tryout. Once the dies have been produced, manufacturing stamps out parts and rivets them together into what's called the screw body, or functional build. Functional build verifies that the parts all fit together. Functional build is an approach used by a number of Japanese automakers in the die development and tryout phase of the design process. The dies are made as close as possible to the original design and then actual parts are stamped out. Parts are screwed together instead of welded to simplify construction. Then manufacturing evaluates the screw body and makes the most economical changes to the dies to achieve a perfect fit. These changes are communicated to product engineering. Styling also reviews the screw body. Similarly, the body assembly and final assembly processes and tools are tested out to make sure they produce the desired vehicle. Thus manufacturing engineering sets the final specifications on parts, which generally need to be within 10 mm of the nominal dimension. Note that at this point manufacturing is driving design changes which are communicated to product engineering, rather than product engineering dictating specifications set in stone which manufacturing engineering must achieve at all costs. The functional build process results in much less design rework and ultimately a higher quality body (Hammett *et al*, 1995).

TRIAL PRODUCTION PHASE

The final phase is trial production. Toyota has two trial production runs, 1A and 2A, conducted in the plant. The trial runs are conducted both on the existing line and in a parallel line. During these runs the final bugs are worked out. A substantial amount of value analysis work is done to drive waste out of the manufacturing process. After each run, changes are implemented as needed. Product engineering has a small group in the plant to help resolve product design issues. On the date specified in the CE concept, Toyota has its "line-off," or start of full production. Toyota managers claim that the line-off date has never been missed.

SUPPLIERS

Consistent with the model of Japanese supplier relations, Toyota involves its suppliers early in the design process, called "design-in" at Toyota. Toyota buys 70 percent of the value of its vehicles from outside companies, much of which is designed by the supplier, so suppliers are critical to Toyota's development process. Ward et al. (1995) describe Toyota's design-in process in detail with a number of case examples.

Toyota's design-in suppliers do their own advanced development relatively independent of Toyota. Since Toyota's vehicles follow a regular 4 year renewal cycle, it's common for a supplier to target its development for a specific Toyota vehicle. During the concept phase, suppliers present their latest technological developments to Toyota engineers. Much like a design show, these presentations generally show how the new part is better than the part Toyota is currently using. They include a fully functional prototype and performance/test data. Some suppliers, especially in suspension parts, may even have the new parts on a Toyota vehicle that engineers can test drive and feel the difference for themselves.

During the system design phase, Toyota uses the information gathered at the suppliers presentation along with the targets in the CE concept and information from Toyota's internal design work (such as styling and engine compartment packing) to create targets for each subsystem or component. The targets are usually based on an existing product, e.g., "we'd like 5 percent better performance, 5 percent less weight, and 10 percent less cost." The information Toyota receives in the supplier presentation is crucial for knowing how aggressive to make the targets. Suppliers who do not do presentations and who do not show improvement from year to year fall into poor favor and will likely lose part of Toyota's business.

Toyota issues its targets and design requirements to suppliers in a Request for Design and Development of Parts (RDDP). The requirements are typically black-box, indicating only interface specifications and performance outcomes, leaving the internal details of part design to the supplier. The RDDP may also request the supplier to include a specific kind of technology, or the new technology they learned about in the supplier presentations. The RDDP for a given part or subsystem may contain one *or more* sets of requirements, asking the supplier to develop a number of parts to different specifications.

Suppliers begin working to achieve the targets and design requirements. During the detail design phase, they propose alternatives for each set of Toyota's requirements, and discuss the relative merits of each alternative. The proposed alternatives are accompanied by test data and analysis, with test data from physical prototypes being preferred. Toyota and the supplier discuss the alternatives, perhaps generating new alternatives or modifications to the proposed one, or perhaps requesting additional development and/or testing. Before Toyota includes a technology in a vehicle, it must be proven technology.

Based on negotiations, Toyota and supplier agree on a specification and Toyota issues the prototype order (which includes the specification). If Toyota is familiar with the technology, it issues "exact specs" for one or two spec prototypes.[5] If Toyota is not familiar with the technology, it issues "tentative specs" and asks the supplier to deliver 3–4 or more spec prototypes. Tentative specs indicate to the supplier approximately where Toyota thinks it wants to set the requirements, but is not confident in the numbers. It then asks the supplier to deliver parts of different designs in this general area so that Toyota can experiment with them and determine the optimal settings. All of these prototypes are destined for the 1S prototype build.

At the 1S prototype stage, Toyota tests suppliers' components and subsystems in the vehicle. Toyota communicates functional and assembly problems to the supplier for correction. These corrections are made as quickly as possible, and must be completed in time for the 2S

[5] Spec prototypes conform to a particular set of specifications, thus multiple spec prototypes means prototypes of different designs, e.g., please deliver 25 prototypes of Design A alternator and 25 prototypes of Design B alternator.

prototype. Up to this point, cost has been discussed, but not settled. Throughout the prototype stages and beyond into production, suppliers are expected to do value analysis/value engineering to decrease the cost of its parts. The final cost and specifications are usually set after 2S and before trial production. The numbers are revised every year throughout the life of the product.

An interesting aspect of Toyota's supplier relations is that Toyota does not treat all its first-tier suppliers equally. The process just described applies to all suppliers who do engineering and design work, but variations can be considerable. A particularly strong and highly competent supplier, such as Nippondenso, may have such advanced product and/or process technology as to render the targets and negotiation process a mere formality. Toyota may mostly buy what the supplier recommends. With less competent suppliers, Toyota may issue targets that are essentially the part specifications. The supplier must strive hard to meet the targets. If they do, the targets become the specifications. If not, the supplier must negotiate with Toyota for less aggressive specifications. In this case, Toyota will not likely ask for more than one spec prototype. Not all Toyota suppliers are design-in suppliers — Toyota has a number of make-to-print suppliers, whereby Toyota designs the part, and sends the print to the supplier who produces it.

SUMMARY

Toyota has a unique approach to product development that violates some of the best practice principles that have become commonplace in the U.S. Their system depends very heavily on the talents of a seemingly "super" individual called the chief engineer who acts as a system integrator and system designer. They also use very disciplined approaches like engineering checklists and a design structure plan to deal with system integration issues up front. Functional engineers reside in their home functions. Functional managers assign them temporarily to design projects, sometimes part-time and sometimes full-time. They work on the "chief engineer's vehicle," but still formally report to their functional managers. They have in-depth knowledge of their function and are expected to spend at least ten years doing engineering work in their function before being assigned any supervisory responsibility. Throughout the process, we saw evidence of what Ward et al. (1995) call "set-based design." Multiple alternatives are identified and considered throughout the process, and alternatives are selected based on actual test results in prototypes.

WHIRLPOOL CONCURRENT ENGINEERING

Whirlpool is one of the world's largest and most successful producers of household appliances. Its main product lines are washers, dryers, dishwashers, refrigerators, air conditioners, and small appliances. This case describes Whirlpool's shift to concurrent engineering in the early 1990s primarily through a complete restructuring of their design process, the use of cross-functional teams for all major projects, and a major change in the way they work with suppliers. It is based primarily on interviews at the St. Joseph Technical Center in November of 1993.

The changing environment for household appliances has made it much more challenging to remain profitable and competitive. Like home electronics, prices have been steadily coming down in real dollars (even before adjusting for inflation) for several decades. Competition is intense domestically among the small number of major players still in the business, and there is always concern about foreign competition, particularly from Asia. Up to this point, Japanese companies have not been major players in the U.S. market, but this could change at any time. The U.S. government has strict safety and other regulations. Any test labs must be AGA and UL approved. Increasingly stringent energy use requirements are demanding a complete evaluation of the mechanics of the product.

To make money in this changing environment, Whirlpool has been experimenting with many organizational approaches to improve quality and reduce costs. In the early 1980s, the emphasis was on organizational experimentation in manufacturing (e.g., participative management, SPC). The goal now is to design products that are what the customer wants in less time and with improved quality and lower manufacturing costs.

Whirlpool laundry and dishwasher product development takes place at its Technical Center in St. Joseph, Michigan. About 200 engineers in this center develop and test new products and are responsible for engineering changes. The center has state-of-the-art facilities for computer-aided design, computer-aided analysis, prototype construction, and product testing. Whirlpool has recently received a good deal of publicity for its rapid prototyping capabilities. Although the technology has been effectively utilized, the product development center only recently began experimenting with new organizational approaches.

Whirlpool's manufacturing is primarily in the midwest. For example, a major facility in Marion, Ohio, makes automatic dryers. Some of the company's new initiatives are aimed at involving manufacturing earlier in the development cycle. However, just as critical for Whirlpool is cementing two linkages that have been tenuous in the past. These are the relationships between design engineering and two other groups — marketing and outside suppliers.

There are several things that make Whirlpool an interesting and distinct case compared to others included in this book. First, it is a home appliance producer and its products are simpler products than some of the other sectors we examine, such as automotive (there are far fewer components and subsystems). A design project often involves changing one part of the product, such as the way information is displayed to the consumer.

Second, Whirlpool did not change its basic organizational structure to facilitate concurrent engineering. Cross-functional teams and a new structured product development process (called Consumer to Consumer, C2C) were overlaid on the existing structure which combines a functional and product organization.

Third, the reward system as it existed was not particularly supportive of concurrent engineering, and it was not changed as the company moved toward cross-functional teamwork. Yet, the lack of rewards for teamwork did not seem to hamper the effort. The significant improvements that resulted illustrate that a wholesale reorganization of organizational structure or the reward system is not always necessary to achieve benefits.

Organizational Structure

Whirlpool's organizational structure is a complex hybrid which combines functional organization, product organization, and brand management. At the highest level, the organization is typically functional. Marketing, procurement, and manufacturing report to separate vice presidents compared to manufacturing and product engineering.

Within the main product development organization at the St. Joseph Technical Center there are four groups — dryer engineering, washer engineering, dishwasher engineering, and product services — each with its own director. Product services is organized into four principal groups: the model shop makes prototypes; computer services supports all computer systems used in engineering; design services includes engineering analysis, drafting (on CAD systems), and materials engineering; and lab services includes technicians who run product tests and the prototype shop.

Design engineers report to the director of a particular type of product such as washers or dryers. Below that, engineers are divided by a combination of electrical versus mechanical systems, and by name brand such as Kenmore, Whirlpool, or Kitchenaid. A rough version of the organization chart for dryer engineering is illustrated in Figure 1. Note that the manager of Whirlpool specializes in mechanical systems, and the manager of Sears products specializes in electronics systems. Design engineers will formally report to a brand manager within a product line, e.g., washer or dryer. In reality, the main dividing line of design engineers is between electrical and mechanical, combined with the product line. An electrical engineer in dryer will typically work on electrical systems within dryers regardless of whether they belong to Whirlpool or Sears. In fact, most systems are similar across these brand names. Thus, the directors of dryer, washer, etc., are very close and there is a good deal of informal swapping of engineers as the need arises. So in this case, the formal organization is not a good representation of actual work practices.

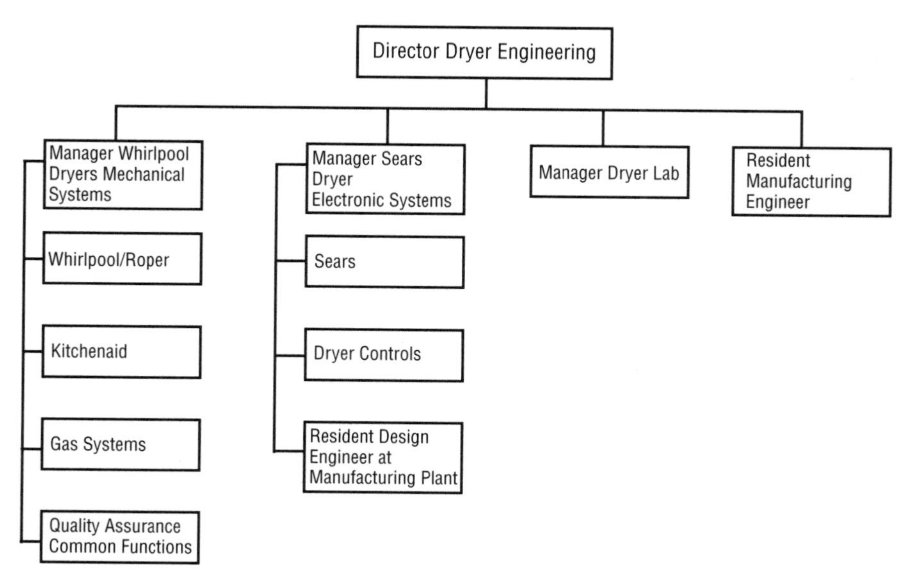

Figure 1 • Organizational Structure for Dryer Engineering

The role of the resident engineers shown on the organization chart is to act as a liaison. A design engineer is resident in the Marion, Ohio, plant to act as a liaison to manufacturing, and a manufacturing process engineer is resident in St. Joseph to link to the design function. The particular engineer resident in Marion in the dryer group had 20 years of manufacturing experience.

This segmented organizational structure would appear to create communication barriers across functional specialties; and in the past it did. But Whirlpool has made significant gains in cross-functional collaboration through changes in its design process and the use of cross-functional teams, without changing this basic organization structure.

The Move to Cross-Functional Teams

Whirlpool's approach to concurrent engineering was mainly through communication and coordination mechanisms. The approach was not implemented in one shot, but rather has evolved over time.

Most of the current initiatives are an outgrowth of Whirlpool's World Wide Excellence System. This is an internal version of the Malcolm Baldridge award. This program has included a great deal of training on modern quality concepts and an internal audit. All functions are encouraged to identify internal customers and find ways to satisfy their customers. There are no prescribed programs, but each unit is encouraged to explore ways of working with their customers, and this has fostered many approaches to cross-functional teamwork.

A turning point in the use of cross-functional teams in product development came when an opportunity arose to attend training on design for manufacturability in 1989. At that time, design and manufacturing representatives were holding design integration meetings. Design engineers said they would attend the DFM training, but only if the manufacturing representatives would also attend. In all, 22 people attended the one week training course in Detroit. A core team from this group looked at Whirlpool products and identified opportunities to apply DFM. They agreed they would focus only on improvements to existing products. They prioritized the list based on expected financial returns.

The result was the assignment of six cross-functional teams to work on these projects, reporting to a steering committee of managers. Each team focused on a different part of the product (e.g., front panel door, wire harness, console). Smaller teams were formed within the larger teams to focus on specific projects (e.g., quick connects, twist mount switches).

The managers on the steering committee made a policy that they would not make decisions but rather would listen to team reports and provide needed resources. The engineering directors stayed out of the loop, also by design. The teams were given complete autonomy (e.g., they picked their own leaders). The result of this DFM initiative was to substantially reduce product cost. And this all happened by turning six cross-functional teams loose.

Building on the success of these teams, the technical center has shifted to a more comprehensive team approach that includes all functions needed for a design project, including marketing, support staff, and key suppliers. There are no longer any teams specially dedicated to DFM. Instead, DFM has become a routine way of developing products.

The new process at Whirlpool utilizes cross-functional teams constructed from departments such as marketing, design engineering, manufacturing, procurement, advanced development, finance, and advanced research. Customers and critical suppliers are also included as part of the team. The exact composition of a development team varies from project to project. The number of representatives from each function is based on the scope and size of the project. Some focused efforts to improve a particular component may have only design engineers, representatives of test labs, and prototype builders. Other broader projects to develop an entirely new product have much more elaborate structures. An example of a very elaborate team structure is illustrated by a washer development project described later in this case (see Figure 2). Most of the team members are collocated at the St. Joseph Technical Center, though supplier representatives and support staff (e.g., finance, marketing) are not.

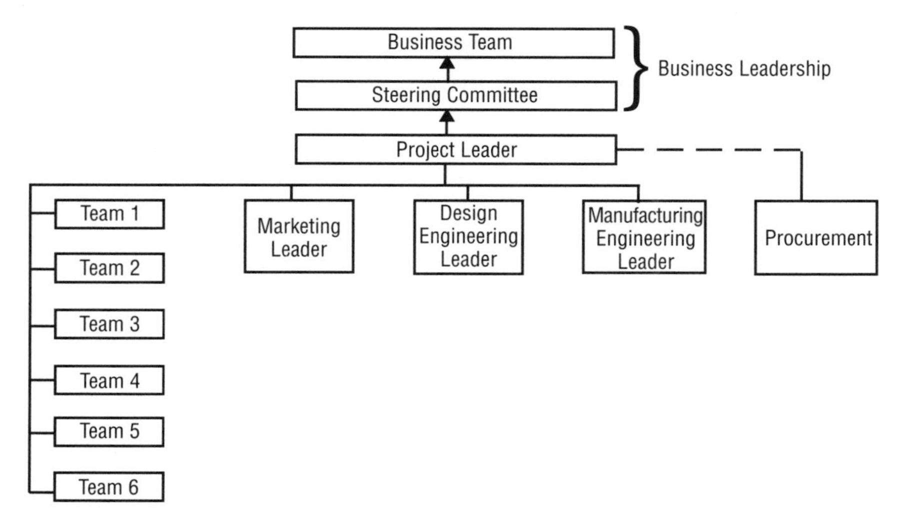

Figure 2 • Team Structure for Next Generation Washer Development Project

Engineers involved in the teams say that the most important changes are the way they work with outside suppliers (via purchasing) and with marketing. Whirlpool had worked hard in the past on design-manufacturing relationships and was quite good at this. But in the past there has been a "wall" between marketing, engineering, and purchasing. Marketing fed customer requirements to engineering without a concept of what those requirements really meant in engineering and manufacturing terms. This led to discussions between marketing and engineering over what is really needed by the customer and what is possible. Now marketing is on the team with engineering performing customer tests and determining customer requirements. Similarly, purchasing in the past was the gatekeeper to suppliers. Engineers might work with outside suppliers, but when it came to getting a quote they had to go through purchasing, who might offer the business to a different supplier who lacked certain technical knowledge. Now purchasing works with suppliers and engineers, and often products are pre-sourced, with the supplier on the design team from the start.

Integrating Suppliers into Product Development

A major part of Whirlpool's change to concurrent engineering is in how it works with parts and material suppliers. In the past, Whirlpool, like most U.S. companies, had an arms length relationship with its suppliers. Whirlpool purchasing dealt with sales managers from the suppliers to get bids and negotiate prices. They were intermediaries that engineers had to go through to work together. Of course, Whirlpool engineers often talked directly to supplier engineers in the design phase, but they could not commit any business, and sometimes good ideas from the suppliers were lost as a lower bidding competitor got the business. This made suppliers naturally reluctant to share ideas and contribute a great deal of their resources to product development that was specific to Whirlpool.

As part of the World Wide Excellence System, purchasing sat down with design engineering and asked how they could better serve engineering. The result was a wholesale change in the way Whirlpool now works with suppliers. Whirlpool procurement looked over their supplier list and has embarked upon a supplier-reduction program. They retained suppliers with the best record for quality, cost, and timeliness, and identified strategic suppliers which could participate in the earliest phases of product development. Now when Whirlpool engineers design a new component, they identify strategic suppliers and invite them to be members of the team. Purchasing participates in the cross-functional teams and can pre-source (i.e., guarantee business to the supplier before the product is even designed).

Supplier involvement has been further extended by giving suppliers responsibility for product testing. The old attitude of the Whirlpool lab managers was that they must test everything, since only they could do it right. It took a long time but they felt it was worth the wait. Now they use suppliers whenever possible for testing. Suppliers use Whirlpool specifications for reliability testing (e.g., time to failure) which Whirlpool knows suppliers can do well. But suppliers are also given responsibility for the more challenging application testing. This might involve shipping a dryer to the supplier so it can test how its part fits and functions in the dryer.

There have been a number of benefits of increased supplier involvement in lab testing. The work load on Whirlpool's lab has been reduced considerably, making it more responsive to the needs of Whirlpool engineers. Supplier expertise is better utilized, and turnaround time on tests has been reduced, which in turn reduced development cycle time. It has reduced some equipment purchase needs, and it gets supplier engineers involved with the application so they see where their part fits. The main challenge has been to treat the supplier as an extension of Whirlpool's own internal capability.

Some lessons learned from involving suppliers in testing include the following.

- Get procurement to identify strategic suppliers who develop long-term sourcing relationships.
- Remove systems that set up adversarial relationships with suppliers.
- Get suppliers to let the test lab work directly with engineers (eliminate the sales intermediary).
- Give suppliers the big picture of design.
- Trust test data from strategic suppliers.
- Do an audit of supplier facilities and expertise.
- Review test standards and expectations with suppliers.

Information/Work Flow

Whirlpool has three distinct systems which determine work flow and its associated information. These are: 1) a Product-by-Product Planning Department which sets project priorities, timetables, and budgets; 2) the Consumer to Consumer (C2C) program management system which manages across the life cycle phases of product development, and 3) a project management system which manages detailed tasks and timetables.

Planning Department

This department coordinates and prioritizes product development and exploratory projects necessary for future products. Individual managers do not set priorities, and manage by exception when the plans cannot be met. An important role of the planning department is to integrate activities across product brands. For example, the planning department may specify the need for a new wire harness. This will typically apply across models. In some cases, a specific feature for a specific brand (e.g., Kenmore) may be needed to distinguish that brand from others.

C2C Structured Development Process

Linked closely to the cross-functional team approach is the C2C (consumer to consumer) process which is a very elaborate structured development process described in two books, each $^3/_4$" thick. It divides all development into a number of distinct phases which have toll gates that must be satisfied before progression to the next phase. The phases are: concept development, advanced product development, full-scale product development, and product introduction and support. Toll gates consist of Idea Screening, Concept Evaluation, Business Evaluation, and Post Audit. The toll gates are points of management review where the project team presents specific, predetermined criteria to upper management for review and approval to the next stage. Tolls are divided into major tolls (those with upper management) and internal project

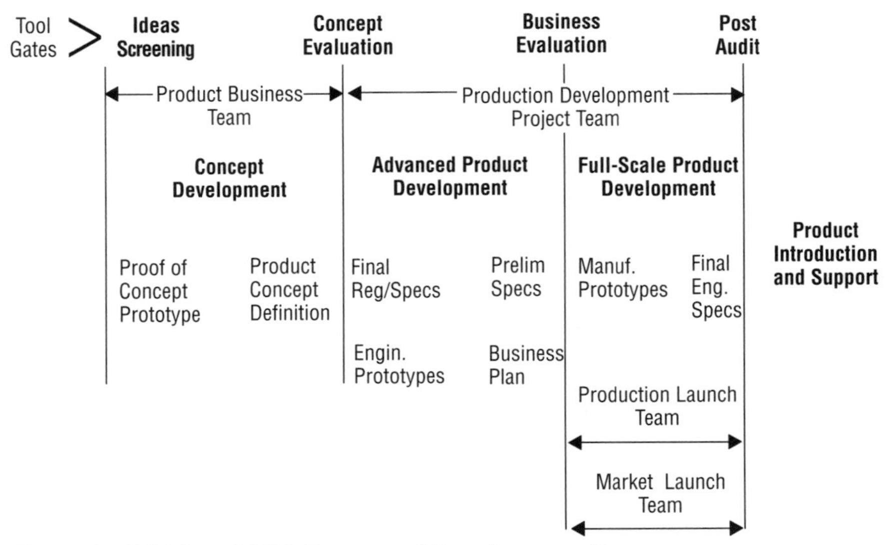

Figure 3 • Whirlpool C2C Structured Development Process

reviews. Major tolls are scheduled by due date, internal tolls are fairly regular and often. Activities that occur within the phases include market analysis, business environment assessment, advanced product and manufacturing concepts, and technology development. Some development projects are approved to span only the early phases of C2C. Figure 3 provides an overview of the highest level of the C2C process.

The application of C2C, which has been introduced within the two years prior to the writing of this case, varies across Whirlpool. C2C has been a major change since it introduced the discipline of business planning using cross-functional teams and required significant training for all participants. Distinct teams appropriate to specific phases were: product business team, product development team, production launch team, and market launch team.

The C2C process has shifted the culture of Whirlpool. In the past, a review consisted of each functional group protecting its interest as well as scrutinizing the work of others. With C2C in place, the cross-functional team wants the whole team to pass through a toll gate. This requires that each individual within the team acts as an interface to their functional management, informing them of progress as well as satisfying the functional interest.

One of the biggest changes brought about by the C2C process is the requirement of a formal business plan before the project can proceed past advanced product development. The business plan forces all functions to reach consensus on the project early in the process. Everyone must sign up or the project will not proceed.

Project Management

The project management system uses personal computer based software which is accessible to anyone in the team as well as management. Detailed tasks and schedules are worked out by subgroups or individuals keeping the overall timeline in mind. Gantt charts are the typical presentation format. Project tracking is reported in a double bar format (planned versus actual).

Reward System

Whirlpool adopted a bonus system tied to company profits long before it was fashionable. Engineers hired out of school typically found that their base pay was a bit lower than their peers, but with the substantial annual bonus their salary surpassed their peers. This had the effect of tying individual goals to the goals of the company.

Historically then, bonus was a straight percent of an individual's salary based only on profit performance. Later in the 1980s the plan was modified so it was partly merit based. The merit component was judged using a management-by-objectives system based on individual performance. Each individual developed objectives for the year and was judged on accomplishment of those objectives. In the 1990s, when Whirlpool began seriously using cross-functional teams for product development it became clear that the MBO system was not particularly supportive of teamwork (the system has since been changed again).

One cross-functional team developed an interesting way to convert part of the individual merit component to a group reward. When they developed their objectives for the year they all agreed that for that particular project every individual on the team would get the same performance appraisal based on team performance. They realized that this project was only one part of their individual objectives and probably would not be the deciding factor in their bonus amount. But this symbolic gesture still led to an enhanced feeling of esprit de corps. And this was a grass roots effort not dictated by management.

What we found most interesting, in general, was the great enthusiasm with which individuals talked about their teams. It was clear that engineers derived great satisfaction from working as part of a team and were excited about their accomplishments. The intrinsic satisfaction we saw seemed to be more important than the performance appraisal system in motivating the group to succeed.

Concurrent Engineering Methods and Tools

Whirlpool utilizes a wide variety of design and analysis methods and tools. These include Story Boarding, Computer Aided Design (CAD — including IGES as an interchange format), Rapid Prototyping, Mold Flow Analysis, Finite Element Analysis (FEA), Progressive Mold Build, Design for Manufacture/Assembly (DFM / DFA), Generating Tooling from Part Databases, Formula Costing, Taguchi, Quality Function Deployment (QFD), Early Supplier Involvement (ESI), Cross-Functional Teams, and Activity Based Costing (ABC).

Each method or tool has been adopted to a varying degree. As an approach demonstrates its usefulness, it becomes embedded into the design process. New engineers are trained in a technique and easily adapt to its approach since they typically have little preconceived notion of how to accomplish a specific design task. Seasoned engineers have to unlearn their previous techniques before they become proficient in a newer technique. As an example, DFM/DFA was introduced several years ago with much attention focused on its application. Today, DFM/DFA is used as the only way to produce such designs, and new engineers hired in cannot understand why anyone would design any other way.

To illustrate the application of CAD methods, combined with a team approach, throughout the design process, a current design flow is compared to an earlier design flow, as shown in Figure 4.

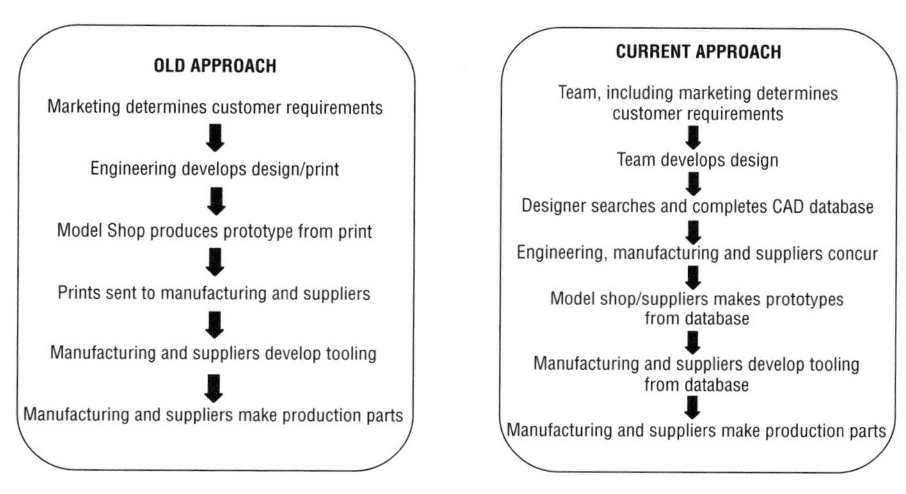

Figure 4 • Use of CAD Methods in the New Development Process

This new process, which integrates teamwork and sophisticated use of CAD, has led to a number of benefits.

- The team feels ownership of the design.
- The final design is developed faster.
- Sales understands and accepts engineering constraints on the design up front.
- Faster design analysis.
- Faster verification of designs with prototypes.
- Fewer design iterations.
- Market research with real parts.

EXAMPLE PROJECTS

NEXT GENERATION WASHER

In part to meet anticipated Department of Energy standards in 1999, Whirlpool was already working on the redesign of its washer. To attack this problem, a complex structure of teams and sub-teams was formed as illustrated in Figure 2. Leaders representing each of the functional groups (marketing, design engineering, manufacturing) were assigned to this project to focus on high-level coordination and ensure that their function delivered. Originally, there were plans to include a financial person assigned full-time to each product team. But there was not enough time needed to justify this.

At a working level, there are subsystem teams that are all cross-functional and have team leaders. Membership might include an electrical engineer, a mechanical engineer, a model maker, a designer, a lab technician, a marketing representative, a manufacturing representative, and representatives from strategic suppliers. Individuals may be members of multiple subsystem teams. For example, the team leader for the enclosure team was also on several other subsystem teams as a member.

A major job of the subsystem team leaders is to manage interfaces. If there is an overall space layout problem (e.g., two subsystems want to use overlapping space within the enclosure), the team leaders get together early to resolve it. If they cannot, they will go up to the marketing leader.

The subsystem team leader has the autonomy to structure the team as he/she wants. For example, the team leader is responsible for bringing in strategic suppliers, scheduling the subsystem development process within the general C2C framework, costing, and recruiting members of teams.

All team members participate in identifying customer requirements and all are invited to participate in consumer use tests. Thus, the consumer is not just an abstraction described in marketing terminology, but represents real people trying out prototypes the team designed. This approach provides some clear advantages over the old system.

- Partitions the product into logical subsystems.
- The subsystem team leader is empowered.
- Cross-fertilization of ideas.
- Develops leadership.
- Provides cross-functional communication.
- All functions go forth through toll gate as a team (as opposed to adversaries).
- Most routine decisions can be made on the spot without having to contact people and wait days or weeks for a response.

The empowerment of team members is critical. In the past, Whirlpool used teams, but all members were representatives of their functions and had to go back to get permission from their bosses. Now the teams are empowered, and the marketing person on the team can voice the marketing view and make most routine decisions on the spot.

Prototyping was also being used in a different way. In the past, prototypes were built to find out what was wrong. The goal for this development project was to use prototypes to verify the concept. The key is learning as much as possible from each prototype round and taking appropriate corrective action immediately. In this way, they hoped to reduce the number of prototype rounds to three. Engineers observed that engineering analysis is helpful, but makes so many assumptions they ultimately can only prove the concept with prototypes.

There were also some rough spots that were being resolved as they went through this new process.

- Managing the design interface between subsystems was better than before, but still a challenge.
- They discovered that this approach demands more people early in the process than had been true in the traditional approach.
- The existing project scheduling methods were well suited to a sequential process, but did not recognize concurrent activities.
- It was challenging to balance the time commitments of all the members who were on other subsystem teams and also had responsibilities on other projects.
- Integration with industrial design was still a weak link. Industrial design tends to think in artistic terms first and feasibility second.

Plastic Tub for Next Generation Washer. The plastic tub was designed by a subsystem team. The team leader was responsible for assembling the team and chose to invite strategic suppliers to be part of the development effort. Thus, the mold maker, resin supplier, and injection molding suppliers were all members of the team from the start. Each supplier provided product expertise and performed the engineering analysis/simulation.

Many of the tasks that in the past had been sequential were done concurrently. For example, finite element analysis was done iteratively as the product was designed and prototypes were built and tested. The FEA results were used to verify test results for each prototype. Steel was purchased before the drawings were even complete.

The subsystem team leader was responsible for the schedule for this sub-project. This meant scheduling about 200 tasks. He maintained a three-month detailed schedule on a moving basis and had more gross schedules farther out in time.

Among the tools used to keep the group progressing and communicating were e-mail over a LAN and groupware (Lotus Notes). Project scheduling was also a critical tool.

As a result of the team approach, test results were available from a prototype washing machine much earlier than usual. They had a tooled part available to test much earlier and then verified the test results through finite element analysis. The results were good parts coming out of the first prototype test. They estimate that 3–4 months were saved.

There was a considerable amount of process learning as this was the first time this group had worked together in this way. They had to figure out the best way of involving different functions (e.g., weekly design reviews, toll gate meetings, etc.). They worked this out as they went. If they erred, they believe it was on the side of involving too many people too early.

The team was collocated in a separate office space from their usual offices. While this was ideal for communication, it had a downside in that the team was perceived by others to be outside the norm. Upper management also found the team approach to be a bit disconcerting at first. They were used to regular reporting of instant results, but the team took their autonomy seriously. Top management had to get used to a relative lack of communication of day-to-day progress. As the team leader described: "we didn't want to rip up the team structure every two weeks to report to upper management." As it was, the team leader acted as the main external interface for the group and found that 80 percent of his time was spent communicating with managers and other teams.

Wire Harness Design. In the past, engineers always assumed they needed to create a brand new wire harness for each new dryer. For this project, engineers, purchasing, and the wire harness supplier worked together to find a lower cost alternative. They discovered that the supplier was equipped to make certain standard length wires cheaper than nonstandard lengths, and had standards for all aspects of the wire harness. By custom designing wire harnesses, Whirlpool was typically outside of those standards.

The solution was to develop a database (in an Excel spreadsheet) of standard parts. The vendor opened its books and provided formula prices for different harnesses composed of different combinations of standard parts. Using this spreadsheet Whirlpool design engineers can look up the cost of everything they design instantly. The engineer can cost out the part without contacting purchasing or the vendor only to play what if games.

The result was a reduction in development lead time and a significant reduction in the cost of wire harnesses. From the supplier's point of view, the use of standard parts reduced machine setup time, inventory, and the time it takes to make samples (it now takes 2 days instead of weeks).

In the past, a high-level person from the wire harness supplier used to come to see Whirlpool purchasing to negotiate on price and features, and the results would be passed down to a working-level engineer. Now the Whirlpool engineer can deal directly with the supplier engineer who can approve the design, and the price is clear to everyone — the salesperson is out of the loop. The job of sales now is to argue for general percentage increases instead of prices of detailed parts.

Quality of Service Meetings. The product service group used an interesting approach to improve the services provided to Whirlpool engineering. They began by holding meetings with engineers to get their input through storyboarding. The process was as follows.

- Customers write on cards the good things about product service (10 minutes) and the things that can be improved (10 minutes). These cards are put on the board and categorized as they are placed.
- The lists of items are read off.
- Customers are asked to place colored dots (reflecting 3,2,1 points) on their biggest concerns.
- The positive and negative points are then summarized and posted.

Cross-functional teams were then formed, including service representatives and customer representatives, and the problems were attacked in order of priority. While this started in the product service group, it was being extended to engineering to see how they can improve service to its customers — manufacturing, procurement, marketing, and services.

In addition to helping to solve specific problems, both sides get a picture of the point of view of the other. The approach has been helpful in building cross-functional relationships. Whirlpool has found that it is critical to get people who actually provide and use the services together — not just their managers.

OVERALL RESULTS

Results to date are encouraging. Overall, it was estimated that concept to production time and cost have been reduced by 25 percent. There are a number of improvements in the process itself. Process learning occurs early in the project cycle, product development time has been reduced, tooling is fabricated correctly on the first pass, quicker and better decisions are made at the team level, and the customer has input into the design, resulting in higher customer satisfaction. Several issues were raised including: more resources are needed early in the project, designing the interfaces between teams could be improved, determining the level and timing of functional resources requires tuning, and interfaces with upper management require redefinition and trust building. Whirlpool engineers also emphasized that the first few times through using this approach are actually slower due to the learning experience.

REFERENCES

Adler, P.S. (1992). "Managing DFM: Learning to Coordinate Product and Process Design," in G. I. Susman, ed., *Integrating Design and Manufacturing for Competitive Advantage.* New York: Oxford University Press.

Adler, P.S. and B. Borys (1996). "Two types of bureaucracy: Enabling and coercive," *Administrative Science Quarterly*, Vol. 41, No. 1, pp. 61–89.

Akao, Y. (1990). *Quality Function Deployment: Integrating Customer Requirements into Product Design.* Cambridge, MA: Productivity Press.

Aldrich, H. (1979). *Organizations and Environments.* Englewood Cliffs, NJ: Prentice-Hall.

Allen, T.J. (1977). *Managing the Flow of Technology*. Cambridge, MA: MIT Press.

Andrews, D.C. and S.K. Stalik (1994). *Business Reengineering.* Englewood Cliffs, NJ: Prentice-Hall.

Asanuma, B. (1985), "The Organization of Parts Purchases in the Japanese Automotive Industry," *Japanese Economic Studies*, Vol. 13, pp. 32–53.

Barkan, P. (1992). "Productivity in the Process of Product Development — An Engineering Perspective," in G.I. Susman, ed., *Integrating Design and Manufacturing for Competitive Advantage.* New York: Oxford University Press, pp. 56–68.

Besterfield, D.H., C. Besterfield-Michna, G.H. Besterfield, and M. Besterfield-Sacre (1995). *Total Quality Management.* Englewood Cliffs, NJ: Prentice-Hall, Ch. 10.

Boothroyd, G. and P. Dewhurst (1989). *Product Design for Assembly.* Wakefield, RI: Boothroyd Dewhurst Inc.

Boothroyd, G., P. Dewhurst, and W. A. Knight (1994). *Product Design for Manufacturing.* New York: Marcel Decker.

Camp, R.C., (1989). *Benchmarking: The Search for Industry Best Practices that Lead to Superior Performance.* Milwaukee, WI: ASQC Quality Press.

Checkland, P. (1981). *Systems Thinking, Systems Practice.* New York: John Wiley & Sons.

Clark, K.B., D. Ellison, T. Fujimoto, and Y. Hyun (1995). "Product development performance in the world auto industry: 1990's update," International Motor Vehicle Program 1995 Annual Sponsors Meeting, Toronto.

Clark, K. B. and T. Fujimoto (1989). "Lead Time in Automobile Product Development Explaining the Japanese Advantage," *Journal of Engineering and Technology Management*, Vol. 6, pp. 25–58.

Clark, K. B. and T. Fujimoto (1991). *Product Development Performance: Strategy, Organization, and Management in the World Auto Industry*. Boston, MA: Harvard Business School Press.

Daft, R. (1995). *Organization Theory and Design*, 5th ed. New York: West Publishing.

Duncan, R. (1979). "What is the Right Organization Structure? Decision Tree Analysis Provides the Answer," *Organizational Dynamics*, Winter.

Fleischer, M. and J.K. Liker (1992). "The Hidden Professionals: Product Designers and their Impact on Design Quality," *IEEE Transactions in Engineering Management*, Vol. 39, pp. 254–264.

Fleischer, M., T. Phelps, D. Arnsdorf., and M. Ensing (1991). *CAD/CAM Data Problems and Costs in the Tool and Die Industry*. Ann Arbor, MI: Industrial Technology Institute.

Fleischer, M., J. White, and D. Carson (1996). *Agile Manufacturing Supply Chains*, Final Report to the Manufacturing Technology Directorate, Wright Laboratory, Air Force Materiel Command, Wright Patterson AFB, Ohio. Ann Arbor, MI: Industrial Technology Institute.

Hackman, J.R. and G.R. Oldham (1980). *Work Redesign*. Reading, MA: Addison-Wesley.

Hammer, M. and J. Champy (1993). *Reengineering the Corporation.* New York: Harper Business.

Hammett, P., W. Hancock, and J. Baron (1995). "Producing a World-Class Automotive Body," in J. Liker, J. Ettlie, and J. Campbell, *Engineered in Japan: Japanese Technology Management Practices.* New York: Oxford University Press.

Harbour and Associates (1996). Reported in *Car and Driver Daily Automotive Update*, May 31, 1996, on http://www.caranddriver.com.

Hauser, R. and D. Clausing (1988). "The House of Quality," *Harvard Business Review*, Vol. 66, May-June, pp. 63–73.

Hiatt, S. R. (1992), *Instructor's Resource Manual to Accompany Organization Theory and Design* by Richard Daft. St. Paul, MN: West Publishing Co.

Himmelfarb, P.A. (1992). *Survival of the Fittest: New Product Development During the 90's.* Englewood Cliffs, NJ: Prentice-Hall.

Iacocca Institute (1991). *21st Century Manufacturing Enterprise Strategy.* Bethlehem, PA: Lehigh University.

Kamath, R.R. and J. K. Liker (1994) "A Second Look at Japanese Product Development," *Harvard Business Review*, Vol. 72, pp. 154–170.

Liker, J.K. (1988). "Survey-guided change in ship design and production: Prospects and limitations," *Journal of Ship Production Research*, Vol. 4, No. 2.

Liker, J.K., P.D. Collins, and F.M. Hull (1995). "Standardization and Flexibility: Test of a hybrid model of concurrent engineering effectiveness," presented at the 1995 Academy of Management Conference, Vancouver, Canada.

Liker, J.K. and M. Fleischer (1992a). "Designers and their Machines: CAD Technology and User Support in the US and Japan," *Communications of the ACM*, Vol. 35, No. 2, February, pp.77–95.

Liker, J. and M . Fleischer (1992b). "Organizational Context Barriers to DFM," in G. I. Susman, ed., *Managing Design and Manufacturing for Competitive Advantage.* New York: Oxford University Press.

Liker, J. K., R. R. Kamath, S. N. Wasti, and M. Nagamachi (1995). "Integrating Suppliers into Fast-Cycle Product Development," in Liker, Ettlie and Campbell, eds., *Engineered in Japan.* New York: Oxford University Press, pp. 152–191.

Liker, J.K., M. Fleischer, and D. Arnsdorf (1992). "Fulfilling the Promises of CAD," *Sloan Management Review*, Spring, pp. 74–86.

Liker, J.K., R. Kamath, N. Wasti, and M. Nagamachi (1996). "Supplier Involvement in Automotive Component Design: Are There Really Large U.S.-Japan Differences?" *Research Policy*, Vol. 25, pp. 59–89.

Majchrzak, A., M. Fleischer, and D. Roitman (1991). *Reference Manual for Performing a HITOP Analysis.* Ann Arbor, MI: Industrial Technology Institute.

Marca, D.A. and C. L. McGowan (1993). *IDEF0/SADT Business Process and Enterprise Modeling.* San Diego, CA: Eclectic Solutions Corporation.

McMillan, J. (1990). "Managing Suppliers: Incentive Systems in Japanese and U.S. Industry," *California Management Review*, Vol. 32, pp. 38–55.

Mintzberg, H. (1983). *Structure in Fives: Designing Effective Organizations.* Englewood Cliffs, NJ: Prentice-Hall.

Nishiguchi, N. (1994). *Strategic Industrial Sourcing: The Japanese Advantage.* New York: Oxford University Press.

Pasmore, W. and J.J. Sherwood (1978). *Sociotechnical Systems: A Sourcebook.* La Jolla, CA: University Associates.

Porter, M. (1980). *Competitive Strategy.* New York: Free Press.

Pryor, L.S. (1989). "Benchmarking: A Self-Improvement Strategy," *The Journal of Business Strategy*, Nov.-Dec.

Richardson, J. (1993). "Parallel Sourcing and Supplier Performance in the Japanese Automobile Industry," *Strategic Management Journal*, Vol. 14, pp. 339–350.

Rummler, G.A. and A.P. Brache (1995). *Improving Performance.* San Francisco, CA: Jossey-Bass.

Sako, M. (1995). "Component Supply Structures in Japan: Myths and Reality of Keiretsu Relationships," *The JAMA Forum*, Vol. 14, No. 2.

Sanderson, S. (1992). "Design for Manufacturing in an Environment of Continuous Change," in G.I. Susman, Ed., *Integrating Design and Manufacturing for Competitive Advantage.* New York: Oxford University Press, pp. 36–55.

Schein, E.H. (1992). *Organizational Culture and Leadership*. San Francisco, CA: Jossey-Bass.

Senge, P. (1990). *The Fifth Discipline: The Art and Practice of the Learning Organization.* New York: Doubleday Currency.

Shigley, J.E. and C.R. Mischke (1989). *Mechanical Engineering Design*, 5th ed. New York: McGraw-Hill.

Suh, N.P. (1990). *The Principles of Design*. New York: Oxford University Press.

Syan, C. and U. Menon (1994). *Concurrent Engineering: Concepts, Implementation and Practice.* New York: Chapman and Hall.

Taguchi, G. (1987). *System of Experimental Design: Engineering Methods to Optimize Quality and Minimize Cost*. Dearborn, MI: American Supplier Institute.

Thompson, J.D. (1967). *Organizations in Action*. New York: McGraw-Hill.

USA Today (1996). April 16.

Ward, A., J. Liker, D. Sobek, and J. Cristiano (1995). "The Second Toyota Paradox: How Delaying Decisions Can Make Better Cars Faster," *Sloan Management Review*, Spring, pp. 43–61.

Wenner, R., J. Pennell, H. Bertrand, and M. Slusarczuk (1988). *The Role of Concurrent Engineering in Weapons System Acquisition*. Alexandria, VA: Institute for Defense Analysis.

Williamson, O. E. (1975). *Markets and Hierarchies: Analysis and Antitrust Implications*. New York: The Free Press.

Womack, J.P. and D.T. Jones (1996). *Lean Thinking*. New York: Simon and Schuster.

Womack, J. P., D. T. Jones, and D. Roos (1990). *The Machine That Changed The World: The Story of Lean Production*. New York: Harper Perennial.

INDEX